Oil and the political economy in the Middle East

MANCHESTER
1824

Manchester University Press

Oil and the political economy in the Middle East

Post-2014 adjustment policies of the Arab Gulf and beyond

Edited by

Martin Beck and Thomas Richter

MANCHESTER UNIVERSITY PRESS

Published by Manchester University Press
Oxford Road, Manchester M13 9PL
www.manchesteruniversitypress.co.uk

British Library Cataloguing-in-Publication Data is available

ISBN 978 1 5261 4909 1 hardback
ISBN 978 1 5261 7186 3 paperback

First published by Manchester University Press in hardback 2021

This edition first published 2023

Typeset by New Best-set Typesetters Ltd

For our loved ones: Martin to Hala (and the flowergirls, Lila and Yasmina), and Thomas to Tuuli, Hanaan, Inken, Jonte, and Elin

Contents

Figures

Tables

Contributors

Amr Adly is an assistant professor at the American University in Cairo. He worked at Carnegie and CDDRL at Stanford University. Amr received his PhD from the European University Institute. He is the author of *Cleft capitalism: The social origins of Egypt's failed market making* (Stanford University Press, 2020) and *State reform and development in the Middle East: The cases of Turkey and Egypt* (Routledge, 2012). He has published in academic journals, including *Geoforum, Business and Politics, Turkish Studies*, and *Middle Eastern Studies*.

Sumaya AlJazeeri is an independent researcher in political economy, finance, and macroeconomics. She covers the Arab Gulf countries in her research and work. Sumaya is a Chevening Scholar and holds an MA in global political economy from the University of Sussex as well as a BA in economics (double majored with political science) from McGill University. She also works as a senior financial equity and economic analyst at an investment bank in the Kingdom of Bahrain, with over ten years of experience in the field.

Riad al Khouri has trained, consulted, lectured, researched, broadcast, and published on political economy, business, and geostrategy over several decades, working in Arabic, English, and French. He holds an MLitt in economics from the University of Oxford, a BA in the same subject from the American University of Beirut, and is a graduate of Ecole Internationale de Genève. Riad is a board member of the Global Challenges Forum Foundation, Geneva, and the principal of the Discover Studies Programme, Amman, among other affiliations.

Said Al-Saqri is the elected president of the Oman Economic Association. He is a researcher and economic adviser and has taught economic and business courses at a number of higher education institutes in Oman, including Sultan Qaboos University. He provides regular public commentary on economic affairs in various media. Said's research interests include the role of natural resources in economic development.

Martin Beck is a professor of modern Middle East studies at the University of Southern Denmark (SDU). His research covers international politics and political economy, in particular Middle Eastern power relations, the Arab–Israeli conflict, regional oil politics, and comparative analysis of rentier states. Martin has published in *Global Policy*, *Middle East Critique*, the *Journal of International Relations and Development*, *Mediterranean Politics*, *European Foreign Affairs Review*, *Democracy and Security*, and the *Journal of Refugee Studies*, among others.

Crystal A. Ennis is a scholar of global political economy and a lecturer at Leiden University. Her research examines the political economy of dependency on hydrocarbon revenue and foreign labour in Gulf economies, and the governance of migration and labour. Crystal has published in *New Political Economy*, *Global Social Policy*, the *International Journal of Middle East Studies*, *Third World Quarterly*, and *Cambridge Review of International Affairs*, among others.

Matthew Gray is a professor at the School of International Liberal Studies, Waseda University, and was previously at the Australian National University. He is the author of *The economy of the Gulf states* (Agenda, 2019), *Global security watch – Saudi Arabia* (Praeger, 2014), *Qatar: Politics and the challenges of development* (Lynne Rienner, 2013), *Conspiracy theories in the Arab world: Sources and politics* (Routledge, 2010), and various journal articles and other pieces on Middle Eastern studies.

Gertjan Hoetjes is currently affiliated to the University of Groningen as a lecturer at the Department of Middle Eastern Studies. In 2020, Gertjan obtained his doctoral degree from the University of Exeter for his PhD thesis on the impact of internet technology on contentious politics in Kuwait and Oman. Besides his interest in state–society relations, he has published a book chapter on Oman's foreign policy towards Iran in 2016 and has a keen interest in political economy.

Mohamad B. Karaki is an associate professor of economics at the Lebanese American University (LAU). He received his PhD and MA from Wayne State University and his BA from the University of Michigan-Dearborn. Before joining the LAU, he held positions as a visiting assistant professor at Oakland University, a lecturer at the University of Michigan-Dearborn, a part-time faculty member at Wayne State University, and a consultant at the World Bank. His research has been published in reputable economics journals.

Robert Mason is a fellow with the Sectarian, Proxies, and De-sectarianisation project at Lancaster University and non-resident fellow at The Arab Gulf States Institute in Washington. Robert was an associate professor and director of the Middle East Studies Center at the American University in Cairo, 2016–19, and was recently a visiting scholar in the Department of Near East Studies at Princeton University. His latest book is *New perspectives on Middle East politics: Economy, society and international relations* (AUC Press, 2021).

Thomas Richter is a senior research fellow at the German Institute for Global and Area Studies (GIGA) in Hamburg, where he works at the GIGA Institute of Middle East Studies. He holds a PhD from the University of Bremen and an MA from the University of Tübingen. Thomas's most recent research relates to structural adjustments and sectoral transformation in the Middle East after the oil price decline in 2014, shrinking civic spaces, and executive personalisation.

Emily Silcock holds an MSc in economics (public policy and development) from the Paris School of Economics and a BA in politics, philosophy, and economics from the University of Oxford. Her research interests focus on the political economy of the Middle East, including the political aspects of aid, the dynamics of peacekeeping in Syria, and public–private dialogue in Jordan, where she has researched and studied extensively.

Karen E. Young is a resident scholar at the American Enterprise Institute (AEI). Her research focuses on the political economy of the Gulf and the wider Middle East. Karen regularly teaches at the US Foreign Service Institute and as a professorial lecturer at George Washington University.

Preface

The basic idea for this book came from the external shock experienced by the Middle East due to the sharp drop in oil prices that started in the middle of 2014. This first wave of oil price declines in the twenty-first century generated enormous pressure on the Arab Gulf states, as well as on Egypt, Jordan, and Lebanon, after over a decade of stunningly high income from hydrocarbons. In this collection, we explore the effects of this price drop as a potential game changer for the political economy of the region with long-lasting consequences. The main reason for our belief that 2014 marked a turning point for the rentier and semi-rentier states in the Middle East is that structural changes in the global energy market in the 2010s make it highly unlikely that the price for a barrel of oil will climb back up above USD 100 in the foreseeable future. In 2020, the COVID-19 pandemic induced another oil price drop. This dramatically revived the pressure on Middle Eastern regimes to launch the adjustment policies that in the present book are analysed in the light of the game-changing oil price decline of 2014. As the editors of this book focusing on the period between 2014 and 2018, we consider ourselves fortunate to be in the position to release original country studies discussing policy adjustments in the immediate years following the price crash of 2014. As the year 2020 deepened a structural change that had started to occur in 2014, the chapters collected in this volume will also shed some light on the deep social, economic, and political changes in the making as triggered by the COVID-19 pandemic. We hope that his book will provide valuable insights for academics, practitioners, and decision makers intrigued by the dynamics of the political economy of the Middle East.

Acknowledgements

The success of a project like the one presented in this edited volume depends first and foremost on the authors for their engagement and patience. We also want to thank the anonymous reviewers who provided us with very helpful critical comments on our book proposal. Gratitude goes furthermore to the four people who kindly provided us with their expertise in editing the manuscript: Rob Byron from Manchester University Press for his encouragement from the very beginning and his careful support throughout the project, Catherine Schwerin for great language editing, Silvia Rojas Castro for an excellent job in formatting the whole manuscript, and also Hala Bejjani for sharing her expertise in transliterating Arabic terms according to the regulations of the *International Journal of Middle East Studies* (IJMES). We are also grateful to several institutions that contributed to the success of this project. The German Institute for Global and Area Studies (GIGA) funded the editing of the book and at the early stage of the project generously financed an authors' workshop in Beirut in March 2019, which the editors of this volume co-organised with the Lebanese Oil and Gas Initiative (LOGI). Kulluna Irada in Beirut supplied us with the premises and logistics. We are grateful to all of them for giving us and the authors the great opportunity to present and discuss our draft papers.

Note on transliteration

The transliteration of Arabic terms follows the regulations of the *International Journal of Middle East Studies* (IJMES). Brand and company names have been rendered as stated on their official websites. Moreover, the editors of this volume have accepted the authors' choice of spelling for the Arabic definite article in personal names.

Abbreviations

ADNOC	Abu Dhabi National Oil Company
Aramco	Saudi Arabian Oil Company
BDL	Banque du Liban
BHD	Bahraini dinar
CapEx	Capital expenditure
CBJ	Central Bank of Jordan
CIS	Commonwealth of Independent States
EDL	Electricité du Liban
EFF	Extended Fund Facility
EGP	Egyptian pound
EGPC	Egyptian General Petroleum Corporation
EITI	Extractive Industries Transparency Initiative
FDI	Foreign direct investment
FGF	Future Generation Fund
GCC	Gulf Cooperation Council
GDP	Gross domestic product
GRF	General Reserve Fund
ILO	International Labour Organization
IMF	International Monetary Fund
IPO	Initial public offering
JCPOA	Joint Comprehensive Plan of Action
JOD	Jordanian dinar
KD	Kuwaiti dinar
KIA	Kuwait Investment Authority
KPI	Key performance indicators
kWh	Kilowatt-hour
LBP	Lebanese pound
LOGI	Lebanese Oil and Gas Initiative
MBS	Muhammad bin Salman
MENA	Middle East and North Africa

NTP	National Transformation Plan
OMR	Omani riyal
OPEC	Organization of the Petroleum Exporting Countries
PIF	Public Investment Fund
QAR	Qatari riyal
QFC	Qatar Financial Centre
QNV	Qatar National Vision 2030
QP	Qatar Petroleum
RPLA	Resource-poor, labour-abundant
RRLP	Resource-rich, labour-poor
RST	Rentier state theory
SABIC	Saudi Basic Industries Corporation
SAGIA	Saudi Arabia General Investment Authority
SAR	Saudi riyal
SCA	Suez Canal Authority
SMEs	Small and medium enterprises
SOEs	State-owned enterprises
SSA	Sub-Saharan Africa
SWF	Sovereign wealth fund
UAE	United Arab Emirates
USD	United States dollar
VAT	Value-added tax

1

Pressured by the decreased price of oil: Post-2014 adjustment policies in the Arab Gulf and beyond

Martin Beck and Thomas Richter

Introduction

The downhill slide in the global price of crude oil, which started mid-2014, has had major repercussions within all the countries of the Middle East, not only for the net oil exporters but also the net oil importers, like Egypt, Jordan, and Lebanon, which are more or less closely connected with the oil-producing countries from the Gulf. After the Arab uprisings of 2010 and 2011, the oil price decline represents a second major shock for the region in the early twenty-first century – one that has imposed constraints, but also constituted opportunities and will do so in the future.

Since the beginning of the latest oil price peak in the mid-2000s, major constraints have arisen due to the generally high share of oil income within state budgets – which is especially true for the Arab monarchies in the Gulf. State spending, which increased heavily after 2010, almost exclusively depends on earnings from the hydrocarbon sector. The decline in the price of oil and its subsequent oscillation on a much lower level have significantly contributed to a relative lack of financial resources (oil rents), which has consequently restricted states' room for manoeuvre with regard to both domestic (e.g. welfare state and economic diversification) and foreign policies (e.g. petrodollar diplomacy). These effects are relevant for Middle Eastern countries beyond the Gulf, too. One reason for this is that some of the non-Gulf Cooperation Council (non-GCC) countries, such as Egypt, also produce oil. Yet, even more important is that smaller oil producers and non-oil producers of the Middle East, in particular Lebanon and Jordan, are structurally dependent on payments from the Arab Gulf – such as loans, direct budget support, investment, and, not least, labour remittances.

At the same time, during periods of declining oil revenues and increasing budget deficits, opportunities emerge: lower government income results in less lavish spending schemes and can potentially strengthen reform-oriented segments within the regime. These groups might eventually start reforming government bureaucracies, seriously tackle the issue of corruption, and even initiate the promotion of job-generating industries with the aim of overcoming prevailing distorted socio-economic structures, institutional deficits, and non-meritocratic habits.

Periods of fiscal crisis are especially exciting for students of comparative political economies. They not only 'provide excellent opportunities to see what really matters in a country's politics' (Moore, 2004: 10), but also condition 'policy outcomes in unexpected ways' (Chaudhry, 1989: 104). In the words of Gourevitch,

> hard times expose strengths and weaknesses to scrutiny, allowing observers to see relationships that are often blurred in prosperous periods, when good times slake the propensity to contest and challenge. (Gourevitch, 1986: 9)

In this chapter we discuss both the empirical importance of policy adjustments after the 2014 oil price decline and academic approaches of political economy apt to analyse these responses. First, we explore the oil price decline of 2014 as a potential game changer for the political economy in the Middle East. We claim that structural changes in the global energy market make it unlikely that oil prices will climb above USD 100 per barrel again in the foreseeable future. We highlight some of the fiscal and budgetary consequences of this for the nine Middle Eastern cases that are investigated in this volume. Second, we outline the most prominent concept for analysing the political economy of the Middle East: rentierism. Third, we scrutinise two major repercussions of decreased oil prices for the political economy in the Middle East. We argue that the predominant context in which policy adjustments take place is, for the six Gulf countries, the change from oil-rent abundance to scarcity and, for the three net oil importers, the significantly reduced energy bill. Fourth, we outline a heuristic framework on how structural changes caused by the oil price decline in 2014 – and in principle also by the 2020 oil price decline induced by the COVID-19 pandemic – could hypothetically translate into policy change. We introduce four domains of adjustment policies: rent-seeking policies, austerity measures, policies of taxation, and structural reform measures. In the final section of this chapter, we highlight the key aspects that are addressed in the nine country studies.

A few remarks on the selection of country studies are expedient. We chose the six members of the GCC – Bahrain, Kuwait, Oman, Qatar, Saudi Arabia, and the United Arab Emirates (UAE) – because no other group of countries in the Middle East and even globally is as dependent on hydrocarbon revenues as they are. These countries are consequently most eligible for exploring the notion that the 2014 oil price decline has triggered a change at the nexus of oil and the political economy in the Middle East. Other regional oil exporters such as Iraq, Libya, and Yemen are omitted because they did not have a functioning, centralised government in the time after the oil price drop of 2014. Algeria is, however, only to a very limited extent integrated into the hydrocarbon centre of the Middle East: the Gulf. We therefore refrained from including this North African country in the comparative exploration of this book. A similar argument can be made with regards to Iran. With Lebanon as the only exception, socio-economic ties between the nine countries selected for this book and the Islamic Republic are rather low. Moreover, as there is intense political and socio-economic interaction and interdependency between Middle Eastern countries, we identified Egypt, Jordan, and Lebanon as those three that are critically dependent on the Arab Gulf through linkages such as political aid, labour remittances, tourism, and private investment. We largely focus on the time after the fall in oil prices in the summer of 2014 until 2018 but also discuss major developments up to early 2020.

The oil price decline of 2014 as a potential game changer for the political economy in the Middle East

The year 2014 marks the beginning of the end of yet another period of high oil prices. The late 1970s were characterised by prices for one barrel of crude oil reaching levels above USD 100 (deflated in 2018 US consumer prices) for the first time. Yet, the 1980s and 1990s witnessed the most severe drop in oil prices in the last century. The year 1998 marked the lowest oil price since 1973. In deflated 2018 USD, the annual average oil price had fallen to less than USD 20. Then a long period of an upsurge in oil prices took its course. Between 1999 and 2011, annual average oil prices increased every year except 2001 and 2009. As a result, crude oil prices more than sextupled from 1999 to 2011, when the annual average price peaked at close to USD 125. From 2012 to mid-2014, it kept up at an outstanding average level of around USD 120 (see Figure 1.1).

For almost fifty years, the price of oil has been subject to repeated fluctuations. However, there is much to suggest that the price decline in 2014 is not cyclical but structural. A return to oil prices of over USD 100 thus remains highly unlikely in the foreseeable future. This is mainly due to technological innovations in hydraulic fracturing – or fracking for short. In this process, a chemically prepared mixture of water and sand is pressed into oil shale at high pressure to extract gas and oil. Fracking in shale oil, the world's largest reserves of which are located in the USA, has meant that the world market can be supplied with sufficient oil at a price of around USD 50 in the medium term. At this price level, exploration costs for new shale oil deposits are covered (Rosenberg, 2019). In addition, due to the

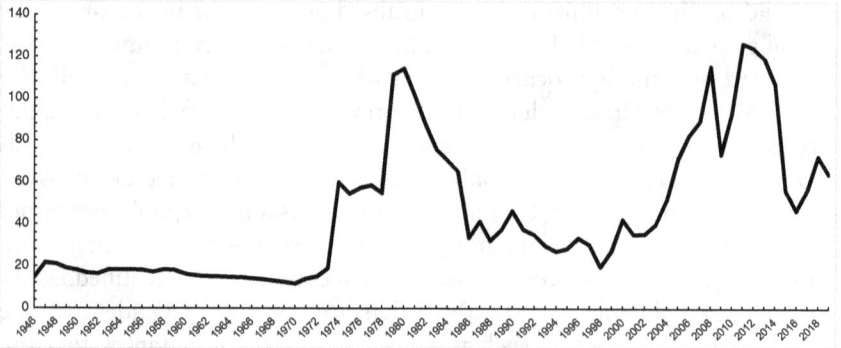

Figure 1.1 Crude oil price per barrel in USD 2019, 1946–2019 (deflated using the consumer price index for the USA). Source: BP (2020b). Note: Between 1945 and 1983, the price of Arabian Light posted at Ras Tanura, and between 1984 and 2018, the dated Brent price (refers to physical cargoes of crude oil in the North Sea).

fracking boom, the USA succeeded in becoming the world's largest oil producer in 2018. In the days of the classic oil price revolution in the 1970s, the price cap for oil was based on the substitution costs for conventional oil, with the largest deposits being in the Gulf region. For the time being, this role is played by shale oil (Beck, 2019).

This shift in production in favour of the USA is by no means sustainable, as they, due to their extremely high production volume in relation to their proved reserves, can produce at the 2019 level for hardly more than another decade. On the other hand, Saudi Arabia would still be able to do so for almost seventy years, and Kuwait even over ninety years (BP, 2020a: 14). However, a return to the dominance of conventional oil in the global energy market is not to be expected for two reasons: first, the deposits of other unconventional and heavy oils are also concentrated in the Americas; second, the demand for hydrocarbons is expected to decline in the medium to the long run due to the energy transition in the making (BP, 2020c).

As a consequence of this development, the oil-rich countries of the Gulf and also some of their net oil-importing Arab neighbours face the epochal challenge of managing economic, social, and political affairs under the condition of significantly reduced levels of oil income. The weight of the inflicted fiscal burden is illustrated in Table 1.1 by the oil-exporting countries' respective break-even oil prices – a notional price on which the national budget is virtually balanced.

Since 2015, the annual break-even oil prices for Bahrain, Oman, and Saudi Arabia have manifestly surpassed the actual annual average oil price

Table 1.1 Fiscal break-even oil prices in USD per barrel, and average oil price annually.

	2011	2012	2013	2014	2015	2016	2017	2018	2019
Bahrain	110.7	119.4	130.4	103.3	118.7	105.7	112.6	118.4	106.3
Kuwait	42.5	49.0	42.5	54.5	47.4	43.4	45.7	53.6	52.6
Oman	77.9	79.8	98.3	94.0	101.7	96.6	96.9	96.7	92.8
Qatar	79.0	63.1	61.9	56.1	52.4	54.0	51.3	48.0	44.9
Saudi Arabia	78.1	77.9	89.0	105.7	94.2	96.4	83.7	88.6	82.6
UAE	93.3	69.9	69.4	91.0	64.7	51.1	62.0	64.1	67.1
Average oil price	107.46	109.45	105.87	96.29	49.49	40.76	52.43	69.78	64.04

Source: IMF (2016b, 2017, 2018, 2019, 2020); average oil price is the annual OPEC basket price (OPEC, 2020).
Note: Fiscal break-even oil prices in 2019 are IMF projections.

shown in the last line of Table 1.1. This indicates a large structural deficit in the state budget. For Oman and Saudi Arabia, this is a direct consequence of the 2014 oil price decline, while in Bahrain this break-even price was above the average annual oil price even before 2014 due to the country's virtual depletion of autonomous oil production. Despite the introduction of effective saving measures, the structural discrepancy between oil income and budgetary needs is still visible in all three countries. Even Qatar and the UAE, often characterised as among the '[t]he uber-rich [rentier] states' (Okruhlik, 2016: 24), felt impelled to take a series of policy measures to adjust their government spending. This was so because in Qatar in 2015 and 2016 and in the UAE from 2015 until 2017 and again in 2019 the respective break-even prices had climbed above the average oil price level. Due to its extraordinarily high production volume per capita, only in Kuwait was the break-even price below average oil price levels for the entire time period.

As presented in Table 1.2, the consequences thereof can be seen in the emerging differences across the Arab Gulf countries with regard to budget balance, debt ratio, and foreign exchange reserve figures as a percentage of gross domestic product (GDP). In Kuwait, Qatar, and the UAE, the discrepancy between state revenue and expenditure has remained at a relatively low, negative level. In fact, Kuwait has seen a small positive budget surplus since 2017, while the debt ratio has risen from 7.5 per cent in 2014 to 15.2 per cent in 2019. In Qatar, the budget turned into modest deficits in 2016 and 2017 only, and the debt ratio increased by about 20 per cent of GDP after 2014 till 2019. Surprisingly, the UAE was hit harder by the fall in oil prices. The federation of emirates has experienced negative budget deficits in all

Table 1.2 Budget balance, debt ratio, and foreign exchange reserves as a percentage of GDP.

	2014	2015	2016	2017	2018	2019
Bahrain						
Budget balance	−1.6	−18.4	−17.6	−14.2	−11.9	−8.0
Debt ratio	44.4	66.0	81.3	88.2	94.7	101.7
Foreign exchange reserves	17.26	10.01	6.75	6.63	4.96	8.57
Egypt						
Budget balance	−12.8	−11.4	−12.0	−10.6	−9.5	−7.6
Debt ratio	85.1	88.5	96.8	103.2	92.7	84.9
Foreign exchange reserves	4.80	4.89	8.62	18.45	16.59	14.38

Table 1.2 Budget balance, debt ratio, and foreign exchange reserves as a percentage of GDP. (Continued)

	2014	2015	2016	2017	2018	2019
Jordan						
Budget balance	−10.3	−5.3	−3.7	−3.7	−4.8	−3.4
Debt ratio	89.0	93.4	93.8	94.3	94.4	94.6
Foreign exchange reserves	44.10	43.49	39.59	38.22	34.47	34.33
Kuwait						
Budget balance	27.1	5.6	0.3	6.3	8.7	6.7
Debt ratio	7.5	4.7	10.0	20.7	14.7	15.2
Foreign exchange reserves	19.81	24.76	28.45	27.93	26.40	29.71
Lebanon						
Budget balance	−6.0	−7.5	−8.9	−8.6	−11.0	−9.8
Debt ratio	133.5	140.9	146.1	149.0	151.0	155.1
Foreign exchange reserves	105.57	97.26	105.47	103.79	92.39	71.10
Oman						
Budget balance	−1.1	−15.9	−21.3	−14.0	−7.9	−6.7
Debt ratio	4.9	15.5	32.7	46.4	53.4	59.9
Foreign exchange reserves	20.13	25.65	30.94	22.79	21.93	20.07
Qatar						
Budget balance	15.3	5.4	−5.2	−2.9	5.3	7.0
Debt ratio	32.3	35.5	46.7	49.8	48.6	53.2
Foreign exchange reserves	21.01	23.04	21.02	8.99	15.86	20.86
Saudi Arabia						
Budget balance	−3.4	−15.8	−17.2	−9.2	−5.9	−6.1
Debt ratio	1.6	5.8	13.1	17.2	19.0	23.2
Foreign exchange reserves	96.82	94.21	83.08	72.30	67.99	63.14
UAE						
Budget balance	5.0	−3.4	−2.0	−1.4	1.2	−1.6
Debt ratio	15.6	18.7	20.2	20.0	19.1	20.1
Foreign exchange reserves	19.46	26.23	23.92	25.26	24.03	25.29

Source: Budget balance and debt ratio: IMF (2017, 2018, 2019); foreign exchange reserves: EIU (2019a, 2019b, 2019c, 2019d, 2019e, 2019f, 2019g, 2019h, 2019i, 2020a, 2020b, 2020c, 2020d, 2020e, 2020f, 2020g, 2020h, 2020i).
Note: Figures for 2019 are IMF projections (budget balance and debt ratio) and Economist Intelligence Unit estimates (foreign exchange reserves).

years since 2014 with 2018 as the only exception when they had a small positive balance of 1.2 per cent. On the other hand, the debt ratio rose only modestly by 5 per cent and foreign exchange reserves slightly increased between 2014 and 2019.

In contrast, since 2015, the budgets of Bahrain, Oman, and Saudi Arabia have exhibited a high deficit of over 10 per cent of GDP; this development is proceeding most dramatically in Bahrain, though. Along with the country's perpetually high budget deficit, its debt ratio has also increased to over 80 per cent of GDP. At the same time, foreign exchange reserves have melted away since 2014. Based on current levels, Bahrain can no longer cover its budget deficit with available currency reserves. Although Oman has had a budget deficit similar to that of Bahrain since 2014, the structural conditions in the sultanate have proven to be more advantageous. There, the debt ratio is rather low at around 35 per cent while foreign exchange reserves account for about one-third of current GDP. Saudi Arabia represents a special case: even though its budget deficit was over 15 per cent in both 2015 and 2016, its debt ratio was registered as being the lowest of all GCC members. In addition, during the oil price boom of the early twenty-first century, Saudi Arabia stockpiled historically large foreign exchange reserves – which will enable it to offset its current budget deficit level over a number of years.

In contrast to the six Gulf countries, macroeconomic data do not signify an immediate impact of the oil price decline on Egypt, Jordan, and Lebanon. At first, looking at the indicators presented in Table 1.2, the three non-Gulf states, all of which are net oil-importing countries, can be said to have weathered the decline in oil prices since 2014 relatively well. Budget deficits have not additionally skyrocketed as in some of the oil countries, and foreign exchange reserves have remained relatively stable – even increased in the case of Egypt. Since 2014, government debt alone has risen further in all three countries from an already very high level since before the oil price drop even started. Only in Lebanon have there been some early warning signals of a looming macroeconomic crisis. The debt ratio has risen from 135.5 per cent of GDP in 2014 to 155.1 per cent in 2019, and foreign exchange reserves have declined by over 30 per cent of GDP despite the fact that Lebanon has benefitted from importing cheaper oil since 2014.

Rentierism and the political economy in the Middle East

Rentierism is a concept which points to the overall importance of income generated by natural resources, in particular hydrocarbon rents, for socio-economic development and politics (Jenkins *et al.*, 2011; Richter, 2019). This concept has developed as the outstanding political economy approach to studying the oil-exporting Middle East and its fringes for decades (for a view departing from rentierism, see Hanieh, 2011). The main idea of economic rents as *unearned* income goes back to the nestors of economics, namely, Adam Smith, David Ricardo, and Karl Marx. Contemporary economics defines rents as a surplus higher than the minimum that the receiver would have otherwise accepted given the availability of alternative opportunities (Buchanan, 1980: 3). Rents typically do not originate from investment or labour – in the capitalist sense of the word – but are generated as the result of natural advantages and organisational skills. Thus, in contrast to entrepreneurs acting on markets, receivers of rents are not under immediate pressure to reinvest their revenues. Typically, rents are therefore at the free disposal of those actors who control access to it. In the Arab Gulf, these are the ruling families (Beck, 2012).

Distinct from classic rent theories in economics, rentierism as an academic concept gained prominence as a result of studying oil-dependent state-building processes in parts of Latin America, mainly Venezuela, which became the first 'petro-state' (Karl, 1997), and – with significance for a whole world region – the Middle East. Mahdavy (1970) pioneered studying the sociopolitical consequences of high oil income using the example of Iran's transformation into an oil-rentier state. His analysis of the monopolisation of revenues from crude oil exports by state institutions, in particular the government, draws attention to two major structural consequences. First, due to the expansion of large public-spending programmes and above-average public-sector growth, the government becomes the dominant actor within the economy. Second, governments of rentier states acquire an unprecedented independence from society. This independence is a result of their ability to expand services and create employment without having to extract resources from the society through taxation. Oil income, as Mahdavy argues, empowers governments 'to bribe pressure groups or to coerce dissidents' (1970: 467). However, he also shows that rentier states tend to suffer from inefficient state bureaucracy and become highly vulnerable to oil price fluctuations, on whose price building they had hardly had any influence in the 1950s and 1960s. This proved to be particularly painful during periods of decreasing oil prices (Mahdavy, 1970: 467).

Ten years later, Delacroix (1980) argued that distributive states – states with large amounts of rents at their disposal – constitute a previously unnoticed outcome of peripheral state-building processes. He points out that in this context the state emerges as the engine of sociopolitical change, which implies that the relevance of class relations for socio-economic and political development diminishes. Thus, state building in distributive states is quite distinct from the European class-based experience. More than fifteen years after Mahdavy's initial observations and in the same year in which

Anderson (1987: 9–10, 14) acknowledged the notion of the rentier state as a major contribution to political science, Beblawi and Luciani (1987) in a seminal edited volume sharpened the debate on rentierism by depicting the various dimensions of the rentier state. Based on the assumption that a rentier economy shapes a rentier mentality, which 'embodies a break in the work-reward causation' (Beblawi, 1987: 52), this kind of non-meritocratic system comes along with the political implication of low political mobilisation. As Luciani (1987) argues, due to the independence of the state from taxes, social groups in the Middle East lack the leverage of Western societies to tame and finally subordinate the state through pressures from below. However, other authors in Beblawi and Luciani (1987) also pinpoint the vulnerabilities of the rentier state. Najmabadi (1987), for instance, explains the downfall of the Shah regime in Iran 1979 by highlighting the alienation of the Iranian rentier state from its own society.

When analysing the 1970s 'oil price revolution' (Schneider, 1983: 101), scholars have highlighted that Middle Eastern rentierism is not confined to oil-exporting states. It eventually spread to the whole region based on petrolism, a system of transnational distribution of oil income across Arab states (Korany, 1986). Petrolism worked through two major channels linking the resource-rich, labour-poor (RRLP: the Gulf monarchies) and resource-poor, labour-abundant (RPLA: e.g. Egypt, Jordan, and Lebanon) countries (Cammett *et al.*, 2015: 27): budget transfers and political aid from the Gulf paid out to the RPLA countries, and, in the opposite direction, labour force migration from the RPLA countries to the oil-producing states. Petrolism resulted in the emergence of so-called semi-rentier states (Beblawi, 1987), among them Egypt, Jordan, and also Lebanon. While these countries do not have oil-rent income in abundance, they receive a mixture of rent income derived from sources such as budget support and aid from the Gulf – often supplemented by financial transfers from the West – the exploitation of other natural resources such as phosphate, levies on the tourism sector, and location rents, which are based on fees for major transport routes such as the Suez Canal (originally Beblawi, 1987: 61; later also Richter and Steiner, 2008; Beck and Hüser, 2015).

Since the 1990s, the scholarly debate on rentierism and the rentier state has become more differentiated and sophisticated (see for recent examples Herb and Lynch, 2019; Yamada and Hertog, 2020). Three main areas of work that are highly relevant for the chapters in this edited volume shall be briefly discussed by way of example.

First, as originally mentioned by Mahdavy (1970: 467), rentier states may use oil income for coercion, too. Ross (2001: 335–6) then attempted to systematically revisit the relevance of repression for the survival of authoritarian rentier states. He argues that the authoritarian regimes of oil rentiers have more financial means available in order to oppress democratic ambitions than authoritarian regimes in resource-poor countries (for a different perspective, see Smith, 2004). Historically, the use of coercive means can be best studied using the examples of the Shah regime in Iran and the rule of Saddam Hussein in Iraq. Yet, as shown by Crystal (2018), in the wake of the Arab uprisings and reinforced by the oil price decline in 2014, coercion also gained in relevance in the Arab Gulf peninsula states.

Second, scholars have attempted to specify the interaction between oil wealth, political institutions, and authoritarianism. As Beblawi highlights, in rentier states only a few control the externally derived rent and this 'would allow them to seize "political power"' (1987: 52). Yet, as Waldner notes, externally derived wealth like oil income does not dictate the establishment of authoritarian regimes (1999: 107). For instance, research from Latin America shows that oil rents crucially supported an elite-negotiated democratisation in Venezuela (Karl, 1997; Dunning, 2008). Thus, historical comparative research comes to the conclusion that the impact of oil wealth depends fundamentally on the quality of political institutions preceding oil and not primarily on whether oil rents represent the majority of government revenues for any given political regime (Smith, 2006: 59).

Third, an elaborated discussion on the developmental effects of oil wealth has emerged that mostly points to the negative consequences of oil rents as a curse rather than a blessing (for a pioneering perspective, see Auty, 1993). Among the many negative effects outlined in this context are Dutch disease and bureaucratic inefficiency. Dutch disease points to the overvaluation of local currencies in oil-exporting countries, which is the result of the inflow of rent income. As a result, the high productivity of oil as the only significant export product damages the productivity rates of all other economic sectors and eventually inhibits export-oriented industrialisation. Bureaucratic inefficiency, already emphasised by Mahdavy (1970: 467) as a negative side-effect of state building based on oil, is – in addition to weak fiscal and macroeconomic institutions – a key hindrance to reaching sufficient levels of economic development within rentier states (Malik, 2019). Yet, Hertog's (2010a, 2010b) contributions highlight that even within the bureaucratic structures of highly inefficient rentier states, profitable and well-managed state-owned enterprises can develop. Contrary to the expectations of mainstream economics, 'islands of efficiency' (Hertog, 2010b: 263) can emerge if market-oriented management is equipped with enough autonomy in its daily operations.

From oil-rent abundance to scarcity in the Gulf

All GCC member states greatly profited from the rise in oil prices in the early twenty-first century. As an outcome, high surpluses benefitted the national budget, and GDP growth figures were markedly good. Many governments began to invest heavily in infrastructure and prestige projects. Gulf states were also able to build up their foreign exchange reserves and to engage in global investments (Hanieh, 2018; Gray, 2019).

Yet, as a consequence of the Arab Spring, the costs of subsidies and expenses in the public sector have increased ever since 2011. As the majority of citizens in the Gulf monarchies are employed in the public sector, their salaries were raised; new positions were created within the security sector and the state bureaucracy, too (Richter, 2017). The goal was to provide the younger generation of citizens with employment opportunities and decent wages. These policies helped the governments to avoid mass protests against their authoritarian rule such as those that occurred in Tunisia in December 2010 and passed through many Arab states in 2011 and beyond. Without the high revenue generated by the sale of hydrocarbon products, the authoritarian Gulf monarchies would not have survived the Arab uprisings so easily; with the exception of Bahrain, the Gulf monarchies faced only short-lived protests (such as in the cases of Kuwait, Oman, and Saudi Arabia) or were even spared from the upheavals in the Arab Middle East (as in the cases of Qatar and the UAE) (Lucas, 2014).

As becomes apparent in Figure 1.2, after flying high in the early 2010s, the basket price of the Organization of the Petroleum Exporting Countries (OPEC) has dropped significantly since summer 2014. The average daily OPEC basket price dropped to USD 41.50 on 13 January 2015 and after a short peak thereafter further to below USD 25 in early 2016. After the OPEC oil basket price partially recovered to above USD 60 per barrel in November 2017 and even reached a level above USD 80 per barrel for a few days in the autumn of 2018, it oscillated in subsequent years in a range between roughly USD 55 and 70. In 2020, however, when demand for hydrocarbons sharply declined as a result of the economic slowdown associated with the COVID-19 pandemic, the OPEC basket price fell below USD 15 in late April before it partially recovered in the next months. From June to September 2020, the oil price regained its footing, reaching a level of around USD 40. Yet, already in 2019, the Gulf states' income from hydrocarbons was almost halved in comparison to the early 2010s, with enormous consequences for the fiscal capacities of the countries.

In the literature on rentierism and the rentier state, different indicators are used to measure the potential impact of oil and hydrocarbon income on socio-economic development and politics. Among them are hydrocarbon production or export values as a percentage of total state revenues or GDP. Yet, none of these indicators measures the true number of rents controlled by governments (Lucas and Richter, 2016) nor do they indicate the financial means available to a government for patronage or coercion in order to manipulate citizens' preferences (Smith, 2017). It is rather the government's

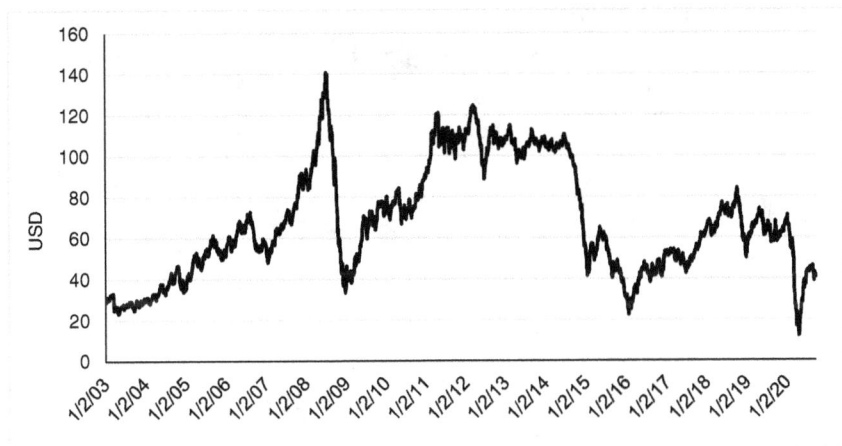

Figure 1.2 OPEC basket price in USD, 2011–20, annual average value based on current prices. Source: OPEC (2020). Note: The OPEC basket price is a weighted average of prices for petroleum blends produced by OPEC members. It was introduced in 2005 and is, as of October 2020, composed of the following variants: Saharan Blend (Algeria), Girassol (Angola), Djeno (Congo), Zafiro (Equatorial Guinea), Rabi Light (Gabon), Iran Heavy (Islamic Republic of Iran), Basra Light (Iraq), Kuwait Export (Kuwait), Es Sider (Libya), Bonny Light (Nigeria), Arab Light (Saudi Arabia), Murban (United Arab Emirates), and Merey (Venezuela) (OPEC, 2020).

hydrocarbon income per capita that indicates best the power of the rentier state. Table 1.3 shows per capita data of natural resource rents for all Gulf countries both for total population and national population, the latter being the subgroup holding citizenship.

Until 2014, most of the authoritarian regimes in the Gulf still had oil revenues in abundance. This enabled them to distribute oil rents through generous subsidies and the provision of financially attractive jobs in the public sector. The figures on oil revenues per capita in Table 1.3 indicate that the era of abundance had come to an end for Bahrain, Oman, and Saudi Arabia, while it continued in the UAE, Kuwait, and Qatar. However, possible long-term effects of the COVID-19 pandemic with regards to a reduced global demand for oil and an accelerated energy transition might eventually terminate hydrocarbon income abundance for all members of the GCC.

A comparison between 2010 and 2015 – for which data on national population is available across all six GCC countries – shows the dramatic decline of rentier leverage per citizen. In 2015, the oil per citizen income in Qatar was only 51 per cent of the 2010 level, 59 per cent in the UAE,

Table 1.3 Natural resource rents per capita across all countries of the GCC (in constant 2010 USD).

Oil rents (in USD) per total/national population		Bahrain	Kuwait	Oman	Qatar	Saudi Arabia	UAE
2004	total population	747	20,768	6,672	22,446	8,264	12,116
	national population	1,334	47,793	9,050	87,801	11,625	n/a
2005	total population	856	27,060	7,782	23,963	9,764	13,913
	national population	1,571	61,913	10,606	n/a	13,797	464,009
2010	total population	690	18,982	6,956	19,153	7,301	7,351
	national population	1,501	49,029	10,808	146,266	10,663	374,048
2014	total population	1,025	20,162	6,063	15,096	8,438	8,881
	national population	2,170	58,327	10,800	n/a	13,507	386,005
2015	total population	549	13,345	3,546	8,989	4,983	5,288
	national population	1,162	39,144	6,454	75,141	8,026	219,783
2016	total population	405	11,355	3,031	7,544	4,134	4,444
	national population	868	33,586	5,592	n/a	6,678	178,225
2017	total population	445	12,371	3,461	9,056	4,781	5,360
	national population	982	36,626	6,446	n/a	7,746	n/a
	in % of total/national population						
2015 at the level of 2010	total population	79.48	70.30	50.98	46.94	68.25	71.94
	national population	77.41	79.84	59.71	51.37	75.27	58.76

Source: GLMM (2020), World Bank (2020a, 2020b).

60 per cent in Oman, 75 per cent in Saudi Arabia, 77 per cent in Bahrain, and 80 per cent in Kuwait.

The end of the era of oil-rent abundance for at least three of the six member states of the GCC is also a consequence of the high annual growth of domestic labour in the labour markets of the more populous Gulf states; Saudi Arabia, but also the UAE and Oman, despite dramatic declines in fertility rates since the 1980s, still show high population growth into the twenty-first century.

Under other socio-economic conditions, this could favour dynamic economic development, but hardly in the Arab Gulf. One reason for this is that the private sector is unattractive for the local labour force due to much lower wages compared to the public sector. On the other hand, the state bureaucracy offers greater prestige and better working conditions; for example, in the form of shorter working hours. While the authoritarian regimes in the Gulf see the economic need to curtail the bloated public sector, they are at the same time under social pressure to make corrections in this regard. This is aggravated by the fact that the population's expectations of the state have risen from generation to generation since the 1970s: governments are forced to spend a large part of their oil revenues on education, which then generates a workforce that is largely directed towards the unproductive public sector (Cammett *et al.*, 2015: 145, 333–42).

Egypt, Jordan, and Lebanon: A significantly reduced energy bill

Egypt (which ceased to be a net energy exporter shortly before the oil price drop in 2014), Jordan, and Lebanon all have benefitted from the oil price decline due to significantly lowered energy bills. As there are hardly any economic transactions free of energy in modern economics, lower energy costs bear the potential to create a boost for the entire economy. At the same time, structures of rentierism are under pressure in the semi-rentier states of Egypt, Jordan, and Lebanon because both pillars of petrolism are diminishing. Due to the 2014 drop in oil prices, Saudi Arabia has fewer financial means available for its petrodollar diplomacy. Also, remittances from the Gulf to Egypt, Jordan, and Lebanon are potentially affected, as the Gulf states are pressurised to reduce their dependence on foreign labour.

Only two of Egypt's rent sources were not hit by the 2014 oil price decline. Due to its location between the Mediterranean and the Red Sea, as well as its historic sites, beaches, and attractive climate conditions, Egypt yields rents from tourism. Moreover, fees for passing the Suez Canal yield a location rent because they indicate a level of earning that highly outweighs the potential income derived from similar investments in an industrial production process (Richter and Steiner, 2008). On the other hand, the drop in oil prices after 2014 constitutes a challenge to Egypt's receipt of two other crucial sources of rent income: political aid payments and workers' remittances, which are also to be considered rents because the recipients of these incomes have not put in labour or made investments for it (Schlumberger, 2010: 245; Abulof, 2017: 67). Since the 1970s, Egypt has heavily relied on budget-to-budget payments, cheap loans, and other forms of political rents from the Arab Gulf, in particular Saudi Arabia. The oil price revolution of the 1970s also enabled the Arab Gulf countries to expand labour immigration. Thus, Egypt became dependent on workers' remittances from the Gulf.

Jordan's only source of rent not affected by the oil price decline of 2014 is its income from phosphate production. Otherwise, Jordan's rentierism has been challenged by the oil price decline in a way similar to the Egyptian case. In the 1970s, Jordan, which was created by the United Kingdom in the 1920s as a state whose king needed to be financially supported in order to maintain his rule, has become dependent on remittances from the Gulf and political aid, particularly from Saudi Arabia but also from Kuwait, even though Amman managed to maintain relatively high political rents from the USA and the EU (Beck and Hüser, 2015).

Lebanon, which has been 'exporting' its labour force en masse in several migration waves since the nineteenth century, has been integrated into the networks of labour migration based on petrolism from the very beginning. Moreover, with the end of the Lebanese Civil War (1975–90), Lebanon became dependent on capital inflows and financial aid from the Arab Gulf, in particular Saudi Arabia, for its neoliberal reconstruction (Baumann, 2016). Lebanon (as well as Jordan) also used the high influx of Syrian refugees in the wake of the Syrian Civil War to acquire a refugee rent by attracting additional foreign aid payments (Tsourapas, 2019).

Rentierism and adjustment policies: A heuristic framework

As orthodox rentier state theory (RST) emphasises, in the first instance a decline in oil-rent income constrains the ruling regime because it disposes over less in rents to stabilise its rule. However, challenges do not only engender *constraints* but can also be taken by the political class as *opportunities* for structural reforms, thereby possibly even paving the way for overcoming rentierism. Initially, the Gulf countries responded to the oil price decline with visionary plans of tackling deficiencies, the Saudi Vision 2030 being just one among several. Yet, there is a long way to go from announcing plans to implement structural change. First, in the past governmental announcements in the Gulf on launching reforms by disseminating development plans and

Table 1.4 Rentierism and adjustment policies.

	Rentierism adjustment policies	Adjustment policies going beyond rentierism
Government revenues	*Rent-seeking policies* Producer cooperation: e.g. OPEC+ Developing the LNG sector	*Taxation policies* Introduction of VAT Introduction of income and corporate taxes
Government expenditures	*Austerity policies* Cutting subsidies Incurring debt Cancellation of investment	*Structural reform measures* Labour market reforms Development projects

Source: Authors' own compilation.

national visions often failed to translate into policy action (e.g. Hvidt, 2013; Ulrichsen, 2016). Similar examples can be found looking at past policy initiatives in Egypt (e.g. Adly, 2020), Jordan (e.g. Moore, 2004), and Lebanon (e.g. Baumann, 2019). Second, it is necessary to discern whether only ad hoc measures are executed, or actual structural change and deep reform are promoted. One example to illustrate the difference are the measures taken against corruption. Purges against corruption are often part of mere populist ad hoc policies or power-consolidating strategies, as seen in the case of Muhammad bin Salman's elimination of potential rivals in 2017 and the following years. Structural reforms, however, would require systematically discouraging rent-seeking opportunities (see Buchanan, 1980), for instance by developing institutions that guarantee an independently audited state budget, which are extremely under-developed in the Arab Gulf, with the exception of Kuwait (AlShehabi, 2017). Another example is that of reorganising the distribution of rents away from providing direct financial support for citizens towards genuinely activating their productive potentials. To frame these differences in a more systematic way, in Table 1.4 we propose an approach that distinguishes adjustment policies with regard to government revenues and expenditures, according to whether the adjustment policies abide by rentierism or go beyond it.

Rent-seeking policies

The Gulf states' most important rent-seeking policy response to the fall of oil prices in 2014 was to intensify cooperation between oil exporters. After an initial phase of inaction and failed attempts to stabilise the price of oil through exporter cooperation, this was achieved under the leadership of Saudi Arabia in 2016 in a similar way to the early 1980s. The members of OPEC were aware that, in view of the greatly increased importance of oil exporters outside the organisation, effective cooperation would only be promising if Russia were involved. A first attempt to launch an OPEC+ in Doha failed in April 2016 due to the resistance of Saudi Arabia, which wanted to prevent Iran, its main competitor for regional power in the Middle East, from benefitting from higher oil prices (Beck, 2016). However, in November 2016, the Vienna Group (also commonly referred to as OPEC+) – consisting of OPEC members plus Russia and other non-members of OPEC such as Bahrain and Oman – reached an agreement on a production quota that would be later renewed in November 2017 and then again in December 2018 (Cohen, 2018). The remarkable upward trend in the OPEC basket price after 2016 – from USD 40.76 per barrel to USD 52.43 per barrel in 2017, and to USD 69.78 per barrel in 2018 (see Figure 1.2) – was certainly caused by different factors, among them socio-economic and political developments such as the ongoing demise of the states of Venezuela and Libya and the USA's decision to impose severe sanctions on Iran and Venezuela. However, there can be hardly any doubt that the resumption of the oil producers' cooperation was a decisive factor in stabilising the oil price. This is remarkable, as all parties involved are facing a prisoner's dilemma: although all members are better off if the cooperative agreement lasts, each and every member has at any time a systemic incentive to exceed its quota. As actors in a prisoner's dilemma have mixed motives over whether or not to cooperate, agreements on production quotas are in principle volatile (Alt *et al.*, 1988: 445–66).

OPEC's post-2014 cooperation has a historical predecessor. However, the production control system established in the early 1980s suffered from the fact that several OPEC members continuously and systematically exceeded their quotas. Since, in addition, oil-producing countries outside the organisation were acting as free riders, Saudi Arabia obtained a majority decision at the end of 1985, which suspended the quota system and forced a price war on producers outside OPEC (Alt *et al.*, 1988: 455–7). As a result, Saudi Arabia in particular has massively expanded its oil production.

However, it is noticeable that the major oil producers in the Middle East leave relatively high shares of their oil in the ground – and this despite the fact that production costs in the Gulf region are extraordinarily low. At the end of 2019, Saudi Arabia, the biggest oil producer in the Gulf and globally the second biggest, had a ratio of proved reserves to production (R/P) of 68.9, whereas the biggest and the third biggest producers in 2019 – the USA and Russia – had R/Ps of 11.1 and 25.5, respectively (BP, 2020a: 14–16). As the relatively high R/P ratios in the Gulf date back to the twentieth century, they indicate that even after the end of the formal quota system in the mid-1980s, the Gulf states continued to 'artificially' depress production in order to prevent a further decline in prices. In doing so, they aimed at earning higher revenues from the sale of oil than they would have been able to do by acting in line with market conditions. Nevertheless, the failure of the cooperation and the flooding of the oil market by the Saudis in the years 1985–86 had serious consequences, as this contributed to a long-lasting decline in prices (see Figure 1.1).

This raises the question whether the production control established in 2016 is sustainable. Similar to OPEC in the 1980s, OPEC+ is in a precarious situation because of the systematic incentive to exceed individual quotas. In addition, Saudi Arabia could be tempted, as it was in Doha in 2016, to exacerbate Iran's economic crisis by accepting the self-harm associated with a termination of production control in the form of a further decline in prices. This becomes all the more likely the more Riyadh perceives the regime in Tehran not only as an adversary but as an enemy that needs to be weakened (Beck, 2020). Yet, there is also a factor that could prevent OPEC+ from failing. In 1985, in order to absorb the failures of other OPEC members as well the increasing production of oil exporters outside OPEC, Saudi Arabia had briefly reduced its oil production to barely more than 2 million barrels per day. In 2018, on the other hand, according to OPEC (2019: 57), Saudi Arabia was producing over 10 million barrels of oil per day. Thus, in the twenty-first century, Riyadh has much greater scope for production cuts than before.

The coronavirus pandemic in 2019–20 caused a sudden extreme drop in demand for oil. Saudi Arabia responded to this crisis in March 2020 by proposing a further reduction in the production quota, but Russia refused. Then, even before the expiry date of the OPEC+ production agreement on 31 March 2020, Riyadh launched a price war in order to bring Russia back to the negotiation table. The immediate result, however, was a price collapse to a level of USD 30 per barrel (Worland, 2020). Yet, on 13 April 2020, pressured and backed by the USA, OPEC+ reached an agreement to cut production, starting in May 2020 (*The Economist*, 2020). The record high reduction of 9.7 million barrels per day from May to July was eased to a still highly significant cut of 7.7 million barrels per day from August 2020 on. In the first three months of the renewed quota system, voluntary additional cuts by Saudi Arabia, the UAE, and Kuwait balanced overproduction committed mainly by Iraq and Nigeria (El Gamal *et al.*, 2020). At the same time, Saudi Minister of Energy Prince Abdulaziz bin Salman publicly announced that he expects all members of OPEC+ to comply to their quota (JPT, 2020).

The Gulf states have quite a number of options for diversifying their rent-seeking activities beyond oil. Qatar is clearly the most important producer of natural gas in the GCC, second in the entire Middle East only to Iran. Qatar's global share of proved reserves is 12.4 per cent, but the 3 per cent that both Saudi Arabia and the UAE held at the end of 2019 are also significant (BP, 2020a: 32). There are also options beyond hydrocarbons to seek additional rent income. For instance, Saudi Arabia launched a programme to attract non-religious tourists.

In contrast to the Arab Gulf states, the two semi-rentier states have never enjoyed rent abundance. Therefore, they were early in exploring ways to increase their rent income and thus look back to a long history of rent-seeking activities. This implies that their room for manoeuvre in terms of unfolding further rent-seeking activities is rather limited. Lebanon is a different case because it possesses potential offshore oil and natural gas deposits. After a long period of inactivity, in February 2020 Lebanon started to explore for hydrocarbons.

Austerity policies

Within all nine countries under scrutiny in this volume, strong influence from government expenditures, mainly for subsidies and social welfare distribution, exists. From the outset, Middle Eastern regimes launched social contracts that included heavily subsidised basic goods such as gasoline, cooking gas, and bread. They also established free public education and health care systems (Loewe and Jawad, 2018). Neoclassical economists have consistently described these state activities as inefficient and harmful (e.g. World Bank, 2003; El-Katiri and Fattouh, 2017). At the same time, social scientists have traced 'Middle Eastern exceptionalism' with regard to the lack of democratisation and low economic performance back to factors that include the historical development of government spending on goods and services, particularly on subsidies and welfare provision funded by oil wealth (e.g. Farsoun, 1988; Ross, 2001).

As for Middle East welfare systems, two major structural differences exist between the three semi-rentier states of Egypt, Jordan, and Lebanon, on the one hand, and the six Gulf rentiers on the other. First, it was only within the former that imperial powers left their footprints, by initialising social welfare systems during the era of colonial penetration. Since British colonial rule on the Arabian Peninsula was mainly meant to safeguard the trade routes with India and British-controlled territory in Iraq and Persia, in comparison to Egypt, Jordan, and Lebanon less colonial social policy legacy exists in the Gulf monarchies. Second, based on the multiplication of rents in the oil-rich Gulf states in the early 1970s, a tremendous difference with regard to the scope and intensity of social welfare and state subsidies emerged. From this high oil income, the GCC states created some of the most lavish subsidy and social welfare systems anywhere in the world (Krane, 2019). Mainly due to limited available financial resources, subsidy and welfare in Egypt, Jordan, and Lebanon remained far more exclusive and mainly confined to only politically relevant social groups (Eibl, 2017). The oil price drop in 2014 put enormous pressure on the expansive welfare and subsidy schemes of the Gulf monarchies. Thus, they have actually been quick to announce the reduction of fuel, energy, and water subsidies. Some actually took their first steps in walking the talk: having some of the lowest petrol prices worldwide, Kuwait and Saudi Arabia raised the price of petrol in 2015 and 2016 by over 100 and 200 per cent, respectively (Krane and Hung, 2016).

An alternative strategy for offsetting a rising budget deficit is to incur debt. As indicated in Table 1.1, the debt-to-GDP ratio has risen since 2014 in most of the countries discussed in this volume. Net oil importers in the Middle East are more experienced with public debt management because they start from a much higher level of debt-to-GDP ratio as a result of accumulating high debt levels in the past. In contrast, with the exception of Bahrain, the only Gulf country with a high debt-to-GDP ratio of 44.4 per cent in 2014, the Arab Gulf states entered the international capital market as newcomers.

With regard to public investment policies, a fundamental structural trap exists during periods of oil price decline. In oil-exporting countries with their dominant public sectors, falling oil prices almost immediately led to the cancellation of previously planned public investments in order to avoid excessive budget deficits because these fiscal measures usually do not spark immediate public opposition (Richter, 2017). This kind of adjustment policy is similar to the austerity policies implemented by oil-producing countries after oil prices dropped in the 1980s (see e.g. Hunter, 1986; Looney, 1994). Often the reduced public investments swiftly cause payment problems within the private sector, which may trigger a more widespread economic recession (Hunter, 1986). Similar effects could be observed in some of the non-oil countries like Egypt, Jordan, and Lebanon during periods of austerity policies, with negative long-term effects for infrastructure and social and economic well-being (Brand, 1992; Hinnebusch, 1993; Baumann, 2019).

Taxation policies

Apart from the payment of zakat – the mandatory contribution in Islam to those in need – there is no taxation on personal income, assets, or corporate gains in the Gulf monarchies. Outside of the hydrocarbon sector, only foreign-owned businesses are levied with a 10 to 20 per cent tax on profits made. Moreover, with the exception of Oman (which uses more general corporate taxation), uniform corporate taxation – which does not distinguish between domestic and foreign owners – is in place only for oil and gas companies. Save for imports (which are subject to a GCC uniform duty rate of 5 per cent) and for parts of the tourism sector, as a general rule no form of goods movement or goods production is taxed (IMF, 2016a).

As a result, the Gulf monarchies have to date served as tax havens, whose benefits for their citizens are possibly under threat due to the drop in the oil price. A first indicator for change, coming with possible far-reaching implications, is the attempt to introduce a value-added tax (VAT) of initially 5 per cent, which was implemented by Saudi Arabia and the UAE in 2018 and by Bahrain one year later. Saudi Arabia increased its VAT to 15 per cent, effective July 2020. Effective VAT collection necessitates documentation of each step of the procurement and the selling process by companies. As for ensuring appropriate payment, implementing VAT collection requires government surveillance of business transactions. The implementation of VAT therefore creates incentives for regimes in the Gulf to deepen their state structures.

In comparison to the GCC countries, in Egypt, Jordan, and Lebanon more complex taxation systems exist. They consist of both direct taxation (of the income and profits of individuals and legal entities) and indirect taxation (mainly VAT and sales tax). In all three countries, the tax system is skewed in favour of capital owners and big business at the expense of the lower income segments of society and the urban poor. This is so for two reasons: first, direct taxation is not progressive, and, second, indirect taxation – which is income insensitive – contributes to the government's total revenues to a much higher degree than direct taxation (Mansour *et al.*, 2015). Moreover, country-specific tax exemptions exist in favour of strategic social groups, for instance the military in Egypt (Momani, 2018) and large enterprises in Jordan (Guscina and Nandwa, 2018; Sidło, 2018). Last but not least, due to the large informal sectors and the limited capacities of tax administrations, tax evasion is a major unmet challenge.

Post-2014 developments in Jordan show how difficult it can be to change the basis and level of taxation. When the government in Amman attempted to alter the taxation system in order to gain support from international organisations like the International Monetary Fund (IMF), it faced massive social resistance (Ali, 2018). With regard to the Gulf, there are two reasons why it does not seem realistic to expect that taxation would lead to the establishment or empowerment of representative institutions (such as parliaments) – thereby constraining the power of ruling dynasties. First, none of the existing regimes and groups within these governments are seriously interested in sharing power, not to mention relinquishing it. Second, no powerful formation of social groups demanding better political representation or even democratisation is in sight.

Structural reform measures

From a labour market perspective, the countries dealt with in this edited volume belong to two different categories: resource-rich, labour-poor (RRLP: the Gulf monarchies) and resource-poor, labour-abundant (RPLA: Egypt, Jordan, and Lebanon) (Cammett *et al.*, 2015: 27). A major traditional pillar of the social contract in the Gulf monarchies is that the state provides its male citizens with jobs in the public sector, including lifetime tenure and far-above-average salaries. This, however, has led to a number of dysfunctionalities. The public sector in the Gulf is bloated, whereas the private sector is unattractive for citizens. Rather than joining the private sector, young graduates prefer a long wait for job openings in the public sector (Cammett *et al.*, 2015: 145). Some of the semi-rentier states suffer from similar disparities. The regime that toppled the Egyptian monarchy in 1952 established a highly complex and inefficient state apparatus. However, downsizing the state bureaucracy is extremely unpopular, as for many the public sector is still the preferred employer (Barsoum, 2018: 775). Another component of labour market dysfunctionality is that high rent income weakens labour-intensive, export-oriented segments of the private sector, which in turn keeps female labour force participation among citizens low (Ross, 2008). The Dutch disease also causes extremely low female labour participation rates in semi-rentier states: Egypt at 23 per cent and Lebanon at 24 per cent reach only about half of the global average, which is at 48 per cent. Jordan at 14 per cent is even further away from this level (World Bank, 2019). Thus, the question arises of whether and how governments have taken the oil price decline in 2014 as an opportunity to curtail the employment of citizens in the public sector, to establish incentives to direct particularly young professionals towards the private sector, and to increase female labour force participation.

In the Arab Gulf, high rent income has facilitated a political economy with a relatively stark influx of foreign labourers in almost all economic segments, because citizens themselves are tied to generous wages and lifetime employment in the state sector – making them less inclined to become engaged in meritocratic work (in the private sector). Labour immigration is a state domain that many Arab countries have filled by advancing the *kafāla* (sponsorship) system, which originated in British attempts to regulate the migrant labour influx to the pearl-diving sector in Bahrain and spread from there to the whole Arab Gulf (AlShehabi, 2019). In Jordan and Lebanon, it only applies to certain segments of the economy – such as care for the elderly and domestic labour – whereas in the GCC countries it is a comprehensive system of employment (Khan and Harroff-Tavel, 2011). A special case is Egypt, where domestic workers from abroad – predominantly from the Horn of Africa and Sudan – are irregular and thus do not even enjoy the minimum protection provided by the *kafāla* system (Thomas, 2010: 999).

The *kafāla* system is tailored to serve the political and socio-economic interests of the citizens at the expense of the migrant workers. This is of utmost importance to the socio-economic systems, particularly of the Arab Gulf, where more than half of the labour force is composed of migrants. The influx of migrant workers is bureaucratically controlled in a way that ensures the recruitment of non-Arabs who lack any means of becoming engaged in politics during their temporary stay there (Khan and Harroff-Tavel, 2011: 302; Hvidt, 2019). To make up for the lack of well-rooted state structures, monitoring and sanctioning costs are shifted to individual employers. Last but not least, the *kafāla* system constitutes a highly asymmetric relationship between employer and employee – and the states often fail to implement what are already low-standard labour law provisions (Khan and Harroff-Tavel, 2011: 298). In other words, the *kafāla* system puts few restrictions on overexploiting the migrant worker labour force.

There are, however, also socio-economic and political dysfunctions beyond the overexploitation of the migrant workers concerned. It has been argued that the Gulf states fail to efficiently utilise particularly highly skilled migrant workers recruited by the *kafāla* system; as they are prohibited from changing employer or seeking permanent residency, they lack the incentive to fully engage with their work (Hvidt, 2019). In the past, some reform efforts by the state bureaucracies to improve the system were blocked by lobby groups (Khan and Harroff-Tavel, 2011: 303). Indeed, increases in productivity are discouraged due to the abundance of cheap labour (Hertog, 2017). However, the most pertinent incentive for the states of the Middle East to reform their respective systems is a political one: at the latest since Qatar won the bid in 2010 to host the 2022 FIFA (Fédération Internationale de Football Association) World Cup, the *kafāla* system has come under scrutiny from global civil society actors and human rights organisations (Khan and Harroff-Tavel, 2011; Pattisson, 2013). Some of them make a pronounced critical analysis of the *kafāla* system, applying the notion of 'contract slavery' to it (cf. Bales, 1999; Jureidini and Moukarbel, 2004). Thus, the Gulf states are under severe pressure to launch reforms of the *kafāla* system. *Kafāla* causes inefficiencies well beyond the Gulf. In Lebanon, for instance, the local low-skilled labour force is prevented from accessing an important segment of the labour market because migrant labour from Asian and African countries takes most of the positions in domestic work, which includes housekeeping but also care-taking of children and the elderly. Thus, the *kafāla* system sets disincentives to establish modern care systems in Lebanon (Beck, 2018).

In addition, the privatisation of state-owned enterprises is becoming a relevant field of policymaking during periods of decreased income levels. Privatisation is often to be considered a more mid-term strategy during such times. However, the successful implementation of large-scale privatisation depends on many factors. Among them are whether domestic or foreign investors are ready to take the risk of investing and the will and the ability of the government to implement privatisation in the face of social and political resistance to it. Last but not least, the assessment of privatisation policies should critically check whether it truly eliminates rent dependency. A historic example of launching a private business sector that rather than reducing rent dependency deepened it is Saudi Arabia's attempt to introduce 'sowing oil rents' in wheat production in the 1980s (Richards and Waterbury, 2008: 163). The protection rate of a policy that, despite all comparative disadvantages, temporarily transferred Saudi Arabia into the world's sixth largest wheat exporter was far above 1,000 per cent. Also, the most spectacular privatisation policy following the oil price decline of 2014, the initial public offering (IPO) of the Saudi Arabian Oil Company (Aramco), barely reduces Saudi's dependency on rent income. As 85 per cent of Aramco is still owned by the Saudi state (Woertz, 2019), the IPO of Aramco is little more than cashing in on future rent generation.

A long-term strategy for governments reacting to the oil price decline consists of the design and implementation of economic or industrial diversification policies. Oil-rentier states as well as net oil importers lack globally competitive economic structures because their economies depend upon very few sectors – in the case of the Gulf monarchies, even upon only one – that are able to hold their own in the world market. Economic diversification therefore aims at addressing this prevailing economic flaw by creating incentives for economic actors to develop technologically advanced exporting industries (Cherif *et al.*, 2016). Economic diversification has been on the agenda of policymakers in all of the countries under scrutiny for decades now (Luciani, 2012; Cammett *et al.*, 2015). However, especially in the course of fiscal crises, the need for diversification becomes more urgent. Yet, such policies will only succeed if a long-term government strategy is applied. This, in turn, requires the political class to be united in its willingness to overcome existing structural obstacles. Last but not least, national entrepreneurs and workers must have the capabilities and motivation to join forces with them.

The decrease in the oil price since 2014 has contributed to an enhanced relevance of foreign direct investment (FDI) in the Arab Gulf states for two reasons: FDI shows promise as a way to make up for the decline of rent income in the balance of payments and may also contribute to the development of new prospering industries. With the exception of Bahrain, whose inward FDI stock reached 90 per cent of its GDP in 2016, FDI inflows in the Arab world in general and among GCC members in particular are rather low in comparison to many other regions of the Global South (Hussein, 2009: 370–1; Al-Tamimi, 2017). Yet, there are significant differences, too: Saudi Arabia and the UAE had inward FDI stocks of above 30 per cent of their GDP in 2016, Oman and Qatar above 20 per cent, whereas Kuwait's level was below 15 per cent (PWC, 2018).

Since 2014, a general trend observable in the Gulf has been the attempt to reverse the downward trend of inward FDI flows, which, after they peaked in 2008 at more than USD 5 billion, did not even reach the level of USD 2 billion in 2015 or 2016. A strategy taken by all GCC countries, albeit in different forms and with varying degrees of intensity, has been to ease restrictions on foreign ownership allowed within each country in order to attract additional FDI (PWC, 2018).

FDI inflow is sensitive to political repercussions, as became conspicuous when, in the wake of Saudi Crown prince and de facto sole ruler Muhammad bin Salman's involvement in the assassination of Jamal Khashoggi, many international companies distanced themselves from the investment conference held in Riyadh in October 2018. In general, the regional political ambitions of some Gulf countries are at odds with their economic and development interests and policies (Young, 2016).

Egypt made many efforts to attract additional FDI, but it frequently had to take setbacks (MEMO, 2019a). Jordan, which had the highest ratio of FDI to GDP in the Arab world in the years prior to the 2014 oil price decline (Méon and Sekkat, 2013: 3), experienced a drop of almost half of its FDI inflow in 2018 compared to the previous year in the wake of social protests against the government's austerity policy (MEMO, 2019b). The change in FDI inflows to Lebanon, which was second to Jordan in terms of the ratio of FDI to GDP before the oil price drop, was highly negative in 2015; yet, this could be more than offset by positive developments in 2016 and 2018 (Nakhoul, 2019).

Country studies

The contributions in this volume aim to shed light on the constraints and opportunities the 2014 oil price drop entailed for nine Middle Eastern countries between 2014 and 2018. The structure of the following chapters complements the logic of case selection as discussed above. We start with the six GCC member states in alphabetical order and proceed with the three oil-importing Arab countries in the Mashriq.

Sumaya AlJazeeri's (2021) analysis highlights the persistence of rentier structures in Bahrain during fiscal crises with the aim of maintaining socio-political stability. By pointing to a neoclassical upgrade of rentierism, she stresses that even though the post-2014 government discourse attempted to present the removal of decades-old customary privileges as structural change, all traditional government structures have largely prevailed. Despite some face-lifting, Bahrain therefore remains a rentier state.

In his analysis on the repercussions of the 2014 oil price drop in Kuwait, Gertjan Hoetjes (2021) highlights how citizens' resistance has effectively frustrated reform initiatives by the ruling regime. Austerity measures such as the imposition of government fees have therefore been effective mainly for migrant workers. In contrast to Kuwaiti citizens who managed to get their voice heard through the most powerful parliament in the Arab Gulf, migrant workers had no political leverage. Due to the country's large fiscal reserves, immediate reform pressure was mitigated and helped the regime to avoid a confrontation with its well-accommodated constituencies.

Crystal Ennis and Said Al-Saqri (2021) highlight that the fiscal adjustment of government spending in Oman does not coalesce steeply during contractions in the oil market. The authors point to the specific difficulty of saving government expenses in a country where economic, social, and political life is deeply linked to oil-dependent and government-initiated development. Effective social pressure has discouraged government attempts to cut subsidies and reform the labour market after the 2014 oil price decline, which indicates that during fiscal crises rentier states are less autonomous vis-à-vis their societies than presented in orthodox rentierism.

Matthew Gray (2021) argues that Qatar has responded to the 2014 oil price drop emphatically, thereby bolstering the nation well by showing how robust the economy is in absorbing shocks. This also helped the regime to be well prepared by coping with the blockade that, among others, Saudi Arabia, the UAE, Bahrain, Egypt, and Jordan imposed on Qatar in June 2017. Backed by huge hydrocarbon reserves, the state's deep reach into society leaves the country rather well equipped for the challenges of rentierism. Yet, there remains the need for structural reforms.

Robert Mason (2021) highlights the intersections and novelties the oil price decline of 2014 has brought forward for Saudi Arabia during a period of transition within the ruling Al Saud family. He portrays Saudi Arabia as being challenged by increasing fiscal constraints caused by a widening budget deficit, on the one hand, and large cohorts of the younger generation entering the job market with the expectation of becoming employed in the highly paid public sector, on the other. As he argues, pure state-led investment programmes could fail to secure lasting growth prospects for Saudi Arabia, which in turn could result in authoritarian upgrading.

In the chapter discussing policy adjustments post-2014 in the UAE, Karen Young (2021) highlights three major policy areas in which promising adjustment initiatives have been launched: fiscal policy, social development policy, and diversification policies. As she points out, the emirates have been more successful at mastering the crisis because domestic affairs in the UAE are deeply shaped by its federal structure – in contrast to the realm of foreign policy, which underwent a centralisation process. Thus, the government promoted policy differentiation, learning, and competition between the seven emirates and set incentives to new policy formulation and experimentation across them.

Amr Adly (2021) highlights Egypt's twisted hydrocarbon dependency. In the light of the overall expectation that Egypt's economy should have benefitted from the oil price drop in 2014, he identifies two elements as being responsible for this not happening. Regulatory and extractive state institutions, which are a prerequisite for coordinating economic diversification and industrial upgrading, are weak. Thus, allocative institutional capacities prevail. In addition to that, continuous petrodollar recycling from the Gulf has offset the positive fiscal impact of declining oil prices.

Riad al Khouri and Emily Silcock (2021) argue that, also for Jordan, yet another net oil importer in the Middle East for which the positive effects of an oil price decline were to be expected, little if any evidence exists that the country has departed from rentierism since 2014. While the benefits of cheap oil post-2014 are dispersed, the Jordanian government has failed to contain the sharp increase in current spending, which mainly represents expenses for subsidies, pensions, social assistance, and wages in the public sector, through which key constituencies are tied to the Hashimite regime.

Mohamad Karaki (2021) analyses the impact of the oil price decline on key macroeconomic variables in Lebanon. He identifies several short-term beneficial impacts, among them an increase in the share of manufacturing in total output and the reduction of the budget deficit as a result of a decline in government allocation towards electricity production. He also discusses the relevance of lower oil prices for the prospects of Lebanon becoming a hydrocarbon producer. He suggests putting in place an upgraded national development strategy that, for instance, promotes the improvement of the current Lebanese system of education in order to avoid the Dutch disease during potentially upcoming periods of exporting natural resources.

In the concluding chapter on oil and the political economy in the Middle East, Martin Beck and Thomas Richter (2021) pose the question of whether rentierism is about to be overcome in the Middle East following the 2014 oil price decline. This question is first addressed by discussing some of the pertinent empirical findings of the book. Second, three theoretical aspects are highlighted: the dominant role of state classes for the evolution of rentier and semi-rentier states, the influence of social and political institutions for adjustment policies, and the specification of rentier state autonomy. Finally, the authors infer that rentierism still prevails.

References

Abulof, U. (2017), '"Can't buy me legitimacy": The elusive stability of Mideast rentier regimes', *Journal of International Relations and Development*, 20:1, 55–79.

Adly, A. (2020), *Cleft capitalism: The social origins of failed market making in Egypt*. Stanford: Stanford University Press.

Adly, A. (2021), 'Egypt's twisted hydrocarbon dependency: A case of persistent semi-rentierism', in M. Beck and T. Richter (eds), *Oil and the political economy in the Middle East: Post-2014 adjustment policies of the Arab Gulf and beyond*. Manchester: Manchester University Press, pp. 164–91.

Ali, D. (2018), 'Jordan, its debt and the mirages of the IMF', *Orient XXI*, 12 November. https://orientxxi.info/magazine/jordan-its-debt-and-the-mirages-of-the-imf,274 5 (accessed 12 January 2019).

AlJazeeri, S. (2021), 'Upgrading towards neoclassical rentier governance: Bahrain's post-2014 oil price decline adjustment', in M. Beck and T. Richter (eds), *Oil and the political economy in the Middle East: Post-2014 adjustment policies of the Arab Gulf and beyond*. Manchester: Manchester University Press, pp. 36–57.

al Khouri, R., and E. Silcock (2021), 'Oil and turmoil: Jordan's adjustment challenges amid local and regional change', in M. Beck and T. Richter (eds), *Oil and the political economy in the Middle East: Post-2014 adjustment policies of the Arab Gulf and beyond*. Manchester: Manchester University Press, pp. 192–212.

AlShehabi, O. H. (2017), 'Show us the money: Oil revenues, undisclosed allocations and accountability in budgets of the GCC states', London School of Economics and Political Science, Kuwait Programme on Development, Governance and Globalisation in the Gulf States, Research Papers, 44. http://eprints.lse.ac.uk/84521/1/show-us-the-money_V1.pdf (accessed 14 December 2019).

AlShehabi, O. H. (2019), 'Policing labour in empire: The modern origins of the kafala sponsorship system in the Gulf Arab States', *British Journal of Middle Eastern Studies*. https://doi.org/10.1080/13530194.2019.1580183 (accessed 14 December 2019).

Alt, J. E., R. L. Calvert, and B. D. Humes (1988), 'Reputation and hegemonic stability: A game-theoretic analysis', *The American Political Science Review*, 82:2, 445–66.

Al-Tamimi, N. (2017), 'World investment report 2017: Arabs still lag', *Arab News*, 9 June. www.arabnews.com/node/1112711 (accessed 9 January 2019).

Anderson, L. (1987), 'The state in the Middle East and North Africa', *Comparative Politics*, 20:1, 1–18.

Auty, R. M. (1993), *Sustaining development in mineral economies: The resource curse thesis*. London: Routledge.

Bales, K. (1999), *Disposable people: New slavery in the global economy*. Berkeley: University of California Press.

Barsoum, G. (2018), 'Egypt's many public administration transitions: Reform vision and implementation challenges', *International Journal of Public Administration*, 41:10, 772–80.

Baumann, H. (2016), *Citizen Hariri: Lebanon's neoliberal reconstruction*. New York: Oxford University Press.

Baumann, H. (2019), 'The causes, nature, and effect of the current crisis of Lebanese capitalism', *Nationalism and Ethnic Politics*, 25:1, 61–77.

Beblawi, H. (1987), 'The rentier state in the Arab world', in H. Beblawi and G. Luciani (eds), *The rentier state*. London: Croom Helm, pp. 49–62.

Beblawi, H., and G. Luciani (1987), 'Introduction', in H. Beblawi and G. Luciani (eds), *The rentier state*. London: Croom Helm, pp. 1–21.

Beck, M. (2012), 'Dynasties', in H. K. Anheier, M. Juergensmeyer, and V. Faessel (eds), *Encyclopedia of global studies*. Thousand Oaks: Sage Publications, pp. 436–9.

Beck, M. (2016), 'Saudi Arabia's foreign policy and the failure of the Doha oil negotiations', E-International Relations, 21 June. https://bit.ly/38F1NEY (accessed 29 June 2019).

Beck, M. (2018), 'Contract slavery? On the political economy of domestic work in Lebanon', E-International Relations, 6 February. https://bit.ly/3tnHfbI (accessed 26 March 2020).

Beck, M. (2019), 'OPEC+ and beyond: How and why oil prices are high', E-International Relations, 24 January. https://bit.ly/3bLaqQ8 (accessed 24 October 2019).

Beck, M. (2020), 'The aggravated struggle for regional power in the Middle East: American allies Saudi Arabia and Israel versus Iran', *Global Policy*, 11:1, 84–92.

Beck, M., and S. Hüser (2015), 'Jordan and the "Arab Spring": No challenge, no change?', *Middle East Critique*, 24:1, 83–97.

Beck, M., and T. Richter (2021), 'Oil and the political economy in the Middle East: Overcoming rentierism?', in M. Beck and T. Richter (eds), *Oil and the political economy in the Middle East: Post-2104 adjustment policies of the Arab Gulf and beyond*. Manchester: Manchester University Press, pp. 1–35.

BP (2020a), 'BP statistical review of world energy 2020'. https://on.bp.com/30J1Gnj (accessed 29 September 2020).

BP (2020b), 'BP statistical review of world energy – all data, 1965–2019'. https://on.bp.com/3rPLTPs (accessed 29 September 2020).

BP (2020c), 'Energy outlook 2020'. https://on.bp.com/3tjunU3 (accessed 29 September 2020).

Brand, L. A. (1992), 'Economic and political liberalization in a rentier economy: The case of the Hashemite Kingdom of Jordan', in I. F. Harik and D. J. Sullivan (eds), *Privatization and liberalization in the Middle East*. Bloomington: Indiana University Press, pp. 167–87.

Buchanan, J. M. (1980), 'Rent seeking and profit seeking', in J. M. Buchanan, R. D. Tollison, and G. Tullock (eds), *Toward a theory of the rent-seeking society*. College Station: Texas A&M University Press, pp. 3–15.

Cammett, M., I. Diwan, A. Richards, and J. Waterbury (2015), *A political economy of the Middle East*. Boulder: Westview Press.

Chaudhry, K. A. (1989), 'The price of wealth: Business and state in labor remittance and oil economies', *International Organization*, 43:1, 101–45.

Cherif, R., F. Hasanov, and M. Zhu (eds) (2016), *Breaking the oil spell: The Gulf falcons' path to diversification*. Washington, DC: International Monetary Fund.

Cohen, A. (2018), 'OPEC is dead, long live OPEC+', *Forbes*, 29 June. https://bit.ly/38DeUq2 (accessed 9 January 2019).

Crystal, J. (2018), 'The securitization of oil and its ramifications in the Gulf', in H. Verhoeven (ed.), *Environmental politics in the Middle East: Local struggles, global connections*. London: C Hurst & Co Publishers Ltd, pp. 75–97.

Delacroix, J. (1980), 'The distributive state in the world system', *Studies in Comparative International Development*, 15:3, 3–21.

Dunning, T. (2008), *Crude democracy: Natural resource wealth and political regimes.* Cambridge: Cambridge University Press.

Eibl, F. (2017), 'Social policies in the Middle East and North Africa', in Palgrave Macmillan (ed.), *The new Palgrave dictionary of economics*. London: Palgrave Macmillan, pp. 1–19.

EIU (2019a), *Country report Bahrain January*. London: Economist Intelligence Unit.

EIU (2019b), *Country report Egypt January*. London: Economist Intelligence Unit.

EIU (2019c), *Country report Jordan January*. London: Economist Intelligence Unit.

EIU (2019d), *Country report Kuwait January*. London: Economist Intelligence Unit.

EIU (2019e), *Country report Lebanon January*. London: Economist Intelligence Unit.

EIU (2019f), *Country report Oman January*. London: Economist Intelligence Unit.

EIU (2019g), *Country report Qatar January*. London: Economist Intelligence Unit.

EIU (2019h), *Country report Saudi Arabia January*. London: Economist Intelligence Unit.

EIU (2019i), *Country report United Arab Emirates January*. London: Economist Intelligence Unit.

EIU (2020a), *Country report Bahrain February*. London: Economist Intelligence Unit.

EIU (2020b), *Country report Egypt February*. London: Economist Intelligence Unit.

EIU (2020c), *Country report Jordan February*. London: Economist Intelligence Unit.

EIU (2020d), *Country report Kuwait February*. London: Economist Intelligence Unit.

EIU (2020e), *Country report Lebanon February*. London: Economist Intelligence Unit.

EIU (2020f), *Country report Oman February*. London: Economist Intelligence Unit.

EIU (2020g), *Country report Qatar January*. London: Economist Intelligence Unit.

EIU (2020h), *Country report Saudi Arabia February*. London: Economist Intelligence Unit.

EIU (2020i), *Country report United Arab Emirates February*. London: Economist Intelligence Unit.

El Gamal, R., A. Lawler, and O. Astakhova (2020), 'OPEC+ presses for compliance with oil cuts', Reuters, 19 August. https://reut.rs/3s0vbgg (accessed 29 September 2020).

El-Katiri, L., and B. Fattouh (2017), 'A brief political economy of energy subsidies in the Middle East and North Africa', *International Development Policy / Revue Internationale de Politique de Développement*, 7:7, 1–26.

Ennis, C. A., and S. Al-Saqri (2021), 'Oil price collapse and the political economy of the post-2014 economic adjustment in the Sultanate of Oman', in M. Beck and T. Richter (eds), *Oil and the political economy in the Middle East: Post-2014 adjustment policies of the Arab Gulf and beyond*. Manchester: Manchester University Press, pp. 79–101.

Farsoun, S. K. (1988), 'Oil, state, and social structure in the Middle East', *Arab Studies Quarterly*, 10:2, 155–75.

GLMM (2020), 'Demographic and economic module'. https://bit.ly/3bMWF3C (accessed 3 March 2020).

Gourevitch, P. (1986), *Politics in hard times: Comparative responses to international economic crises*. Ithaca: Cornell University Press.

Gray, M. (2019), *The economy of the Gulf states*. Newcastle upon Tyne: Agenda Publishing Limited.

Gray, M. (2021), 'Qatar: Leadership transition, regional crisis, and the imperatives for reform', in M. Beck and T. Richter (eds), *Oil and the political economy in*

    the Middle East: Post-2014 adjustment policies of the Arab Gulf and beyond.
    Manchester: Manchester University Press, pp. 102–23.
Guscina, A., and B. Nandwa (2018), 'How the Middle East and Central Asian
    countries can reduce debt and preserve growth', International Monetary Fund,
    Country Focus, 13 November. https://bit.ly/3ctQK2t (accessed 12 January 2019).
Hanieh, A. (2011), *Capitalism and class in the Gulf Arab states.* New York: Palgrave
    Macmillan.
Hanieh, A. (2018), *Money, markets, and monarchies: The Gulf Cooperation Council
    and the political economy of the contemporary Middle East.* Cambridge: Cambridge
    University Press.
Herb, M., and M. Lynch (eds) (2019), 'The politics of rentier states in the Gulf',
    POMEPS Studies, 33. https://bit.ly/3e5xuZS (accessed 2 February 2019).
Hertog, S. (2010a), 'Defying the resource curse: Explaining successful state-owned
    enterprises in rentier states', *World Politics,* 62:2, 261–301.
Hertog, S. (2010b), *Princes, brokers, and bureaucrats: Oil and the state in Saudi
    Arabia.* Ithaca: Cornell University Press.
Hertog, S. (2017), 'Making wealth sharing more efficient in high-rent countries:
    The citizens' income', *Energy Transitions,* 1:7, 2–14.
Hinnebusch, R. A. (1993), 'The politics of economic reform in Egypt', *Third World
    Quarterly,* 14:1, 159–71.
Hoetjes, G. (2021), 'Stalled reform: The resilience of rentierism in Kuwait', in
    M. Beck and T. Richter (eds), *Oil and the political economy in the Middle
    East: Post-2014 adjustment policies of the Arab Gulf and beyond.* Manchester:
    Manchester University Press, pp. 58–78.
Hunter, S. T. (1986), 'The Gulf economic crisis and its social and political conse-
    quences', *Middle East Journal,* 40:4, 593–613.
Hussein, M. A. (2009), 'Impacts of foreign direct investment on economic growth in
    the Gulf Cooperation Council (GCC) countries', *International Review of Business
    Research Papers,* 5:3, 362–76.
Hvidt, M. (2013), 'Economic diversification in GCC Countries: Past record and future
    trends', London School of Economics and Political Science, Kuwait Programme
    on Development, Governance and Globalisation in the Gulf States, Research
    Papers, 27. https://bit.ly/3tdLxT4 (accessed 10 January 2019).
Hvidt, M. (2019), 'Exploring the nexus between highly-skilled migrants, the
    kafala system, and development in the UAE', *Journal of Arabian Studies,* 9:1,
    75–91.
IMF (2016a), 'Diversifying government revenue in the GCC: Next steps', International
    Monetary Fund, GCC Tax Policy Paper, 26 October. www.imf.org/external/np/
    pp/eng/2016/102616.pdf (accessed 17 April 2020).
IMF (2016b), 'Regional economic outlook: Middle East and Central Asia (statistical
    appendix)', May. https://bit.ly/3eDQGjb (accessed 10 January 2019).
IMF (2017), 'Regional economic outlook: Middle East and Central Asia (statistical
    appendix)', October. https://bit.ly/3rNtYsP (accessed 10 January 2019).
IMF (2018), 'Regional economic outlook: Middle East and Central Asia (statistical
    appendix)', November. https://bit.ly/3rOpLVO (accessed 10 January 2019).
IMF (2019), 'Regional economic outlook: Middle East and Central Asia (statistical
    appendix)', October. https://bit.ly/2OP6g0G (accessed 27 January 2020).
IMF (2020), 'Regional economic outlook: Middle East and Central Asia (statistical
    appendix)', April. https://bit.ly/3cvNGTw (accessed 16 April 2020).

Jenkins, J. C., K. Meyer, M. Costello, and H. Aly (2011), 'International rentierism in the Middle East and North Africa, 1971–2008', *International Area Studies Review*, 14:3, 3–31.

JPT (2020), 'Oil jumps after Saudi oil minister presses OPEC+ for better compliance with production cuts', *Journal of Petroleum Technology*. https://bit.ly/2PZb5Wj (accessed 29 September 2020).

Jureidini, R., and N. Moukarbel (2004), 'Female Sri Lankan domestic workers in Lebanon: A case of "contract slavery"?', *Journal of Ethnic and Migration Studies*, 30:4, 581–607.

Karaki, M. B. (2021), 'Lower oil prices since 2014: Good news or bad news for the Lebanese economy?', in M. Beck and T. Richter (eds), *Oil and the political economy in the Middle East: Post-2014 adjustment policies of the Arab Gulf and beyond*. Manchester: Manchester University Press, pp. 213–36.

Karl, T. L. (1997), *The paradox of plenty: Oil booms and petro-states*. Berkeley: University of California Press.

Khan, A., and H. Harroff-Tavel (2011), 'Reforming the *kafala*: Challenges and opportunities in moving forward', *Asian and Pacific Migration Journal*, 20:3–4, 293–313.

Korany, B. (1986), 'Political petrolism and contemporary Arab politics, 1967–1983', *Journal of Asian and African Studies*, 21:1–2, 66–80.

Krane, J. (2019), *Energy kingdoms: Oil and political survival in the Persian Gulf*. New York: Columbia University Press.

Krane, J., and S. Y. Hung (2016), 'Energy subsidy reform in the Persian Gulf: The end of the big oil giveaway', James A. Baker III Institute for Public Policy of Rice University, Issue Brief. https://bit.ly/3tkZu1D (accessed 17 April 2020).

Loewe, M., and R. Jawad (2018), 'Introducing social protection in the Middle East and North Africa: Prospects for a new social contract?', *International Social Security Review*, 71:2, 3–18.

Looney, R. (1994), *Manpower policies and development in the Persian Gulf region*. Westport: Praeger.

Lucas, R. E. (2014), 'The Persian Gulf monarchies and the Arab Spring', in M. Kamrava (ed.), *Beyond the Arab Spring*. London: C Hurst & Co Publishers Ltd, pp. 313–40.

Lucas, V., and T. Richter (2016), 'State hydrocarbon rents, authoritarian survival and the onset of democracy: Evidence from a new dataset', *Research & Politics*, 3:3. https://doi.org/10.1177/2053168016666110 (accessed 12 January 2019).

Luciani, G. (1987), 'Allocation vs. production state', in H. Beblawi and G. Luciani (eds), *The rentier state*. London: Croom Helm, pp. 63–84.

Luciani, G. (ed.) (2012), *Resources blessed: Diversification and the Gulf development model*. Berlin: Gerlach Press.

Mahdavy, H. (1970), 'Patterns and problems of economic development in rentier states: The case of Iran', in M. A. Cook (ed.), *Studies in the economic history of the Middle East: From the rise of Islam to the present day*. New York: Oxford University Press, pp. 428–67.

Malik, A. (2019), 'The political economy of macroeconomic policy in resource-rich Arab economies', in K. Mohaddes, J. B. Nugent, and H. Selim (eds), *Institutions and macroeconomic policies in resource-rich Arab economies*. Oxford: Oxford University Press, pp. 17–51.

Mansour, M., P. Mitra, C. A. Sdralevich, and A. Jewell (2015), 'Fair taxation in the Middle East and North Africa', International Monetary Fund Staff Discussion Notes, 2 September. https://bit.ly/2OR1PTc (accessed 12 January 2019).

Mason, R. (2021), 'The nexus between state-led economic reform programmes, security, and reputation damage in the Kingdom of Saudi Arabia', in M. Beck and T. Richter (eds), *Oil and the political economy in the Middle East: Post-2014 adjustment policies of the Arab Gulf and beyond.* Manchester: Manchester University Press, pp. 124–44.

MEMO (2019a), 'Egypt's foreign direct investment falls by 24.4%'. https://bit.ly/2PZylDD (accessed 12 March 2020).

MEMO (2019b), 'Report: Foreign investment in Jordan down 52.6% in 2018'. https://bit.ly/3glpOFP (accessed 12 March 2020).

Méon, P.-G., and K. Sekkat (2013), 'The impact of foreign direct investment in Arab countries', Economic Research Forum, Policy Perspective, 9. https://bit.ly/3rNhPE5 (accessed 13 January 2019).

Momani, B. (2018), 'Egypt's IMF program: Assessing the political economy challenges', Brookings, 30 January. https://brook.gs/3lhtmsR (accessed 12 January 2019).

Moore, P. W. (2004), *Doing business in the Middle East: Politics and economic crisis in Jordan and Kuwait.* Cambridge: Cambridge University Press.

Najmabadi, A. (1987), 'Depoliticisation of a rentier state: The case of Pahlavi Iran', in H. Beblawi and G. Luciani (eds), *The rentier state.* London: Croom Helm, pp. 211–27.

Nakhoul, S. (2019), 'Foreign investment inflows up 14 percent to $2.9 billion', BusinessNews.com.lb, 17 June. https://bit.ly/3bMHQOz (accessed 12 March 2020).

Okruhlik, G. (2016), 'Rethinking the politics of distributive states', in K. Selvik and B. O. Utvik (eds), *Oil states in the new Middle East: Uprisings and stability.* Abingdon: Routledge, pp. 18–38.

OPEC (2019), 'Monthly oil market report February 2019', 12 February. https://bit.ly/3viKdAa (accessed 13 April 2020).

OPEC (2020), 'OPEC basket price'. www.opec.org/opec_web/en/data_graphs/40.htm (accessed 2 October 2020).

Pattisson, P. (2013), 'Revealed: Qatar's world cup "slaves"', *The Guardian*, 25 September. https://bit.ly/2NoRjCg (accessed 10 January 2019).

PWC (2018), 'Foreign investment: GCC raises ownership limits to catalyse diversification', Middle East Economy Watch. https://pwc.to/3vplbQ1 (accessed 10 January 2019).

Richards, A., and J. Waterbury (2008), *A political economy of the Middle East.* Boulder: Westview Press.

Richter, T. (2017), 'Structural reform in the Arab Gulf states – Limited influence of the G20', GIGA Focus Middle East, 3. https://bit.ly/3qO3CWm (accessed 18 April 2020).

Richter, T. (2019), 'Oil and the rentier state in the Middle East', in R. Hinnebusch and J. Gani (eds), *The Routledge handbook to the Middle East and North African state and states system.* London: Routledge, pp. 225–37.

Richter, T., and C. Steiner (2008), 'Politics, economics and tourism development in Egypt: Insights into the sectoral transformations of a neo-patrimonial rentier-state', *Third World Quarterly*, 29:5, 935–55.

Rosenberg, E. (2019), 'Can fracking survive at $50 a barrel?', Investopedia, 25 June. https://bit.ly/2OAJgmd (accessed 29 October 2019).

Ross, M. L. (2001), 'Does oil hinder democracy?', *World Politics*, 53:3, 325–61.

Ross, M. L. (2008), 'Oil, Islam, and women', *American Political Science Review*, 102:1, 107–23.

Schlumberger, O. (2010), 'Opening old bottles in search of new wine: On non-democratic legitimacy in the Middle East', *Middle East Critique*, 19:3, 233–50.

Schneider, S. A. (1983), *The oil price revolution*. Baltimore: Johns Hopkins University Press.

Sidło, K. (2018), 'No taxation without representation? On tax reforms in Jordan', Medium, 3 September. https://bit.ly/3eFDJ8S (accessed 12 January 2019).

Smith, B. (2004), 'Oil wealth and regime survival in the developing world, 1960–1999', *American Journal of Political Science*, 48:2, 232–46.

Smith, B. (2006), 'The wrong kind of crisis: Why oil booms and busts rarely lead to authoritarian breakdown', *Studies in Comparative International Development*, 40:4, 55–76.

Smith, B. (2017), 'Resource wealth as rent leverage: Rethinking the oil–stability nexus', *Conflict Management and Peace Science*, 34:6, 597–617.

The Economist (2020), 'A historic OPEC+ deal to curb oil output faces many obstacles', 13 April. https://econ.st/3rRM9NZ (accessed 15 April 2020).

Thomas, C. (2010), 'Migrant domestic workers in Egypt: A case study of the economic family in global context', *American Journal of Comparative Law*, 58:4, 987–1022.

Tsourapas, G. (2019), 'The Syrian refugee crisis and foreign policy decision-making in Jordan, Lebanon, and Turkey', *Journal of Global Security Studies*, 4:4, 464–81.

Ulrichsen, K. C. (2016), 'Economic diversification plans: Challenges and prospects for Gulf policymakers', The Arab Gulf States Institute in Washington, Policy Paper, 2. https://bit.ly/2OR2pAc (accessed 3 November 2016).

Waldner, D. (1999), *State building and late development*. Ithaca: Cornell University Press.

Woertz, E. (2019), 'Aramco goes public: The Saudi diversification conundrum', GIGA Focus Middle East, November. https://bit.ly/3ctr0Du (accessed 18 April 2020).

Worland, J. (2020), 'Answers to six key questions about the oil price collapse', *Time*, 11 March. https://bit.ly/3qOaGC4 (accessed 16 March 2020).

World Bank (2003), *Trade, investment, and development in the Middle East and North Africa: Engaging with the world*. Washington, DC: World Bank.

World Bank (2019), 'Labor force participation rate, female (% of female population ages 15+) (modeled ILO estimate)', World Bank Data. https://bit.ly/30G75LU (accessed 26 March 2020).

World Bank (2020a), 'World development indicators: Total natural resources rents (% of GDP) (NY.GDP.TOTL.RT.ZS)', World Bank Data. https://bit.ly/3tiRT3F (accessed 3 March 2012).

World Bank (2020b), 'World development indicators: Total population (SP.POP.TOTL)', World Bank Data. https://bit.ly/3qMrbih (accessed 3 March 2012).

Yamada, M., and S. Hertog (2020), 'Introduction: Revisiting rentierism – with a short note by Giacomo Luciani', *British Journal of Middle Eastern Studies*, 47:1, 1–5.

Young, K. E. (2016), 'Gulf states are torn between economic sense and military ambition', The Conversation, 19 September. https://bit.ly/3qQo9JI (accessed 10 January 2019).

Young, K. E. (2021), 'Federal benefits: How federalism encourages economic diversification in the United Arab Emirates', in M. Beck and T. Richter (eds), *Oil and the political economy in the Middle East: Post-2014 adjustment policies of the Arab Gulf and beyond*. Manchester: Manchester University Press, pp. 145–63.

2

Upgrading towards neoclassical rentier governance: Bahrain's post-2014 oil price decline adjustment

Sumaya AlJazeeri

Introduction

What has been witnessed since 2014 goes beyond the steep decline in oil prices and marks a critical interval of unprecedented change in political-economic policymaking within the Gulf Cooperation Council (GCC). Although this decline in oil prices was preceded by at least five other episodes from the 1980s (Baffes *et al.*, 2015), it is this last episode that has initiated a discussion about the viability of rentierism and whether this interval of change could mark a substantial divergence away from it. At the core of this discussion is how the magnitude of economic and fiscal repercussions of low oil prices has ushered in a series of fiscal actions that go against the classical understanding of rentier state theory (RST). Moreover, this discussion prompted questions about the extent to which these changes are core fiscal reforms signalling the 'demise of rentier governance' or are merely a 'coordinated course of corrections' (Krane, 2018: 24) in rentier fiscal policy.

A significant amount of literature has long argued that Bahrain, along with other GCC countries, is a classical rentier state based on Beblawi and Luciani's (1987) prominent account. While agreeing broadly with this classification, this chapter emphasises that the intricate political-economic structure of Bahrain's rentier economy has historically been and continues to be governed by 'political fiscal management' rather than 'oil wealth management'. Hence, the driver of fiscal policy is not the oil price movement, although this directly impacts the government's oil revenues. Rather it is political considerations, ultimately serving to maintain the 'social contract' founded long ago based on wealth distribution in return for quiescence and stability through allegiance and co-optation. These political considerations continue to serve as the backbone of today's fiscal policies, whereby the reallocation of oil rents resulting from fiscal pressure has been associated with mitigating measures, taking into account the importance of maintaining social stability.

Existing research has shown that fiscal policy in Bahrain is not strictly driven by oil price volatility (El-Enbaby and Selim, 2018). This is reflected in the inconsistency in fiscal policy since the 1980s, where at times expenditure has been procyclical and in other periods countercyclical (Krane and Monaldi, 2017). More specifically, six major episodes of oil price declines can be observed, including the latest one occurring in 2014. The first episode was in 1985–86, associated with a significantly increased supply in oil by the Organization of the Petroleum Exporting Countries (OPEC), resulting in oil price decline. The second episode was between 1990 and 1991 following the oil price spike triggered by the First Gulf War in the Middle East. The third episode was in 1998 following the 1997 OPEC expansion of production, and the fourth episode was in 2001 following the 11 September terrorist attacks in the USA and intensified by the dotcom bubble. The fifth episode occurred following the 2008–9 financial crisis resulting from a severe contraction in global consumer demand, and the final episode was in the second half of 2014.

Strikingly, it is the most recent episode that has complicated the fiscal challenges of Bahrain during a politically critical period. Nonetheless, it also served as a catalyst for a fiscal upgrade that has been increasingly pressing, providing the government with the political cover and mandate for updating the existing social contract to better reconcile with the requirements of modernising an oil economy. This cover would have otherwise been politically unviable and socially rejected while fiscal balances were in surplus (Krane and Monaldi, 2017). Therefore, constrained by its fiscal challenges and the need to make key changes in its economic model away from its dependence on oil, the government embarked on engaging citizens in financing the widening deficits through the introduction of low-level value-added taxation, increasing fees for government services, and the reduction of especially energy and utility subsidies.

These adjustment policies, as argued in this chapter, revolved around the reallocation of rents by streamlining policies based on patronage to sustain rentier governance for the long term (Krane and Monaldi, 2017) with the aim of maintaining the existing political-economic structure. Hence the means of governance have been upgraded, while 'the state or the government, being the principal rentier in the economy, plays the crucial role of the prime mover of the economic activity' (Beblawi, 1987: 53) and fiscal management. In this context, government expenditure and projects have remained mainly state capitalist (Gray, 2011) and reliant on oil rents since 2014. More importantly, weak institutions and inefficient governance, which are characteristic of rentier states, have not been substantially reformed despite a rising rhetoric of the need for deep (fiscal) reform adopted by the government in Bahrain. Therefore, this chapter argues that the so-called reforms put in place since 2014 remain within the scope of rentier governance rather than substantially diverging from it. This thesis emphasises that what has been implemented thus far are adjustment policies in order to upgrade rentier governance in a neoclassical way, which falls short of true reform, as the latter would constitute the internalising of new fiscal norms stemming from an organic systemic process of reforming institutions and the form of governance.

This chapter proceeds with a description of the political-economic structure which best characterises Bahrain based on the existing literature. The second section conducts a historical comparative discussion of the two most similar episodes of oil price declines (1980s vs post-2014), mainly focusing on the consequential fiscal policy outcomes in light of these episodes. The third section then analyses the fiscal outcomes following 2014 and the nature of the measures undertaken. This section goes through the reasons why the adjustment policies implemented thus far do not substantially diverge away from the existing structures of rentier state governance. It also discusses the government's rhetoric of reform and whether this has translated into actual steps towards structural and institutional reforms that would limit the scope of the rentier state's control. The final section concludes and draws a broader picture of the current political economy of Bahrain post-2014.

Bahrain's political economy: A historical perspective

Bahrain was the first country among the GCC to discover oil in 1931. Today, however, it is the least endowed, with limited oil reserves compared to its oil-rich neighbours. Between 1935 (since oil production commenced in Bahrain) and 1965 'about 50 per cent of the country's oil resources had been depleted' (Askari, 2013: 36). In 1963 in particular, Saudi Arabia's Aramco discovered a new oil field – the offshore field of Abu Saʿfa, which Bahrain shares with Saudi Arabia. Both Bahrain and Saudi Arabia negotiated a production-sharing agreement in which Aramco would be given full operational control while Bahrain would receive 50 per cent of the revenues from the Saudi government without incurring any of the production costs (Askari, 2013). During the early years of oil discovery, Bahrain was under British rule. However, it gained independence in 1971 at the time of the 'oil boom'. The de facto state following independence was a continuation of the rule of the Al Khalifa family which existed prior to British imperial rule and is referred to as the Al Khalifa Sheikhdom. Under British rule, and more specifically since the discovery of oil in 1931, the citizens' resentment was largely directed at imperialist companies, and the imperial state, principally the United Kingdom (Askari, 2013). During this period, foreigners were key players and major determinants of the political economy of oil in the Gulf (Askari, 2013).

However, since the end of British rule in Bahrain and in the entire Arab Gulf, the people's resentment or demand for improved welfare and tangible benefits from their countries' oil wealth became redirected towards the existing rulers. It is at this point in time, which coincided with the oil boom of the 1970s, that high oil rents facilitated the easiest route in policymaking, which is to redistribute or reallocate the wealth through the provision of subsidised inflexible goods and services to the population and public employment for nationals, which was the way to secure citizens' approval and gain legitimacy (Askari, 2013).

Accordingly, placing the expenditure boom in the 1970s within this context, the impact it had on the oil states fits perfectly within the definition of classic RST. The oil boom of the 1970s generated a spike in public expenditure. However, it was not the abundance of oil rents which drove the spending boom, but rather the fact that, due to political-economic considerations 'in the aftermath of the price increase of 1973–74, popular expectations in the Persian Gulf also exploded' (Askari, 2013: 74). Therefore, public expenditure was seen within the context of reallocation – whereby all the 'rentier state' needs is an expenditure policy (Luciani, 1990: 76). This process of distributing oil rents within society took place as the government funded its oil-driven economic development through a massive public investment programme in infrastructure, utilities, and basic industries, in addition to funding a generous welfare system. This initially resulted in a period of rapid growth in non-oil activities. In light of this, Bahrain's public expenditure, being in line with its neighbouring GCC countries, began to increase at elevated levels. However, what distinguished Bahrain early on is the fact that it had very limited oil reserves, and by 1970 it was relying mostly on the Abu Sa'fa field, which it shares with Saudi Arabia; however, without operational control to it. In the first decade of operations at Abu Sa'fa, production was about 44 million barrels, with nearly half of Bahrain's oil revenues coming from Abu Sa'fa. Bahrain has been receiving 50 per cent of the revenues for most of the time, except during the period between 1996 and 2004, when it received all revenues from Abu Sa'fa's production (Askari, 2013). This reliance on the Abu Sa'fa field has subjected Bahrain to a different type of volatility in income growth that is impacted by changes to the terms of agreement, thereby making Bahrain dependent on Saudi Arabia as the bigger and more powerful party in the agreement.

At these elevated levels of expenditure, and with the reality of limited oil reserves, this prompted Bahrain to pursue an economic diversification strategy from the early 1970s, encompassing a wide range of industrial developments including ship repair at its ports and aluminium manufacturing in its flagship company Aluminium Bahrain (ALBA). In addition, Bahrain became a financial hub, taking over from Lebanon, which went through a devastating civil war between 1975 and the late 1980s. Additionally, Bahrain also became the most industrialised GCC country by the 1980s (Hvidt, 2013). The backbone of gross domestic product (GDP) expenditure continued to be by and large government expenditure, government investments in state-owned enterprises (ALBA, the Bahrain Petroleum Company also known as BAPCO), and banks (National Bank of Bahrain, Bank of Bahrain and Kuwait now known as BBK) to name a few. In what follows, this chapter will look further into historical data and government policies in reaction to oil price changes to better gauge how Bahrain as a 'rentier state' has structured its fiscal policies at different periods, mainly related to significant episodes of oil price decline.

Fiscal adjustments during oil boom and bust

Historically, Bahrain posted deficits in twenty-one out of twenty-nine years between 1990 and 2018. However, the movement in expenditure has been mostly volatile and negatively correlated with oil prices. El-Enbaby and Selim's empirical study on the correlation between expenditure and oil prices has demonstrated that 'real spending shocks are negatively correlated with oil price shocks' (2018: 4). Official data published, plotted in Figure 2.1 and Figure 2.2, supports the finding that the relation between oil prices and deficit is not cyclical at all times.

During periods of very low oil prices during the 1990s, Bahrain's deficits were lower than during recent years. Additionally, data following the 1994 social-political unrest show that public expenditure rose sharply despite oil prices being significantly low due to public wages increasing from 48 per cent of total spending in 1994 to 54 per cent in 1995 and 1996 (El-Enbaby and Selim, 2018: 11). In 1997 through 1998, Bahrain's budget deficit widened due to a significant spike in expenditure, which increased by 40 per cent despite oil prices remaining at low levels since the oil price revolution of the early 1970s. Furthermore, later in 1999, following the passing of Bahrain's then ruler Shaykh Isa bin Salman Al Khalifa and the accession of his son King Hamad to the throne, Bahrain's fiscal expenditure increased again due to the introduction of new, generous, public gestures of patronage. These gestures included the provision of free electricity for around 10,000 low-income families, cutting higher education fees at Bahrain's national university by 80 per cent, reducing import tariffs on vehicles, and offering cost of living grants for single-parent families and orphans (Wright, 2006).

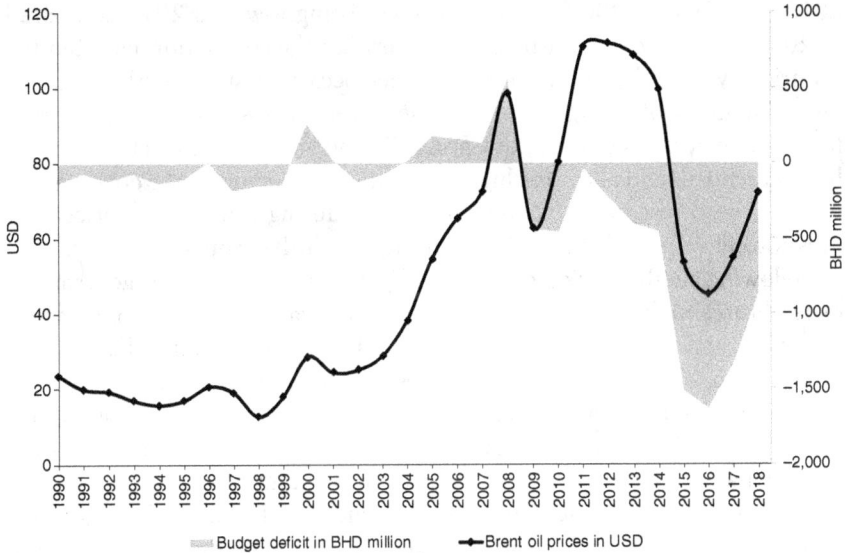

Figure 2.1 Bahrain's deficits since the 1990s (in BHD billion). Source: Ministry of Finance (2008–20) and IMF (1991–2019). Author's calculation.

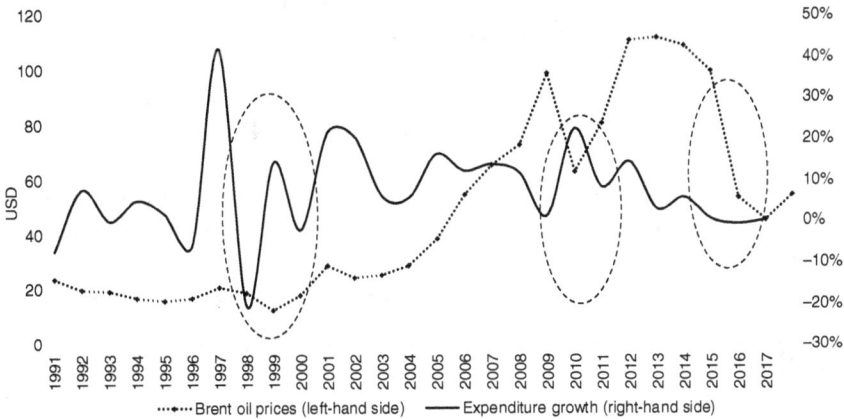

Figure 2.2 Bahrain's budget expenditure growth vs oil price movements. Source: Ministry of Finance (2008–20) and IMF (1991–2019). Author's calculation.

Another significant spike in expenditure was witnessed after the financial crisis of 2008, in which the state played a role in stimulating the economy at a time of distress. Higher spending on projects saw the country's capital expenditure (CapEx) – also referred to as investment expenditure – on projects, infrastructure, and long-term assets increasing by 66 per cent in

2010 compared to 2009 despite oil prices being lower in 2010 compared to 2008. However, it must be pointed out that the correlation between the oil price cycle and CapEx on many other occasions was mostly negative, where at times of rising oil prices, the government's current expenditure tended to increase as opposed to CapEx (Fasano and Wang, 2001). Intuitively, high oil rents should allow for higher investment spending by the government, which otherwise could be compromised on during times of oil price suppression. However, this has not been the case in Bahrain since 1990.

Following another period of social and political unrest in 2011, government expenditure, mainly current expenditure on wages in the public sector, subsidies, and public goods and services, has been countercyclical. Large increases in wages, transfers, and subsidies have been observed at times where oil price movements were volatile and have taken a steep downturn since mid-2014. Considering that the political and social situation in the country has remained unstable after 2011 throughout 2014, the government of Bahrain did not cut back on current expenditure. Rather, Bahrain's deficit spending was substantially fortified after 2014 despite the steep decline in oil prices. The country's elevated levels of expenditure translated into unprecedented deficits, affirming again a countercyclical relationship between the oil price movement and expenditure growth. Hence it can be argued that government expenditure in Bahrain is mostly stimulated by political causes rather than oil price movements.

Policy adjustment since 2014

The plunge in oil prices of 2014 complicated Bahrain's political-economic situation with squeezed fiscal balances[1] and an unstable political-economic environment following the Arab uprisings of 2011. Bahrain in particular was clearly the most affected GCC country during this period due to the political and social unrest which crippled the economy at a time when fiscal imbalances were unfolding, and debt levels were growing exponentially.

Bahrain's debt has grown significantly, increasing from 25 per cent of GDP in 2010 to 85 per cent of GDP in 2017, reaching a total of Bahraini dinar (BHD) 11.5 billion in 2018 (USD 30.5 billion). This amount excludes external debt. Debt-servicing costs have therefore been moving in an upward direction, reaching 22 per cent of total revenue and accounting for 5 per cent of total debt as of 2018.

Compared to previous periods of oil busts, this time fiscal adjustment was different. The oil bust which extended from the mid-1980s until 2003 was paralleled with social and political unrest met by the government, leaving domestic subsidies and benefits untouched while extending various public gestures of benevolence and patronage (see Table 2.4, which provides a summary of most government decisions, policies, and adjustments that fall within the patronage concept and welfare financing). In contrast to that, the oil bust post-2014 was met, although not immediately, with a reduction in capital expenditure (see Table 2.1), a reduction in non-wage recurrent spending (including energy subsidies) (see Table 2.2), and the introduction of new forms of taxation (value-added tax [VAT], excise taxes) in addition to an increase in government fees and penalties.

The substantial change in course of action with unprecedented measures that have been contrary to the fiscal responses during historical oil busts has rendered the assumptions of RST somewhat inconsistent with recent events. In the light of this finding, this chapter emphasises the attempts to reconcile the change in policies with the unchanged form of rentier governance in Bahrain. After having established that, historically, fiscal policymaking has for the most part been in line with the social contract stipulations of rentier governance, this section discusses the continuation of a fiscal policy following 2014's oil price decline, which, as argued in this chapter, did not run against the basic rationale of rentier governance.

Several key elements seem to be worth discussing in this context. First, the adjustments implemented have not changed the core political-economic and fiscal structure of Bahrain, which remains largely dependent on oil rents in its fiscal balance and in its contribution to the country's GDP. Oil revenue continues to be the main source of government revenue, and a key factor in sustaining elevated levels of welfare expenditure despite the rationalisation of oil rents. In this context, what is meant by the rationalisation of rents is the process of efficient, or conscious, mobilisation of oil rents in a way that promotes economic growth, diversification, and movement away from oil-focused production. However, despite these efforts, this so-called movement away from oil continues to be largely dependent on oil rents and is therefore more accurately described as oil-driven economic modernisation. This is evident in Bahrain's fiscal performance over the last decade, whereby Bahrain's government expenditure to GDP has actually been slightly increasing. Although Bahrain's oil GDP (which is the monetary value of the gross domestic produc-tion of oil and gas – or hydrocarbon-related activities and sectors) accounts for an average of 20 per cent of GDP (cf. Table 2.3), the oil industry remains a key sector, which has spillover effects over other non-oil-related sectors. Hence Bahrain's non-oil GDP (which represents the total value of domestic production from non-oil and gas sectors), including sectors like manufacturing, construction, transportation, retail, financial, and real estate, is largely and directly impacted by government expenditure, or indirectly by spillover effects from demand stimulated by projects initiated and funded by the government.

Table 2.1 Bahrain's historic expenditure breakdown (in BHD million).

	2008	2009	2010	2011	2012	2013	2014	2015	2016	2017	2018	2019	2020
Capital expenditure (in BHD million)	508	462	767	441	737	477	448	444	411	355	350	200	200
Total expenditure (in BHD million)	1,552	1,692	1,868	2,412	2,524	2,877	3,096	3,116	3,121	3,182	3,253	3,312	3,347

Source: Ministry of Finance (2008–20). Note: Figures for 2008–18 are based on actuals from final accounts, whereas figures for 2019 and 2020 are based on budgeted figures.

Table 2.2 Subsidies as a percentage of current expenditure in Bahrain.

	2013	2014	2015	2016	2017	2018	2019	2020
Subsidies as % of current expenditure	22	21	24	21	19	17	18	18
Wages and other current expenditure (%)	78	79	76	79	81	83	82	82

Source: Ministry of Finance (2008–20). Note: Figures for 2008–18 are based on actuals from final accounts, whereas figures for 2019 and 2020 are based on budgeted figures.

Table 2.3 Bahrain's private vs government expenditure as percentage of GDP.

	2009	2010	2011	2012	2013	2014	2015	2016	2017
Private expenditure as % of GDP	41	41	39	38	41	40	45	45	42
Government expenditure as % of GDP	14	13	14	15	16	16	18	17	17

Source: Ministry of Finance (2008–20). Note: Figures for 2008–18 are based on actuals from final accounts, whereas figures for 2019 and 2020 are based on budgeted figures.

Second, a closer look at some of the policies implemented and royal decrees issued (for an overview, see Table 2.4) shows that overall consistency in policies has been lacking. Rather, many policies seem to have been implemented as ad hoc measures, taking precedence over long-term reform measures that would be truly reformist in nature. The latter would signal a clear shift in fiscal policy towards lowering the population's dependence on the distribution of oil rents vis-à-vis subsidies and various allowances and freely provided public goods and services. Hence, when, on the one hand, the government of Bahrain implements policies (e.g. lowering subsidies) that seem to stem from fiscal pressure, while, on the other, following them with measures that offset or mitigate the impact of these policies on the population in general, and nationals in particular, this signals inconsistency and a degree of arbitrariness in the fiscal policymaking. It does not show a process of transformation or core institutional changes. The key drivers of fiscal policy have been to reach social stability and mitigate impacts of unemployment, low wages, or higher costs of living, all of which are crucial for the stability of the regime. Whether it be with regard to lowering subsidies

Table 2.4 Policy adjustments in Bahrain since the early 2000s.

Regulatory timeline	Policies aimed at increasing non-oil revenue and lowering fiscal burdens	Policy reversal or policies with countering effects
Pre-2008	Unemployment insurance law; labour law restructuring Establishing Tamkeen: labour fund to support Bahraini employees and private-sector institutions	Lowering utilities bills (mainly electricity and water bills) for 10,000 low-income families in 1999, cutting higher education fees at University of Bahrain, reducing import tariffs, and offering cost-of-living grants
Post-2008	Gulf Air (national carrier – to be privatised)	Royal public gesture: BHD 1,000 (USD 2,650) ordered to be distributed to 13,265 Bahraini families registered within the Ministry of Social Development 1,000 houses to be built for impoverished Bahraini families at a cost of BHD 4 million Cost-of-living allowance announced to combat inflationary pressure on low-income families
2015	Hike in traffic violation penalties and fees	
2016	First fuel price hike; Electricity Water Authority tariff monthly charges – first increase; increase in expats' healthcare fees Highest-ranking public jobs (cabinet ministers) saw a 20 per cent salary cut, bonuses, and other allowances given to public servants were removed	
2017	Mumtalakat (Bahrain's investment arm) announced it would be distributing BHD 20 million (USD 53 million) of its profits to the national budget for two years	Allowances and overtime pay to public servants removed in 2016 – reversed in April 2017 Debt ceiling raised to BHD 13 million

Table 2.4 Policy adjustments in Bahrain since the early 2000s. (Continued)

Regulatory timeline	Policies aimed at increasing non-oil revenue and lowering fiscal burdens	Policy reversal or policies with countering effects
2018	Excise tax implemented; second fuel hike in January Fiscal consolidation programme announced to lower current expenditure	
2019	5 per cent VAT implemented	Bahrain scraps subsidy reform plan to merge meat subsidies and living allowance into one package, which allegedly was to reduce overall subsidy spending
	Early retirement law passed	Gulf Air privatisation scrapped off the plan
	Reduction in water and electricity subsidies to continue	NEP (National Employment Programme) – increased compensation fees – which add to the government's current expenditure bill
	NEP programme announced: increase in parallel Bahrainisation programme fees and amendment of the unemployment insurance draft law	Financing the early retirement support programme: BHD 230 million (approximately USD 610 million) approved to be moved to the public account from the surplus of the insurance against unemployment programme Legislative efforts ongoing to increase minimum social benefit payments paid to Bahraini citizens

Source: Author's compilation.

in the late 1990s or conducting public gestures of benevolence intended mainly to mitigate social and economic distress by extending cost of living allowances and benefits to low-income families, in many cases, fiscal policy was implemented following critical periods in Bahrain's political history, more specifically during the 1990s, post-2011, and again in 2014. Bahrain was challenged by a myriad of deep social and political conflicts from the 1990s onwards (Jones, 2017), beginning with political protests and instability in 1994, when opposing factions demanded the restoration of the National Assembly, and moving to the early 2000s, especially in 2011. Overall, this has resulted in a fiscal policy which largely depends on the reallocation of rents. As a lasting effect of this, the government has tended to overspend, which in the long term has eroded Bahrain's fiscal position.

The character of these problems is displayed in the regulatory timeline compiled in Table 2.4. This overview compares policies intent on improving the fiscal balances with those policies contradicting fiscal consolidation. For each year, the table shows two columns, one highlighting the policies aimed mainly at lowering oil dependence and increasing economic diversity, while the second column portrays contradictory measures pointing to fiscal policies which reversed some of the policies aiming at mitigating oil dependence. The compilation shows the lack of a clear agenda for reforming the country's socio-economic policymaking. For instance, the government implemented key policies, such as energy price reforms, but they were followed by policies characterised by counter effects, such as the introduction of living allowances or social benefits for low-income families. Furthermore, while the government allocated BHD 961 million (USD 2.55 billion) in 2014 for oil and gas subsidies, from 2017 onwards, it reduced that budget item to zero. In 2014, the cost-of-living allowance paid to low-income families was allocated a budget of BHD 105 million (USD 278 million), which increased to BHD 128 million (USD 340 million) in the budget allocation for 2019. This shows that policies can best be described as reactive and ad hoc resulting from a combination of social-political and economic-fiscal pressure. They are mainly aimed at alleviating those pressures, while maintaining the status quo of the Bahraini rentier state.

Third, the steps taken recently (e.g. subsidy reform, higher and new taxation) are best placed in a context of the reallocation of benefits or customary privileges given to all citizens. These customary privileges are handed out in the form of welfare financing through subsidised goods and services, the extension of various citizen benefits and allowances, and zero taxation. Hence, when the government embarks on lowering energy subsidies, thereby increasing the regulated prices of fuel (which is also referred to as energy price reforms), it takes away those long-held, customary privileges from both the citizens and residents. However, when it follows this with measures to mitigate the impact of higher energy prices on its citizens, for instance by introducing allowances such as the 'cost-of-living allowance', it reinstates the privileges taken from the whole populace to a certain faction of its citizens, namely, those with lower incomes. Therefore, the reallocation of benefits to low-income citizens and other areas of need within the government's fiscal finances epitomises the reallocative role of a rentier government. Based on this, the patronage system remains unchanged in principle (Krane, 2019), whereby the existing rentier ruling body continues to sustain control over the type, form, and direction of welfare financing while maintaining certain customary privileges provided to the most sensitive segments of its citizens (Tsai, 2018). In other words, policy adjustments implemented cannot take place without prudence and mitigating policies that preserve the patronage structure, which is sensitive to the social welfare of most impacted citizens.

Fourth, the nature of fiscal reforms – though unprecedented – has been misconstrued as inflexible (Krane, 2019). Based on this assertion, reforming subsidies cannot go without upsetting stability. However, as we have seen, the oil price decline in 2014 came at a politically challenging time for Bahrain. Hence, embarking on these fiscal changes at a critical time would have seemed irrational and in contradiction to some of the claims made in the literature on the rentier state. However, considering that these adjustments did take place, one should argue that they can best be considered as a form of taking back certain benefits in a time of need and reallocating them. This solves the problematics discussed in the rentier literature, which argues that social contracts are rigid. The social acquiescence of these policy adjustments has been largely influenced by social caution amid stronger state control and coercion, in which citizens have no choice but to accept and absorb these policy adjustments without making demands (Krane, 2019).

Fifth, these fiscal reforms have come at a time where younger ruling elites such as the Crown prince of Bahrain, the amir of Qatar, and the Crown prince of Saudi Arabia have been constantly advised by international organisations like the International Monetary Fund (IMF) to reduce oil consumption, ration government spending, and speed fiscal consolidation. At a time when oil rents were declining significantly – albeit at different levels across the GCC – while overall domestic consumption of oil and gas remains high and the total demand for social welfare is increasing, this constellation is particularly worrying for these younger rulers, putting the fiscal underpinnings of the rentier state at risk. Therefore, embarking on fiscal adjustment policies by resorting to only streamlining existing patronage structures is arguably necessary to sustain rent-based governance for the future. In other words, this can be considered as a form of updating or correcting the trajectory that rent-based governance has been moving along for the past three decades rather than jettisoning it (Krane and Monaldi, 2017; Krane, 2018).

Overall, the current political-economic structure in Bahrain still strongly exhibits characteristics of the classical rentier state whereby the political and social constraints make it difficult to diverge away from the rentier mentality of wealth distribution and patronage. Hence, rentierism remains the backbone of governance, albeit in a neoclassical way that reconciles the existing demands of the political-social construct with the requirements of a modern rentier economy in an increasingly challenging fiscal time amid the ongoing suppression of oil prices and the global movement away from oil.

Is there change beyond rhetoric?

In 2008, Bahrain launched its 'Bahrain Vision 2030', which lays out the government's plan for the next decades to achieve economic transformation and diversification. This vision was succeeded by a national development strategy (2015–18) that proposed actions aimed at improving the business and investment environment in Bahrain. Among various aspects of economic diversification and regulatory changes, Bahrain's Vision 2030 postulated that 'the sustainability of government finances is strengthened by reducing dependence on oil revenues to fund current expenditures' (Government of Bahrain, 2008: 17).

Data from Bahrain's final accounts show that oil revenue as a percentage of total revenue has indeed declined. While in 2008 oil revenue accounted for 85 per cent of total revenue, in 2018 it accounted for 82 per cent. However, recurrent expenditure, which refers to all government expenses related to wage bills, payments made on goods and services, subsidies, and debt-service costs (Ministry of Finance 2008–20) but excludes capital costs, as a percentage of oil revenue has been on the rise. Total recurrent expenditure[2] as a percentage of oil revenue in 2008 was 68 per cent, while it grew exponentially to 146 per cent of total oil revenues in 2018, and 91 per cent of total expenditure during the same year. The result has been widening budget deficits. To cover these deficits, Bahrain's debt has grown exponentially since 2008 at a compounded annual growth rate of 32 per cent, reaching BHD 11.5 billion (USD 30.5 billion) in 2018.

Although theoretically resorting to debt issuance has created a new source of government funding other than oil rents, in practical terms debt has not replaced oil revenues. Indebtedness offers something that can best be described as local fiscal-budget anaesthesia, which allows the government to finance its inelastic recurrent expenditure. In fact, resorting to debt issuance has more than ever increased Bahrain's oil dependence because oil income is the main source of insurance against debt default in addition to regional support from the country's oil-rich neighbours. The ballooning debt of Bahrain's government has therefore reaffirmed the significance of oil rents while complicating fiscal challenges and increasing the urgency of fiscal regulatory reform.

In April 2018, Bahrain announced that off its west coast it had discovered its largest oil and gas field since 1932, estimated to contain at least 80 billion barrels of tight oil.[3] The announcement was arguably politically timed to give the necessary reassurance that Bahrain's oil wealth is far from depleted and that it could therefore afford to sustain its welfare-rentier state. However, the time lag between the discovery and the actual extraction of oil, assuming the process to be feasible, could take years. Moreover, the timing of the oil discovery was arguably more of a political-financial assurance to international creditors and markets, given that it had to cancel a planned sale of international conventional bonds as investors demanded higher yields in March 2018, just before the announcement (Barbuscia, 2018a). Despite this, international creditors and credit-rating agencies continued to be concerned about Bahrain's ability to manage its sovereign credit risk and to pay off its long-term debt. Increasing speculation and pressure resulted in a huge sell-off of conventional papers in Bahrain's bond market in June 2018 as European and US investors began dumping the debt (Barbuscia, 2018b). Furthermore, due to creditor scepticism and unfavourable demand, which reflected low confidence in Bahrain's financial strength despite its oil discovery announcement, Bahrain had to redeem USD 750 million in Islamic bonds in November 2018.

Later in October of that year, Bahrain's oil-rich neighbours (Saudi Arabia, United Arab Emirates, and Kuwait) announced a USD 10 billion aid package. This was allegedly to help the Bahraini government meet some of its maturing debt requirements and to support the Central Bank after its net foreign reserves hit a low in June 2018. Overall the help package provided additional reassurance to international creditors and investors that Bahrain would not immediately default on its debt obligations (Barbuscia, 2018a).

In light of this, and as a condition for the financial aid given, Bahrain had to reassure its GCC peers that it will embark on serious structural fiscal and economic reforms to alleviate its fiscal pressure. This resulted in the launch of Bahrain's Fiscal Balance Programme (for more details, see Government of Bahrain (2018)), a programme which outlined Bahrain's way forward to have long-term fiscal sustainability and a balanced budget by 2022. Nonetheless, much of the rhetoric and language used in the document remains at headline and broad levels, for example employing such general formulations as announcing the intention to reduce government's operational expenditure or to simplify government processes with the aim of increasing non-oil revenues (Government of Bahrain, 2018). Compared to Bahrain's Vision 2030 document, it seems that the targets set in 2008 are being reiterated in 2018, as in the 2008 vision document, an example headline states: 'transform public sector human resource management, review and adjust government organisation and processes to streamline them' (Government of Bahrain, 2018: 18).

The key takeaway from this is that ten years later the core governing system has not changed, and the way forward remains ambiguous. While the rhetoric of reform is elevated, it remains within broad headlines, without transparently discussing the road map to achieve a balanced budget. It also comes at a time where government actions, such as the announcement of the newfound oil discovery, potentially contradicts this rhetoric.

Beyond adjustments: Has the institutional structure of governance changed?

Inefficient structures of governance are key characteristics of a rentier state, as outlined in the literature. This inefficiency is attributed to the dependence on oil, which tends to skew 'the institutional development of the state because oil rents weaken agencies of restraint' (Karl, 2004: 666). In a rentier state like Bahrain, fiscal policy became challenging, as weak institutions have made it easy to 'use current spending as a channel for the redistribution of oil rents' (El-Enbaby and Selim, 2018: 3) without accountability. These institutions – ranging from budgetary institutions, ministries, and government investment arms to weak legislative bodies like the parliament and the *majlis al-shūrā*[4] to audit institutions such as the National Audit Office – have remained unchanged so far, with the same structures in place before and after the 2014 oil price plunge.

For example, while emphasis was set on strengthening the private-sector institutions in Bahrain's Fiscal Balance programme, it did not set a plan for reforming the existing governance structures. Rather it vaguely addressed cases of financial irregularity, financial leakages, or corruption highlighted in the National Audit Office Report – a report which is issued on a yearly basis by Bahrain's National Audit Office. However, the report of the Council of Representatives (2018) does not outline any measures of accountability but merely highlights violations and inefficiencies in the general performance of government institutions. In addition, in January 2018, Bahrain's *majlis al-shūrā* rejected a legislative motion passed by the elected lower house which would have obliged the National Audit Office to report those actors implicated by the report as being involved in any corruption and mismanagement of public funds to the public prosecutor's office. Had this law been passed, it would have been a major leap forward in reforming the existing governance institutions by achieving core structural reform that would have led to higher transparency, less corruption in the system, and thereby less exhaustion of funds.

It is from the unchanged, inefficient, and weak institutional structures that the adjustment policies implemented, ranging from lowering subsidies to increasing taxation and fees, have been passed. Although across all existing institutions the rhetoric has changed, now emphasising fiscal consolidation or highlighting the eradication of corruption, existing institutions have failed to overhaul the current political-economic form of governance that is based on the legacy of inefficient rentierism. On the contrary, what has taken place so far is that fiscal corrections have been introduced only to revive the fiscal balances rather than to focus on building and centrally setting the general directions for the desired changes in the overall institutional structures.

Conclusion

The discussion brought forward in this chapter suggests that recent reforms in Bahrain do not mark a divergence from the legacies of rentier governance. Based on the foundations laid by Beblawi and Luciani (1987) while incorporating the neoclassical understandings offered by Krane (2018, 2019) and Yamada (2019), four key elements are highlighted as being relevant in the context of upgrading the contemporary Bahraini rentier state.

First, the reforms implemented since 2014 have not changed the core political-economic structure of Bahrain's rentier state. Oil income continues to be the main source of government revenue, despite the rationalisation of its spending, whereas the government continues to be the prime mover and allocative force with regard to most economic activities.

Second, a closer look at royal decrees issued about key decisions and measures implemented pre-2014 compared to post-2014 shows that consistency in policies has been lacking. Fiscal policy adjustments in Bahrain are generally ad hoc and often reversed by mitigating decisions. The steps taken more recently (e.g. subsidy reductions, increasing tariffs, VAT introduction, along with other new forms of taxation) should be rhetorically placed in the context of taking away 'customary privileges' given to citizens when rents were high in the past. Although unprecedented, these steps have been misconstrued and understood as a clear divergence away from classical rentierism, whereby lowering welfare financing is equivalent to taking away the basic rights of citizens to free or subsidised inflexible public goods as argued in the RST. This is so because the classic rentier state approach treated the provision of inflexible public goods like subsidies as part of a social contract, which is rendered difficult to change since the right of citizens to receive these subsidies and welfare financing is a function of their allegiance.

However, beyond the theoretical assumption, when placing these public goods in a more flexible context whereby subsidies are customary privileges and not basic rights, then adjusting them or removing them can be established with or without citizen approval, co-optation, and acceptance. Accordingly, the government can single-handedly update the social contract, while the rentier form of governance remains intact. The argument laid out here is not so much about a change of discourse in classifying these public goods and services as inflexible basic rights. It is more about establishing a better understanding of the context in which we believe Bahrain's rentier governance is best placed and understood. Hence, in so doing, the rhetoric of adjusting the fiscal measures does not challenge RST, which has best defined the governing systems of countries like Bahrain.

Third, recent fiscal reforms have come at a time when younger ruling elites have been pressured by international monetary organisations to reduce domestic oil consumption, amid rising global environmental concerns, the introduction of electrical motorisation, a slowing global economy, and depleting levels of oil in Bahrain, while overall domestic consumption of oil and gas has been on the rise. This is particularly worrying for the ruling elite, as the economic underpinnings of the rentier state are at risk. Hence, this makes clear their greater emphasis on streamlining the long-existing customary privileges given to the population to sustain rentier governance for the future. Therefore, recent reforms in Bahrain also appear to be a form of updating rent-based governance rather than jettisoning it (Krane, 2018).

Finally, the rhetoric of fiscal reform used by the government that gained salience in the 2010s remains within broad and general headlines when talking about change but lacks a transparent and precise agenda. Further, the backbone to fiscal consolidation and real fiscal reform is having strong institutions. Hence, the potential reform of the existing institutional setting would entail a significant divergence from the present political-economic structure of the current rentier state in Bahrain. However, the evidence for and the discussion of the role of the National Audit Office and the legislative bodies show that these institutions remain unchanged and weak, contradicting the discourse of reform on achieving fiscal consolidation and streamlining government expenditure. Without the government embarking on actual structural reforms, its rhetoric remains in the realm of aspirations rather than signalling actual policy change.

Therefore, this chapter concludes that Bahrain's political economy continues to be in status quo albeit with updates. The adjustment policies implemented so far do not rise to the level of substantial systemic reform and are merely policy corrections or updates along neoclassical lines. These corrections are a 'facelift' rather than an actual complete transformation of political-economic governance. Today, Bahrain's key source of income in addition to oil rents is debt. While subsidies are being reduced, debt is rising. This reflects Luciani's (1990) point about reallocation, but with a modern notion. In Bahrain, rents have now been reallocated from the citizens to creditors and back to citizens. With Bahrain's debt-servicing costs rising, this has come at the expense of citizens, who have seen their welfare privileges narrowing. The amount the government saves from its fiscal adjustments has been largely directed towards funding its debt-servicing costs. Hence oil rents have now been partially redirected towards creditors financing the ongoing deficits due to the inelasticity of the government's welfare financing. This in itself is what is referred to as the ultimate modernising and updating of the 'oil wealth distribution lines'.

Notes

1 Bahrain's current expenditure grew at double-digit levels on average from 2009 to 2014; however, it slowed significantly post-2014 (Ministry of Finance, 2008–20).
2 Recurrent expenditure of Bahrain includes wages (manpower), expenditure on services, expenditure on consumables, expenditure on assets and maintenance, budget transfers, grants, and repayment of loans and interest (Ministry of Finance, 2008–20).
3 Tight oil is a form of light crude oil held in shale deep below the earth's surface, which is extracted with hydraulic fracturing, or fracking, using deep horizontal wells. For more on Bahrain's shale oil discovery and reserves, see Barrington and Barbuscia (2019).
4 Bahrain's *majlis al-shūrā* is the 'Consultative Council' which represents the upper house in Bahrain's National Assembly. The council comprises 40 members, appointed by the king. Bahrain's lower house, the Council of Representatives, also known as the *nuwwāb*, comprises 40 elected members. For legislation to pass, it has to be passed by both councils.

References

Askari, H. (2013), *Collaborative colonialism: The political economy of oil in the Persian Gulf*. New York: Palgrave Macmillan.

Baffes, J., M. Kose, F. Ohnsorge, and M. Stocker (2015), 'The great plunge in oil prices: Causes, consequences, and policy responses', World Bank, Policy Research Note, 1. https://bit.ly/3eFFLpw (accessed 27 July 2019).

Barbuscia, D. (2018a), 'Bahrain does not plan new dollar bond issue this year', Reuters, 9 October. https://reut.rs/2ZGYY1f (accessed 27 July 2019).

Barbuscia, D. (2018b), 'Bahrain bond-sukuk spread balloons as investors disagree on credit outlook', Reuters, 25 June. https://reut.rs/3kcFACP (accessed 27 July 2019).

Barrington, L., and D. Barbuscia (2019), 'Bahrain talking to U.S. oil companies about tight oil deal', Reuters, 26 February. https://reut.rs/3pJ2BhM (accessed 27 July 2019).

Beblawi, H. (1987), 'The rentier state in the Arab world', in H. Beblawi and G. Luciani (eds), *The rentier state*. London: Croom Helm, pp. 49–62.

Beblawi, H., and G. Luciani (eds) (1987), *The rentier state*. London: Croom Helm.

Council of Representatives (2018), 'al-Taqrir al-Sanawi 2017/2018 Diwan al-Raqaba al-Maliyya wa-l-Idariyya' [Council of Financial and Administrative Audit 2017/18], Kingdom of Bahrain Council of Representatives Speaker's Office, 26 December. https://bit.ly/3pJ3iYq (accessed 16 February 2020).

El-Enbaby, H., and H. Selim (2018), 'Fiscal outcomes in Bahrain: Oil price volatility, fiscal institutions or politics?', Economic Research Forum, Working Papers, 1234, October. https://bit.ly/3vr4kMM (accessed 27 July 2019).

Fasano, U., and Q. Wang (2001), 'Fiscal expenditure policy and non-oil economic growth: Evidence from GCC countries', International Monetary Fund, Working Paper, 195. www.imf.org/external/pubs/ft/wp/2001/wp01195.pdf (accessed 17 February 2020).

Government of Bahrain (2008), 'From regional pioneer to global contender: The economic vision 2030 for Bahrain', Bahrain's e-Government Portal. https://bit.ly/2MdllYV (accessed 27 July 2019).

Government of Bahrain (2018), 'Fiscal balance program', Ministry of Finance and National Economy. www.mofne.gov.bh/fbp_en.pdf (accessed 28 February 2020).

Gray, M. (2011), 'A theory of "late rentierism" in the Arab states of the Gulf', Georgetown University Center for International and Regional Studies, Occasional Paper, 7. https://bit.ly/3qRlK1s (accessed 17 February 2020).

Hvidt, M. (2013), 'Economic diversification in GCC countries: Past record and future trends', London School of Economics and Political Science, Kuwait Programme on Development, Governance and Globalisation in the Gulf States, Research Papers, 27. http://eprints.lse.ac.uk/55252/ (accessed 17 February 2020).

IMF (1991–2019), 'World economic outlook database'. https://bit.ly/3toQRDp (accessed 11 August 2019).

Jones, M. O. (2017), 'History of Bahrain', in C. Matthews (ed.), *The Middle East and North Africa 2017*. London: Routledge. https://www.researchgate.net/publication/315774944_History_of_Bahrain (accessed 11 February 2020).

Karl, T. L. (2004), 'Oil-led development – social, political and economic consequences', in C. Cleveland (ed.), *Encyclopedia of energy*. San Diego: Elsevier, pp. 661–72.

Krane, J. (2018), 'Political enablers of energy subsidy reform in Middle Eastern oil exporters', *Nature Energy*, 3:7, 547–52.

Krane, J. (2019), 'Subsidy reform and tax increases in the rentier Middle East: The politics of rentier states in the Gulf', *POMEPS Studies*, 33, 18–24. https://bit.ly/38FzRkf (accessed 19 July 2019).

Krane, J., and F. Monaldi (2017), 'Oil prices, political instability, and energy subsidy reform in MENA oil exporters', James A. Baker III Institute for Public Policy of Rice University. https://bit.ly/3nDatlj (accessed 19 July 2019).

Luciani, G. (1990), 'Allocation vs. production states: A theoretical framework', in G. Luciani (ed.), *The Arab state*. London: Routledge, pp. 65–84.

Ministry of Finance (2008–20), 'Final accounts of Bahrain', Ministry of Finance and National Economy. www.mofne.gov.bh/FinancialFramework.aspx (accessed 17 February 2021).

Tsai, I.-T. (2018), 'Political economy of energy policy reforms in the Gulf Cooperation Council: Implications of paradigm change in the rentier social contract', *Energy Research & Social Science*, 41, 89–96.

Wright, S. M. (2006), 'Generational change and elite driven reform in the Kingdom of Bahrain', University of Durham, Institute for Middle Eastern and Islamic Studies, Working Paper, 7. http://dro.dur.ac.uk/456/ (accessed 17 February 2020).

Yamada, M. (2019), 'Exploring why institutional upgrading is not so easy in rentier states: The politics of rentier states in the Gulf', *POMEPS Studies*, 33, 13–17. https://bit.ly/2OAEEfQ (accessed 19 July 2019).

3

Stalled reform: The resilience of rentierism in Kuwait

Gertjan Hoetjes

Introduction[1]

The distribution of oil rents still plays a significant role in the Kuwaiti economy. According to figures of the World Bank released in 2019, oil rents contributed to 36.6 per cent of Kuwait's total gross domestic product (GDP) in 2017, which is higher than that of any of the other Gulf Cooperation Council (GCC) states (World Bank, 2019). The decline in global crude oil prices that commenced in 2014 has stimulated the Kuwaiti government to implement new initiatives to reduce state spending, particularly revolving around the removal of subsidies on fuel and utilities in order to tackle growing deficits and ensure the sustainability of fiscal expenditure in the long term. At the same time, the decline in oil prices has spurred plans for economic diversification, encapsulated in the 'NewKuwait' vision. However, these initiatives which aim to reduce fiscal expenditure and the country's dependence on oil have faced public resistance. This resistance has been particularly articulated in the National Assembly, a popularly elected institution with significant legislative power that exercises checks on the power of the Kuwaiti government.

This chapter analyses the resilience of rentierism in Kuwait, conceptualised as the dependence of the domestic economy on external rents, and the central role of the state in the allocation of wealth to the population. Recognising the importance of pre-oil social formations in Kuwait in the allocation of rents and the structure of the political economy, the chapter will first examine the impact of rents on state–society relations in Kuwait. This will be followed by an overview of the process of modern state formation in the country. After that, this chapter scrutinises previous episodes of fiscal adjustment in the mid-1980s and the period after the Iraqi invasion of Kuwait, followed by an analysis of the limited fiscal adjustment and efforts towards economic diversification that were initiated before and after the drop in oil prices in 2014.

Pre-oil social formations and rent distribution in Kuwait

Following the start of the first crude oil export in 1946, external rents obtained from the sale of the country's oil resources have been central to Kuwait's economy. The state that has evolved since the late 1940s has played a pivotal role in the distribution of these rents, with different social groups making claims to them to gain access to the country's oil wealth. Specific historical contingencies such as pre-oil social formations and institutional configurations have determined access to the oil rents and affected the response of the Kuwaiti government to the structural decline in the price of oil since 2014.

Pre-oil social formations still have a significant impact on Kuwait's political economy and access to the country's oil wealth. The private sector, which has been projected by the amir to 'lead the economy' (S. A. J. Al-Sabah, 2019) in the country's plans for economic diversification, has been historically dominated by the merchant class. This class emerged when the city-state was established in the early eighteenth century by a group of families from the Sunni Najd-based 'Anaiza confederation. This group of families, which included the Al Sabah family, took the name Bani Utub when they arrived in Kuwait. The Bani Utub have formed the backbone of an oligarchic merchant class that emerged throughout the centuries, which also included several Shi'i families. The merchants controlled the means of production in Kuwait, managed trade and imports, as well as the pearl ships, and dominated the financial sector. They offered employment to a lower class engaged in fishing, pearl diving, and shipbuilding (Al-Nakib, 2016: 22, 75), while also providing social services to the inhabitants of Kuwait. The ruling Al Sabah family depended on the voluntary donations provided by the merchants to sustain their reign. In return, they shied away from introducing taxes (Herb, 1999: 69).

The influence of the merchants in the decision-making process was undermined by Shaykh Mubarak (r. 1896–1915), who signed an agreement with Great Britain in 1899 that gradually enhanced the autonomy of the ruling family vis-à-vis the merchants. This was followed by the oil concession Shaykh Ahmad (r. 1921–50) obtained from the British in 1934 and the large-scale exploitation of oil that was initiated in the late 1940s. Meanwhile, Sunni merchants maintained a strong corporate sense through shared economic interests and intermarriage (Crystal, 1995: 37), which was symbolised with their effective political opposition against the ruler in 1909, 1921, and 1938. This strong corporate sense encouraged Abdullah Salim (r. 1950–65), Kuwait's first amir, to co-opt them during his reign, with the merchants benefitting from the sanctioning of land grabbing, their access to lucrative government contracts, and the benefits they obtained from the restrictions on foreign ownership of companies in Kuwait that allowed them to operate as import agents. Many of the family-based companies that initially benefitted from this state largesse managed to develop into massive conglomerates with activities in multiple sectors (Hanieh, 2011: 68). This has sustained the social hierarchy that had been established in the pre-oil era, with the merchant class together with members of the ruling family having privileged access to external rents through their access to government contracts and their dominant position in the private sector (Crystal, 1995: 7–37). Simultaneously, they have been the main beneficiaries of the willingness of the government to shore up the private sector after the 1976–77 stock market crash, the crash of an offshore market, known as the Suq al-Manakh, in 1982, and the financial crisis in 2008 (Nosova, 2016: 74). Their privileged access to the rents has been a source of tension in Kuwaiti society and has fostered claims for redistribution among disadvantaged social groups (Beaugrand, 2019: 59), complicating attempts towards fiscal reform following the drop in crude oil prices in 2014.

Modern state formation and the National Assembly

The historical political opposition of the merchant class against the ruling family stimulated a process of modern state formation in Kuwait that has been central to the institutional structure in which the current debate about fiscal adjustment is embedded. In 1938, Sunni merchants mobilised successfully and forced the ruler, Shaykh Ahmad, to consent to the establishment of an elected National Legislative Council. This council, elected by the heads of 150 merchant families, had significant legislative power and sought to 'control all the state's income and expenditure' (Salih, 1992: 68–77) including those originating from the oil concession, in order to fund social reforms and build a modern administrative system.

Although the legislative council was disbanded by the ruler after six months, it fostered the perception of 'citizens' ownership' (Beaugrand, 2019: 56) of oil revenues and, together with the increase in oil revenues, stimulated the development of a modern state through the creation of new ministries. These ministries, headed by members of the ruling family, have been central to the 'organisation and management of the distribution of wealth' (Khalaf and Hammoud, 1987: 351), establishing a modern infrastructure and providing lavish welfare services and jobs in the public sector to Kuwaiti nationals. Central to this is the allocative function of the state, which has guaranteed Kuwaitis free access to health care, education, and low-income housing, as well as marriage and housing loans (Crystal, 1995: 79). It has even committed itself to full employment of its citizens through Law No. 18 of 1960, which 'stipulates that every citizen has the right to a job in the public sector' (Nosova, 2016: 83).

Despite the limited fiscal sustainability of the continued provision of these benefits due to the structural decline in the price of crude oil, it has been very difficult for the government to make any cuts to these provisions. An important safeguard for the privileges of Kuwait nationals has been the National Assembly, an elected legislative institution[2] which has to pass every law before it can be sanctioned by the amir (A. S. Al-Sabah, 1962: 13). The institution was founded shortly after the country's independence in 1961, when the Iraqi Prime Minister Abd al-Karim Qasim threatened to seize Kuwait. In order to boost popular support in the face of this threat to the external security of Kuwait, Amir Abdullah Salim promised to the population new public institutions and a constitution. The constitution was promulgated in 1962 and established the National Assembly, for which the first elections took place in 1963 (Herb, 2016: 16).

This legislative institution has reduced the autonomy of the Kuwaiti government from social pressures, forcing it to seek consensus with members of parliament (MPs) in order to pass legislation. This has further fostered the perception of the 'citizens' ownership' of the external rents, with MPs representing different social groups using the National Assembly to stake their claim to access the oil rents. While this allows the ruling family to 'divide and rule' through establishing constantly shifting parliamentary alliances, it has also enhanced accountability, as the National Assembly has been granted the power to remove ministers and the prime minister through a vote of no confidence (Herb, 2014: 50–66).

This accountability has been further enshrined in the 1962 Constitution, which regards all natural resources as property of the state and commits the latter to the 'preservation and proper exploitation' of them (A. S. Al-Sabah, 1962: 4). Realising the exhaustibility of the country's oil resources and the importance of intergenerational solidarity, Shaykh Abdullah Salim already pressed for economic diversification in the 1950s. In 1953, he issued a decree that established the Kuwait Investment Authority (KIA), thereby establishing the first sovereign wealth fund (SWF) in the region. The KIA manages both the General Reserve Fund (GRF) and the Future Generation Fund (FGF). The GRF was founded in 1960 and could be considered as a 'government holding fund for revenues and assets' (Shehabi, 2015: 24), as it manages the annual state revenues. Recorded budget surpluses and revenues from KIA's assets originating from its ownership of the Kuwait Fund for Arabic Economic Development, the Kuwait Petroleum Company, and its investments abroad are channelled to the GRF. As a result, the GRF plays a crucial role in smoothening short-term fiscal deficits and thus helping the country to deal with the volatility of oil prices.

The FGF was established in 1976 in order to provide an income for future generations. The government is required by law to transfer 10 per cent of the annual revenues to the FGF (Shehabi, 2015: 25). The FGF and GRF have enabled the government to deal with large account surpluses, avoiding a huge increase in the exchange rate (James, 2017). In order to ensure the proper management of the country's SWFs, the KIA is monitored by the state's Audit Bureau, which reports directly to the parliament (Sartawi, 2012: 103). In addition, the head of the KIA can be questioned by the parliament (Young, 2019: 48).

Previous periods of fiscal adjustment

The KIA was established shortly before Kuwait was hit by a small economic recession in the mid-1950s. Similar to the contemporary debate about fiscal adjustment, according to Zahlan this recession generated 'a generally widespread feeling that Kuwait belonged to the Kuwaitis first and foremost, particularly when there were cutbacks in expenditure' (Zahlan, 1989: 35). This feeling of entitlement among Kuwaiti nationals was further spurred in the 1950s, as new regulations were implemented that required employers to discharge foreigners before nationals and exempted Kuwaiti university graduates from mandatory entrance examinations to enter the public sector (Zahlan, 1989: 35–7). This was combined with the introduction of a new Nationality Law in 1959, which overturned the 1948 Nationality Law. While the latter granted Kuwaiti citizenship to children of Muslim and Arab fathers born in Kuwait, the 1959 Nationality Law only allowed the sons of Kuwaiti fathers to pass Kuwaiti citizenship (Longva, 2005: 121) and restricted the total number of annual nationalisations to fifty (Longva, 1997: 48). The 1959 Nationality Law has enhanced the importance of exclusive 'Kuwaiti' origin to gain access to the country's oil rents and participate in the country's National Assembly (Longva, 2006: 181). At the same time, it has fostered nationalist sentiments among the population that justify the exclusion of expatriates and the *bidūn* (people with no state affiliation who claim entitlement to Kuwaiti nationality) to the country's oil wealth, which particularly gain popularity when the privileges of Kuwaiti nationals are threatened.

During later periods of lower oil revenues and economic crises, the SWFs established in the 1950s and 1970s provided a buffer that helped to mitigate the effects of economic recessions. Through their effectiveness they have provided a sense of fiscal security, which could explain the limited urge lawmakers feel to address the fiscal deficits that started to occur in Kuwait after 2016. In the 1980s, significant investment returns derived from the

SWFs started to exceed oil revenues from 1986 onwards (Shehabi, 2015: 25), partly compensating for the loss of government revenue as a result of the drop in oil prices because of the 1980s oil glut. This helped to limit cuts to social services and benefits (El-Katiri *et al.*, 2011: 9).

The reserves accumulated in the GRF and FGF proved again to be crucial in the 1990s to responding to the fiscal crisis caused by the Iraqi invasion of Kuwait. The invasion had a devastating impact on Kuwait's economy due to the significant damage to the country's petroleum industry, infrastructure, and environment and the loss of economic output. Meanwhile, Kuwait had accumulated significant new debts, as it owed the countries that participated in the military coalition to liberate the country an estimated sum of USD 20 billion. To respond to this, a massive reconstruction programme was initiated, with the country spending approximately USD 40–50 billion of its accumulated savings in the GRF and FGF. Other sources for funding were the international debt market, with the government raising the debt ceiling to USD 33 billion, and a small amount of the war reparations Kuwait received from Iraq (Shehabi, 2015: 1–23).

Privatisation and the shifting role of the Kuwaiti state

The strains on the budget caused by the loss of economic output and accumulated debt after the liberation of Kuwait in 1991 encouraged a reconsideration of the role of the state in the economy. The government embarked on neoliberal reforms, with the privatisation of state-owned enterprises (SOEs) and the removal of restrictions on foreign direct investment (FDI). These have formed the prelude to the reform programme advocated by the current government, which has sought to boost the private sector in Kuwait, reduce government subsidies, and enhance FDI inflows.

After the liberation from Iraq, Kuwait was the first of the GCC states to initiate a privatisation programme (Nosova, 2016: 216). This programme sought to reduce the influence of SOEs, which played important roles in the provision of credit and subsidised products to Kuwaiti nationals, the industrialisation of the country, and the execution of large infrastructural projects. This privatisation was encouraged by the fiscal deficits and resulted in the offloading of several loss-making SOEs through initial public offerings (IPOs), closed bidding contests, and open auctions (Sartawi, 2012: 9–99).

At the same time, attempts were made by the Kuwaiti government to improve the investment climate for foreign companies. This has been deemed necessary to import new technology that could foster productivity and help to provide attractive high-skilled jobs to Kuwaiti nationals in the private sector. A free-trade zone was established in 1999, and a law was introduced in 2000 allowing foreign ownership of shares in companies listed on the Kuwaiti Stock Exchange, a law which was extended eight years later to allow foreign ownership of non-listed shareholding companies.

Despite these attempts, Kuwait has struggled to receive FDI, as it attracted the lowest amount of FDI among the GCC states between 2000 and 2011 (Nosova, 2016: 39, 225). Investment has been deterred by the aforementioned economic nationalist policies implemented in the 1950s that have provided preference to local companies in public tenders. In addition, further impediments have been intensive red tape, the corruption scandals that have plagued the country (Narayan, 2018), and the sensitivity of land acquisitions. The latter have been highly politicised since the land grab during the 1950s, which has made the National Assembly increasingly suspicious of contracts between the government and private parties that involve land. As a result, the government only leases land for a limited period of time to the private sector. This happens through a build–operate–transfer (BOT) mechanism, through which a private entity constructs and operates a facility for a limited period and eventually transfers it back to the government (Herb, 2009: 386).

Political stagnation and princely struggles

A further impediment to the neoliberal reform agenda is posed by growing infighting among members of the ruling family, which has contributed to an increase in tensions between the government and the parliament, leading to frequent cabinet reshuffles and several dissolutions of parliament. Because of the high ages of the current amir and Crown prince, several contenders among the ruling family have tried to position themselves in anticipation of a future succession. Particularly the bitter rivalry between Shaykh Nasser al-Muhammad and Shaykh Ahmad al-Fahd, who are both nephews of the amir, cast a shadow over the Kuwaiti political scene from 2006 until 2014. The former served as a prime minister after Shaykh Sabah Al-Ahmad became amir in 2006 until his resignation in 2011, while the latter served in several ministerial positions from 2003 until 2011. Both Shaykh Nasser al-Muhammad and Shaykh Ahmad al-Fahd used allies in the parliament, in Kuwaiti media outlets, and on the Internet to influence the public debate and discredit each other. Their struggle played out in parliament, with an increasing number of MPs submitting interpellations to ministers, a practice which can potentially lead to a motion of no confidence. This resulted in a situation in which the amir suspended eight governments between 2006 and 2012, while five parliaments were dissolved, three by the amir and two by the constitutional court.

The infighting between these leading members of the ruling family emerged while transformations were unfolding at a social level. While in the first thirty years of independence tribal segments of the population tended to side with the ruling family, this constellation changed during the end of the 1990s. As a result of growing educational access, members of the previously marginalised tribes enhanced their political consciousness and started to criticise the privileged access of an elite formed by princes and merchants to the country's oil wealth (Alnajjar and Selvik, 2015: 117). Their antagonism was further fostered with the government vigorously enforcing the prohibition of tribal primaries in 2008 (Azoulay and Beaugrand, 2015) and the adoption of a law in the same year that forbade the use of public land for the building of private constructions, which particularly affected the tribal areas in Kuwait. These government actions and a growing anti-tribal discourse in the expanding media realm shifted loyalty of a large segment of the tribes to the opposition. This led to the election of a parliament dominated by opposition members in 2008, which effectively pressured the government to almost double allowances and public-sector salaries (Nosova, 2016: 201).

The stiff opposition the government faced in parliament coincided with the proliferation of youth movements, which effectively used the opportunities offered by the Internet to agitate against political corruption in the country, declining social services, and failure of the government to adhere to the country's constitution. These youth movements were effective in organising the 2006 street protests that eventually persuaded parliament to approve electoral reform. In November 2011, youth movements organised protests that helped to force the resignation of Prime Minister Shaykh Nasser al-Muhammad, while in the autumn of 2012 they initiated unprecedented large street protests against the electoral reform (reversing some of the gains made by the protestors in 2006) enforced by the amir.

However, these protests did not persuade the amir to make a U-turn. Following an opposition boycott of the parliamentary elections in December 2012, which led to a historically low turnout, a new parliament was elected, which mainly consisted of pro-government MPs. Meanwhile, the government resorted to authoritarian measures in order to stymie the large opposition movement that had been created between 2011 and 2012 (Alnajjar and Selvik, 2015: 119). To appease the population, social spending was further increased. During the height of the Arab Spring protests in February 2011, the government gave every Kuwaiti citizen a cash handout of USD 3,500 and over a year of free food (Thompson, 2011). In 2012, public-sector employees saw a 25 per cent increase in their salaries (OBG, 2014). A year later, a USD 2.6 billion debt relief plan was agreed with the parliament that covered the personal loans of Kuwaitis who were struggling to repay the interest and debt they had accumulated on loans from commercial banks (Westall, 2013). In exchange for this measure advocated by lawmakers in their election campaigns, the parliament approved measures in line with the government's neoliberal reform agenda. In 2013, MPs approved a new FDI law that lifted some restrictions on foreign ownership of local companies (OBG, 2017) and supported the establishment of the Kuwait Direct Investment Promotion Authority (KDIPA), which helps foreign companies to apply for an FDI licence (Nosova, 2018: 39).

Fiscal adjustment since 2014

The tensions described above in the form of succession struggles, the emancipation of tribal segments in Kuwaiti society, and the social protests initiated by youth movements provided the backdrop to the political and social context in which the drop in oil prices occurred as of mid-2014. As a result of the country's loss of state revenue from oil, Kuwait recorded its first financial deficit in the fiscal year 2015–16, which has widened over the years (Carvalho *et al.*, 2017: 5). The deficit is not only caused by declining crude oil prices, but also the result of a decline in oil production, as Kuwait agreed to commit to the Organization of the Petroleum Exporting Countries (OPEC) production cuts of 3 per cent in 2017 (Meredith and Huang, 2019). At the same time, fiscal expenditure has been rising because of an increase in capital spending to stimulate economic diversification (IMF, 2019a: 10) and failure to significantly diminish current expenditure after it had tripled between 2007 and 2014 (OBG, 2014). This current expenditure primarily consists of the public-sector wage bill, which constitutes about 55 per cent of total fiscal expenditure. In addition, government subsidies, which include fuel, electricity, water, and food staples, account for more than 20 per cent of fiscal expenditure (Carvalho *et al.*, 2017: 6).

   In order to finance the deficits, the government has drawn money from the GRF, similar to the previous episode of fiscal adjustment after the Iraqi invasion, when approximately USD 40 to 50 billion of Kuwait's investment portfolio was spent (Shehabi, 2015: 12). According to Moody's, assets of the GRF have declined by Kuwaiti dinars (KD) 14.7 billion (USD 48 billion) from 2015–16 (the first fiscal year in which a deficit was recorded) until the fiscal year of 2018–19. To reduce its dependence on the GRF, the Kuwaiti government tapped into the global debt market in 2017 with a successful USD 8 billion debt sale (Barbuscia, 2019). However, a repeat of this is unlikely, inasmuch as the government withdrew a new debt law that would raise the debt ceiling to KD 25 billion (USD 81.6 billion) in March 2018. This amid stiff opposition of MPs, who challenged the necessity of borrowing from the financial markets because of concerns about the misappropriation of public funds by several public institutions (Izzak, 2018). They have urged the government to find alternatives, forcing the latter to use the GRF for its financing needs (IMF, 2019a: 13).

The utilisation of the GRF has put the availability of liquid assets to finance Kuwait's fiscal deficits under strain, with Moody's projecting their depletion within the next four years under current rates (Barbuscia, 2019). This has created new pressures on the KIA to enhance returns. In January 2017, then managing director Bader al-Saad announced that KIA would increase the total assets managed in-house from 2 per cent to 8 per cent at the expense of the investment in public funds, which have recorded diminishing returns and have been hit by increasing market volatility (James, 2017). The KIA has particularly eyed private investment into infrastructure by investing in projects for power distribution, seaports and airports in the United Kingdom, Spain, and Australia. In addition, more investments are projected into private equity and real estate (Lacqua and Schatzker, 2017).

The change in strategy has been initiated while the activities of the KIA have come under increased scrutiny from the National Assembly. In May 2015, ten staff members were recalled from the fund's London office because of accusations from a committee of MPs over accounting irregularities and 'claims that employees deliberately made bad investments' (Tomlinson *et al.*, 2015). Concerns among MPs about the functioning of the KIA increased further in 2016, when the fund conceded that it had recorded losses of around USD 1 billion throughout the fiscal year 2015–16 (Kuwait News Agency, 2017) and in March 2018, when it was announced that the KIA had failed to recoup almost USD 706 million, as companies it had invested in went bankrupt (*Arab Times*, 2018). Managing Director Bader al-Saad was replaced in February 2017 by Farouk Bastaki, with the former being blamed by Finance Minister Nayef al-Hajraf for allowing 'improper investments' (*Times Kuwait*, 2019) amid accusations levelled by MPs that senior officials had used the funds of the KIA to invest in projects they were personally involved in (Izzak, 2019a).

Stalled subsidy reform

The reduction in government revenue since the drop in oil prices in 2014 has also given impetus to the reform of subsidies in Kuwait. Less than a year before the drop in crude oil prices, the Kuwaiti government had already announced its intention to review subsidies on fuel and electricity, which were estimated to account for around 22 per cent of the fiscal budget (Azoulay and Wells, 2014). The subsidies introduced following the start of large-scale oil exports in the 1950s had the objective of making basic goods such as electricity, water, and petrol affordable to every Kuwaiti (Sartawi, 2012: 94). Over the years, tight regulation has kept the prices for diesel, petrol, and electricity far below international and regional prices, resulting in a situation in which the production costs of electricity exceed the tariffs for electricity usage (IMF, 2015: 5–6). This necessitates the transfer of government funds to utility companies in order to compensate them for their operations. The subsidy system has stimulated extensive household consumption, reduced the incentive for energy efficiency, and sustained a regressive system in which wealthier households that use more electricity and petrol benefit more from the system than less wealthy households (Eltony, 2007: 80).

Following the review, the Kuwaiti Ministry of Finance announced in May 2014 that subsidies on fuel, electricity, and water had to be reduced by at least 15 per cent in two years (Thomas, 2014). The drop in oil prices later that year gave further impetus to subsidy reform. In January 2015, diesel prices in Kuwait were hiked from USD 0.69 to USD 1.32 per gallon, with the exception of 'businesses with heavy demand' (Krane, 2019: 141). In April 2016, the National Assembly approved a bill sponsored by the government which more than doubled electricity and water charges for expatriates, the commercial sector, and government institutions. Kuwaiti nationals and the industrial and agricultural sectors were exempted from the increase (Izzak, 2016a). In August 2016, the Kuwaiti cabinet announced that it would raise the price of petrol, with standard petrol going up by 41 per cent, with a government committee revising petrol prices every three months (Agence France-Presse, 2016).

In comparison to other GCC states the prices for petrol and diesel are still modest, as the price of petrol in Kuwait is the lowest among the GCC states (see Table 3.1), while the price of diesel is the second lowest among the GCC states (after those of Saudi Arabia). Nevertheless, the rise in petrol prices has been deeply unpopular among Kuwaiti nationals. In response to the petrol prices, members of the National Assembly threatened to grill Prime Minister Shaykh Jaber al-Mubarak al-Hamad Sabah and Finance Minister Anas Saleh (Izzak, 2016b). These threats eventually encouraged the amir to dissolve the National Assembly in October 2016 (Weiner, 2016). The elections in December 2016 resulted in the resurgence of the Kuwaiti opposition (of which the majority had boycotted the previous elections) in the National Assembly, as they won twenty-four out of the fifty seats, benefitting from widespread discontent over subsidy reform. Soon after the elections, members of the Kuwaiti opposition promised to act as a safeguard against government austerity measures and roll back some of the subsidy reforms (Martin, 2017).

Table 3.1 Prices of petrol and diesel in GCC states (November 2019).

	Gasoline (USD per litre)	Diesel (USD per litre)
Bahrain	0.53	0.42
Kuwait	0.36	0.38
Oman	0.56	0.62
Qatar	0.52	0.51
Saudi Arabia	0.55	0.13
UAE	0.57	0.65

Source: Valev (2019).

Imposing austerity upon non-nationals

The resistance against the reform of the subsidy system shows the difficulty for the Kuwaiti government to reconfigure the relations that were established between the state and citizens following the start of large-scale oil exploitation, which enabled the emergence of an oil welfare state. This is also exemplified by the hostility of MPs against taxation that could directly impact nationals. Although the Kuwaiti government is a signatory to the GCC value-added tax (VAT) agreement, the introduction of this tax has been delayed until 2021 by the Budget Committee of the parliament (Ghantous and Torchia, 2018). Instead, a decision was taken to accelerate the implementation of excise taxes on tobacco and sugary drinks, which were expected to have been introduced in 2020 (Saadi, 2019). The introduction of taxes is further inhibited by the lack of extractive capacity, with the Kuwaiti tax department currently lacking administrative infrastructural and information technology (IT) systems to enable wider tax collection (Carvalho *et al.*, 2017: 19).

The failure to implement new taxes has been combined with growing calls to introduce new measures for tax expatriates. While Qatar (Goncalves, 2019) and the United Arab Emirates (Rizvi, 2019) started to issue permanent residency schemes for a limited number of expats in 2019, Kuwait has witnessed a growing anti-expat discourse relating to concerns about the large number of non-nationals in the country. Particularly vocal has been the MP Safa al-Hashem, who made Kuwaiti headlines by calling for a tax on expats 'for the air they breathe in Kuwait and for walking on the streets' (Izzak, 2019b). While these statements were partly made to make media headlines, they also relate to a broader disquiet about the failure of the government to reduce the dependency on foreign labour. Despite government efforts towards labour nationalisation from 1978 onwards (Nosova, 2016: 84), expatriates currently comprise 83 per cent of the total labour force in

Kuwait. They are mainly employed in the private sector, earn low wages, and work on flexible labour contracts, allowing their employers to respond quickly to economic downturns and keep prices low. This has reduced the incentive for Kuwaitis to join the private sector, as in many instances they could earn higher wages in the public sector for comparable levels of education, while benefitting from inflexible contracts that are offered in the latter (Shehabi, 2019: 13).

Initiatives towards labour nationalisation have so far been unsuccessful, with the government unable to counter the large-scale visa trading in Kuwait (*Arab Times*, 2017). Alongside broader calls to address demographic imbalances, there has particularly been pressure upon the government by a broad alliance including both Islamist and liberal MPs to replace expatriates in the public sector with Kuwaiti nationals. This is in response to public demand for more public-sector jobs. However, this substitution has proved to be difficult, with particularly the health care and education sectors (which employ most of the expats in the public sector) struggling to find a sufficient number of qualified Kuwaiti nationals (Toumi, 2018).

Despite the dependency of the private sector on flexible and cheap expatriate labour, the need for specific skills not offered by Kuwaiti nationals, and the convenience expatriates provide to nearly every Kuwaiti household as maids, there have been long-standing calls for the rectification of Kuwait's 'demographic imbalance'. The argument in favour of this is that the presence of a large number of expatriates puts strains on public services, has fostered traffic congestion, and ignited fears for the 'cultural survival' (Longva, 1997: 17–44) of the nationals, as they have become a minority in their country. The need for fiscal adjustment has fostered policies to ensure a growing contribution of the non-national population to public services in the country, further inhibiting their indirect access to the country's oil wealth.

In 2017, the Ministry of Health in Kuwait increased health care fees for expatriates at public hospitals for the first time in more than twenty years, alongside doubling the annual health care insurance fees from KD 50 (USD 165) to KD 130 (USD 430). Health care fees were further increased in 2019 in order to 'ease congestion at public hospitals' and allow 'government-run clinics to take in more patients' (Nagraj, 2019). At the same time, the government has taken steps to segregate health care and limit access of expatriates to public hospitals and clinics. Central to this has been the establishment of the Kuwait Health Assurance Company (Dhaman) in 2015. This company is structured through a public–private partnership (PPP) that involves the Kuwaiti private conglomerate Arabi Holding Group, the KIA, the Public Institution for Social Security (a Kuwaiti public pension fund), and a listing on the stock market (*Kuwait Times*, 2018). Dhaman has started the construction of primary health care centres and hospitals to provide health care to the expatriate population (Garcia, 2019), profiting from the increased demand for private medical services the forced segregation will induce. The step towards segregation in health care provision has opened up a new market for private investors, coming at the expense of non-nationals, who are subject to the commodification of health care services (Hanieh, 2018: 227–8).

Besides the severe hikes in the costs of health care, expatriates have also faced significant increases in charges for work permits. In 2016, the charge for a first-time work permit was increased from KD 2 (USD 6.5) to KD 50 (USD 163.2), while renewal of the permit was increased from KD 2 (USD 6.5) to KD 10 (USD 32.5) (Izzak, 2016a). Furthermore, the Kuwaiti parliament has been discussing proposals to levy a remittances tax of 5 per cent. These proposals have been opposed by the Kuwait Chamber for Commerce and Industry, whose members rely heavily on low-wage expat labour (Goncalves, 2019). It was also criticised by the Kuwaiti government and the governor of the Central Bank because of fears it would create a black market for remittances (Nagraj, 2019). The debate over the introduction of remittance taxes reveals the divide between the business sector and a new generation of political entrepreneurs from middle-class backgrounds that mainly represent public-sector workers (Azoulay, 2013: 69–71). While the former attempts to maintain access to foreign labour to reduce costs and have flexibility, the latter advocates for an increase in the contribution of the private sector in the form of employment for nationals and enhancing the sector's (indirect) contribution to revenues.

'NewKuwait' and economic diversification in Kuwait

Competing interests between the business sector and political entrepreneurs from Kuwait's salaried middle class have also impacted the country's plans for economic diversification. The drop in oil price has been used by the government to push forward its plans for economic diversification. Soon after entering office in December 2016, the new Kuwaiti cabinet unveiled its vision entitled 'NewKuwait'. This ambitious vision aims to transform Kuwait into a regional financial, logistical, and cultural hub by 2035, profiting from Kuwait's position on the Chinese Belt and Road Initiative and its geographical proximity to Iran and Iraq. The vision gained more prominence after the eldest son of the amir, Shaykh Nasser Sabah, threw his political weight behind it in March 2018 through his position as first deputy prime minister and potential contender for the position of Crown prince. The infrastructural projects that encompass 'NewKuwait' are centred on the development of five islands (Bubyan, Uarba, Failaka, Miskan, and Auha) and the erection of a new city, Madinat al-Harir (City of Silk) (*Gulf States Newsletter*, 2018: 9).

However, the implementation of these projects will be difficult, particularly because of concerns among MPs that these large infrastructural projects mainly benefit merchant families (Nosova, 2016: 207). Furthermore, political opposition has emerged in parliament about the powers that should be attributed to the new General Authority of the City of Silk, Bubyan Island, and Port Mubarak, with concerns that the draft laws submitted for this new authority subvert parliamentary oversight and stand in contradiction to the Kuwaiti constitution (Diwan, 2018). In addition, Islamist MPs have slammed plans to allow the consumption of alcohol in the new city, which would be in breach of the Islamic norms that apply in the country (Tarzi, 2019).

Beyond the glossy headlines, 'NewKuwait' is also projected to increase the contribution of the private sector to GDP, reinvigorating the aforementioned neoliberal reforms that were initiated in the 1990s. The vision seeks to privatise SOEs and launch PPPs to increase private investment in public infrastructure. Private-sector growth and FDI is further sought by lifting restrictions imposed by burdensome bureaucratic procedures (Carvalho *et al.*, 2017: 12). These procedures have allowed individuals with formal or informal access to certain state resources to prey on businesses, and they privilege established companies which are able to use their connections to keep out potential competitors (Nosova, 2016: 90). Reform of the bureaucracy is deemed crucial to creating a more transparent and competitive business environment and for small and medium enterprises (SMEs) to flourish given the restraints extensive red tape currently imposes on the establishment of new businesses (Carvalho *et al.*, 2017: 12). However, reform is likely to face resistance from actors that benefit from the opaque bureaucratic system.

Adjustment policies since 2017: A shift towards piecemeal reform

Prospects for structural fiscal reform have become dimmer since the resurgence of the Kuwaiti opposition after the parliamentary elections in 2016. The election campaigns that were initiated after the dissolution of parliament in October 2016 revealed widespread public opposition against the austerity measures implemented by the government, particularly against the removal of the subsidies. In their election campaigns, opposition candidates particularly took issue with the large expenses on overseas medical allowances, which according to them have cost KD 760 million a year (approximately USD 2.5 billion) and have primarily benefitted people with ties to government ministers. The Kuwaiti opposition, which includes Islamists and tribal, liberal, and independent candidates, called upon the government during the election campaign to put an end to these practices before making cuts to the allowances and subsidies provided to Kuwaiti nationals (Hagagy, 2016).

Following the election of a new National Assembly in December 2016, the opposition resurged by winning twenty-four of the fifty seats, overturning the largely pro-government parliament elected in June 2013. Given the large-scale opposition to the austerity measures, the government has scaled back its ambitions for reform. It has instead concentrated on ad hoc measures that could be implemented to boost non-oil revenues, such as enhancing fees for government services, imposing penalties on businesses who do not meet the quotas on Kuwaitisation, and improving revenue collection. Furthermore, it has encouraged the streamlining of government spending by improving procurement, rationalising capital expenditure, closing loopholes in the administration of various transfers, and limiting grants provided to priority sectors (IMF, 2019b). The Kuwaiti government has aimed to reduce spending on public-sector employment by streamlining allowances and rationalising employment benefits (IMF, 2018: 12). The shift towards piecemeal reform is illustrated by the new budget for the new tax year starting in April 2019, in which no significant efforts are made to reduce the public-sector wage bill and further reform the subsidy system.

This gradual reform is further encouraged by rising global oil prices, resulting in higher-than-expected revenue and fiscal expenditure below the projected budget (Carvalho *et al.*, 2017: 21). This has resulted in a new budget for the tax year 2019–20 in which expenses on subsidies and public-sector wages have not been significantly lowered (MacDonald, 2019). Tensions with parliament as exemplified by the resignation of four ministers in December 2018 (IMF, 2019a: 4) further highlight the tightrope the government has to walk not to antagonise the opposition, limiting the scope for large-scale reform and the necessity to seek consensus.

Conclusion

Historical contingencies with the establishment of an extensive welfare state in the 1950s, social protections provided by the 1962 Constitution, and a relatively powerful parliament could explain Kuwait's high level of current expenditure on public-sector employment and subsidies, distinguishing Kuwait from other GCC states which have relied on a higher level of capital spending. In comparison with other GCC states, Kuwait has been slow to react to the decrease in global crude oil prices and shied away from structural reforms. The government has primarily concentrated on subsidy reform in order to reduce expenditure rather than the introduction of new taxes or reducing the public-sector wage bill.

The subsidy reform has ignited widespread discontent, particularly the rise of petrol prices in September 2016. This has created tension between the National Assembly and the Kuwaiti government, with the latter eventually resorting to a piecemeal reform of public expenditure in order to address declining government revenues. This has avoided the implementation of structural change to reconfigure the relations between the state and its citizens, as so far measures to reduce the fiscal budget have primarily affected non-nationals.

While it will eventually be necessary to make the budget more sustainable in the long term, this is not in the interest of a large salaried middle class in Kuwait dependent on public-sector employment and profiting from subsidies on fuel and utilities. Meanwhile, the prospect for structural change through successful economic diversification and non-oil sector growth are limited, as large-scale infrastructure projects are difficult to implement given concerns among the Kuwaiti opposition that they primarily benefit the merchant elite.

Reforms targeted to lift certain bureaucratic procedures are likely to face opposition from government officials and established businesses who benefit from the cronyism that is fostered through the extensive red tape, while efforts to encourage SMEs and attract more FDI are doomed to fail without an overhaul of the bureaucracy. At the same time, the increase in crude oil prices since 2018 and the cushion provided by GRF have been other factors that have reduced the appetite to renegotiate the social contract between citizens and the state. This has led to a situation in which Kuwait is stuck in transition, with none of the social groups being willing to reduce its access to the oil rents.

Notes

1 The author would like to thank Durham University for its generous funding for him to attend the DAVO Congress in Hamburg in October 2019, during which he presented the draft of this chapter. At the time, the author was employed as a teaching fellow in Middle East politics at the School of Government and International Affairs, Durham University.
2 Before 2005, only Kuwaiti men were eligible to vote and stand for election for the National Assembly.

References

Agence France-Presse (2016), 'Kuwait announces "ultra" petrol price rise of 83% – to 55 cents a litre', *The Guardian*, 2 August. https://bit.ly/3rPQnFM (accessed 13 January 2020).

Alnajjar, G., and K. Selvik (2015), 'Kuwait: The politics of crisis', in K. Selvik and B. O. Utvik (eds), *Oil states in the new Middle East: Uprisings and stability*. London: Routledge, pp. 105–24.

Al-Nakib, F. (2016), *Kuwait transformed: A history of oil and urban life*. Stanford: Stanford University Press.

Al-Sabah, A. S. (1962), 'Kuwaiti Constitution'. https://www.wipo.int/edocs/lexdocs/laws/en/kw/kw004en.pdf (accessed 12 July 2019).

Al-Sabah, S. A. J. (2019), 'Kuwait Vision 2035 "New Kuwait"', New Kuwait Summit. https://newkuwaitsummit.com/ (accessed 9 November 2019).

Arab Times (2017), 'Al-Hashim reiterates intention to call for imposing taxes on expats', 3 February. https://bit.ly/2OOmebn (accessed 13 January 2020).

Arab Times (2018), 'Kuwait Investment Authority fails to "recoup" almost $706 million in investments', 28 March. https://bit.ly/3cCI5uO (accessed 13 January 2020).

Azoulay, R. (2013), 'The politics of Shi'i merchants in Kuwait', in S. Hertog, G. Luciani, and M. Valeri (eds), *Business politics in the Middle East*. London: Hurst & Company, pp. 67–100.

Azoulay, R., and M. Wells (2014), 'Contesting welfare state politics in Kuwait', *Middle East Report*, 272, 43–7.

Azoulay, R., and C. Beaugrand (2015), 'Limits of political clientelism: Elites' struggles in Kuwait fragmenting politics', *International Journal of Archaeology and Social Sciences in the Arabian Peninsula*, 4. http://journals.openedition.org/cy/2827 (accessed 9 November 2019).

Barbuscia, D. (2019), 'Mideast-debt: Kuwait, Oman wealth funds being depleted to finance deficits', Reuters, 17 June. https://reut.rs/2Nqjqky (accessed 12 July 2019).

Beaugrand, C. (2019), 'Oil metonym, citizens' entitlement, and rent maximizing: Reflections on the specificity of Kuwait', in M. Herb and M. Lynch (eds), *The politics of rentier states in the Gulf*. Washington, DC: Project on Middle East Political Science, pp. 29–33.

Carvalho, A., J. Youssef, J. Ghosn, and L. Talih (2017), 'Kuwait in transition: Towards a post-oil economy', Tri International Consulting Group. https://bit.ly/37D8IxG (accessed 14 July 2019).

Crystal, J. (1995), *Oil and politics in the Gulf: Rulers and merchants in Kuwait and Qatar*. Cambridge: Cambridge University Press.

Diwan, K. S. (2018). 'Kuwait's MbS: The reform agenda of Nasser Sabah al-Ahmed al-Sabah', The Arab Gulf States Institute in Washington, 16 April. https://bit.ly/3qOg08w (accessed 8 November 2019).

El-Katiri, L., B. Fattouh, and P. Segal (2011), 'Anatomy of an oil-based welfare state: Rent distribution in Kuwait', London School of Economics and Political Science, Kuwait Programme on Development, Governance and Globalisation in the Gulf States, Research Papers, 13. https://bit.ly/3eJEXye (accessed 12 July 2019).

Eltony, M. N. (2007), 'The economic development experience of Kuwait: Some useful lessons', *Journal of Economic and Administrative Sciences*, 23:1, 77–102.

Garcia, B. (2019), 'Dhaman opens first primary clinic for expats in Hawally', *Kuwait Times*, 11 November. https://tinyurl.com/33ykdn3d (accessed 13 January 2020).

Ghantous, G., and A. Torchia (2018), 'Kuwait to postpone VAT implementation to 2021, says parliament committee', Reuters, 16 May. https://reut.rs/37X2f15 (accessed 12 July 2019).

Goncalves, P. (2019), 'Qatar starts issuing permanent residency visas to expats', International Investment, 1 February. https://tinyurl.com/3dcekcks (accessed 13 July 2019).

Gulf States Newsletter (2018), 'Succession in Kuwait: Emir's son Nasser Sabah in the ascendancy', 5 July. https://bit.ly/3pHlLEx (accessed 12 July 2019).

Hagagy, A. (2016), 'Analysis – Kuwait's anti-austerity lawmakers threaten reform plans', Reuters, 5 December. https://tinyurl.com/dsfhpksx (accessed 8 November 2019).

Hanieh, A. (2011), *Capitalism and class in the Gulf Arab states*. Basingstoke: Palgrave Macmillan.

Hanieh, A. (2018), *Money, markets and monarchies*. Cambridge: Cambridge University Press.

Herb, M. (1999), *All in the family: Absolutism, revolution, and democracy in the Middle Eastern monarchies*. Albany: State University of New York Press.

Herb, M. (2009), 'A nation of bureaucrats: Political participation and economic diversification in Kuwait and the United Arab Emirates', *International Journal of Middle East Studies*, 41:3, 375–95.

Herb, M. (2014), *The wages of oil: Parliaments and economic development in Kuwait and the UAE*. Ithaca: Cornell University Press.

Herb, M. (2016), 'The origins of Kuwait's National Assembly', London School of Economics and Political Science, Kuwait Programme on Development, Governance and Globalisation in the Gulf States, Research Papers, 32. http://eprints.lse.ac.uk/65693/1/39_MichaelHerb.pdf (accessed 12 July 2019).

IMF (2015), 'Kuwait: Selected issues', Country Report, 328. https://bit.ly/3h6uQq7 (accessed 12 July 2019).

IMF (2018), 'Kuwait: Selected issues', Country Report, 22. https://bit.ly/3xD1Lbw (accessed 12 July 2019).

IMF (2019a), 'Kuwait: Article IV consultation', Country Report, 95. www.imf.org/~/media/Files/Publications/CR/2019/1KWTEA2019001.ashx (accessed 15 July 2019).

IMF (2019b), 'Kuwait: Staff concluding statement of the 2018 Article IV mission', 28 January. https://tinyurl.com/6za4s6bb (accessed 28 February 2019).

Izzak, B. (2016a), 'Work permits, transfer fees raised up to 50 KDs – All Kuwaitis, farms, industries exempt from new power tariffs', *Kuwait Times*, 19 April. https://tinyurl.com/ah58bhk3 (accessed 13 January 2020).

Izzak, B. (2016b), 'MPs to file requests to grill PM, finance minister', *Kuwait Times*, 12 October. https://tinyurl.com/37p8p7f7 (accessed 13 January 2020).

Izzak, B. (2018), 'MP says no need to borrow, highlights bribery scandal', *Kuwait Times*, 27 March. https://tinyurl.com/6vxt9z83 (accessed 13 January 2020).

Izzak, B. (2019a), 'Minister warned over KIA; call to forgive dead citizens' debts', *Kuwait Times*, 20 January. https://tinyurl.com/24dvmjr2 (accessed 13 January 2020).

Izzak, B. (2019b), 'MP Safa Al-Hashem calls for taxing expatriates; Dallal slams maid recruitment offices', *Kuwait Times*, 5 February. https://tinyurl.com/y7uwwn8y (accessed 13 January 2020).

James, R. (2017), 'What's happening at Kuwait Investment Authority?', Private Equity International, 20 November. https://tinyurl.com/hx24wj4m (accessed 12 July 2019).

Khalaf, S., and H. Hammoud (1987), 'The emergence of the oil welfare state: The case of Kuwait', *Dialectical Anthropology*, 12:3, 343–57.

Krane, J. (2019), *Energy kingdoms: Oil and political survival in the Persian Gulf.* New York: Columbia University Press.

Kuwait News Agency (2017), 'Kuwait Investment Authority denies USD six bln losses', 22 May. https://tinyurl.com/34xbaxp2 (accessed 14 November 2019).

Kuwait Times (2018), 'Dhaman signs partnership agreement with IBM to establish advanced healthcare system', 26 March. https://tinyurl.com/3kt76mvd (accessed 13 January 2020).

Lacqua, F., and E. Schatzker (2017), 'KIA's Al-Saad: Take more risk to maintain returns', Bloomberg, 18 January. https://tinyurl.com/28uxc9t4 (accessed 24 February 2021).

Longva, A. N. (1997), *Walls built on sand: Migration, exclusion, and society in Kuwait.* Boulder: Westview Press.

Longva, A. N. (2005), 'Neither autocracy nor democracy but ethnocracy: Citizens, expatriates and the socio-political system in Kuwait', in P. Dresch and J. P. Piscatori (eds), *Monarchies and nations: Globalisation and identity in the Arab states of the Gulf.* London: I.B. Tauris, pp. 114–35.

Longva, A. N. (2006), 'Nationalism in pre-modern guise: The discourse on Hadhar and Badu in Kuwait', *International Journal of Middle East Studies*, 38:2, 171–87.

MacDonald, F. (2019), 'Kuwait budget has smaller deficit, lacks engines for change', Bloomberg, 21 January. https://bloom.bg/3eJtqig (accessed 28 February 2019).

Martin, G. (2017), 'A look at the state of Kuwait's political landscape', Fair Observer, 4 February. https://tinyurl.com/7uhcu5da (accessed 12 July 2019).

Meredith, S., and E. Huang (2019), 'OPEC agrees to extend current production cuts by 9 months', CNBC, 1 July. https://tinyurl.com/y59683j9 (accessed 15 July 2019).

Nagraj, A. (2019), 'Kuwait further hikes healthcare fees for expats', Gulf Business, 17 April. https://gulfbusiness.com/kuwait-hikes-healthcare-fees-expats/ (accessed 14 July 2019).

Narayan, P. (2018), 'The Middle East petro state with the most riding on rising oil prices isn't Saudi Arabia – it is Kuwait', CNBC, 13 July. https://cnb.cx/2P4lajP (accessed 1 March 2019).

Nosova, A. (2016), 'The merchant elite and parliamentary politics in Kuwait: The dynamics of business political participation in a rentier state'. PhD dissertation, London School of Economics.

Nosova, A. (2018), 'Private sector and economic diversification in Kuwait', in A. Mishrif and Y. Al Balushi (eds), *Economic diversification in the Gulf region, Volume I: The private sector as an engine of growth.* Singapore: Palgrave Macmillan, pp. 27–44.

OBG (2014), 'Kuwait targets subsidy reforms', 17 November. https://bit.ly/3udPKHw (accessed 7 November 2019).

OBG (2017), 'Kuwait's recent tax changes and regulations'. https://bit.ly/3eNPc4j (accessed 1 March 2019).

Rizvi, A. (2019), 'New permanent residency scheme will attract gold standard of young talent', *The National*, 22 May. https://tinyurl.com/p4rbt4zn (accessed 13 January 2020).

Saadi, D. (2019), 'Kuwait's economy set to accelerate after returning to growth in 2018, IMF says', *The National*, 29 January. https://tinyurl.com/9e78be7e (accessed 13 January 2020).

Salih, K. O. (1992), 'The 1938 Kuwait legislative council', *Middle Eastern Studies*, 28:1, 66–100.

Sartawi, M. (2012), 'State-owned enterprises in Kuwait: History and recent developments', in A. Amico (ed.), *Towards new arrangements for state ownership in the Middle East and North Africa*. Paris: OECD Publishing, pp. 93–116.

Shehabi, M. (2015), 'An extraordinary recovery: Kuwait following the Gulf War', The University of Western Australia, Working Papers, 15–20. https://ideas.repec.org/p/uwa/wpaper/15-20.html (accessed 13 November 2019).

Shehabi, M. (2019), 'Diversification in Gulf hydrocarbon economies and interactions with energy subsidy reform: Lessons from Kuwait', The Oxford Institute for Energy, OIES Paper, MEP 23. https://bit.ly/3tbuQax (accessed 14 July 2019).

Tarzi, N. (2019), 'Kuwait's modernist drive faces political setbacks', *The Arab Weekly*, 5 May. https://tinyurl.com/yk5c76z8 (accessed 8 November 2019).

Thomas, B. (2014), 'Kuwait subsidy cuts "will not hurt low-income earners"', Arabian Business, 18 May. https://tinyurl.com/43vvt4pk (accessed 13 January 2020).

Thompson, B. (2011), 'Kuwait's citizens get $3,500 cash handouts and free food', BBC News, 12 February. https://tinyurl.com/96ea5zjp (accessed 14 November 2019).

Times Kuwait (2019), 'Kuwait finance minister grilled in parliament', 11 June. https://bit.ly/334XVK7 (accessed 14 November 2019).

Tomlinson, H., M. Waller, and A. Dawood (2015), 'Kuwait fund staff made "deliberately bad" deals', *The Times*, 14 May. https://tinyurl.com/brmah7af (accessed 13 January 2020).

Toumi, H. (2018), 'Replacing public sector expats "too difficult"', *Gulf News*, 20 May. https://tinyurl.com/28dsad5s (accessed 13 January 2020).

Valev, N. (2019), 'Global petrol prices', Global Petrol Prices.com. https://www.globalpetrolprices.com/ (accessed 12 November 2019).

Weiner, S. (2016), 'Kuwait's emir just dissolved the country's parliament. Here's why', *The Washington Post*, 17 October. https://wapo.st/3qSWF7m (accessed 13 January 2020).

Westall, S. (2013), 'Kuwait's parliament approves personal debt relief law', Reuters, 9 April. https://tinyurl.com/y8zj3dyr (accessed 10 November 2019).

World Bank (2019), 'Oil rents (% of GDP)', World Bank Data. https://data.worldbank.org/indicator/ny.gdp.petr.rt.zs (accessed 27 February 2019).

Young, K. E. (2019), 'What's yours is mine: Gulf SWFs as a barometer of state–society relations', in M. Herb and M. Lynch (eds), *The politics of rentier states in the Gulf*. Washington, DC: Project on Middle East Political Science, pp. 44–50.

Zahlan, R. S. (1989), *The making of the modern Gulf states: Kuwait, Bahrain, Qatar, the United Arab Emirates, and Oman*. London: Unwin Hyman.

4

Oil price collapse and the political economy of the post-2014 economic adjustment in the Sultanate of Oman

Crystal A. Ennis and Said Al-Saqri

Introduction

If the onset of the oil age radically transformed economic life in Oman, its decline promises economic ruptures as well. Since the 1970s, the state and the economy have grown dependent on black gold. Not only is oil the main driver of economic growth and government revenue, but government expenditure drives the non-oil economy. Thus, since late 2014, when the global oil market readjusted to lower levels, austerity pressures have mounted. This is not the first time the economy has experienced such economic gravities. Commodities markets are notoriously volatile. How can we understand government economic responses to the oil downturn since 2014 in Oman?

This chapter addresses this question and finds that economic adjustments mirror oil market fluctuations, but fiscal restraint does not adjust as steeply during contractions in the oil market. This illustrates the difficulty in cutting expenditure in spaces where both state and economic life are deeply tied to each other and to oil-supported government spending. Over the next pages, we examine this pattern, with specific attention to policy measures since 2014. First, we introduce the structure of the economy and the role of oil and government expenditure. Alongside this, we discuss government policy initiatives to respond to fiscal pressures and towards the goals of diversifying the economy away from hydrocarbons. The core of the chapter examines a selection of these economic policies and initiatives: government borrowing, labour market regulation, subsidy reduction, the *tanfīdh* programme, and changes in taxation regimes.

Given that the economic downturn and its implications are still unfolding, the policy measures and trends we discuss are still in flux. That said, some patterns are observable. The first pattern is how dependent the economy remains on oil. The entrenchment of policymaking and the limitations on fiscal policy are tied up with the oscillations of the global oil market. The second observation is the impact of social pressure on policy choices. We argue that the government is trapped between socio-economic demands and fiscal realities – a disruption in one can lead to undesirable instability in the other. If the government makes policy changes in reaction to social pressure during a weakened fiscal position, it can damage its financial position and may risk having to revalue the currency peg. However, if the government undertakes austerity measures, these can cause more social instability. The state is in essence caught between a rock and a hard place.

When oil flows, the economy flows

The advent of the oil age restructured developmental possibilities, the government's funding and expenditure structure, and the types of economic opportunities available for society (Al-Yousef, 1996; Valeri, 2009). It facilitated rapid economic growth and change, but also a structural dependence on oil that has proven difficult to loosen. Such a condition is common among rentier states in the global economy (Beblawi and Luciani, 1987; Gray, 2011; Ahmed and Al-Saqri, 2012; Ennis, 2015).

Rentier economies include states characterised by oil abundance, measured by the ratio of per capita oil revenue to all government revenues, and oil dependence, measured by the percentage value of resource exports to gross domestic product (GDP) or the percentage of resource exports to government revenue (Luciani, 1990; Kropf, 2010; Herb, 2014: 11–15). By both definitions, Oman is defined as a rentier economy. Herb categorises it as a 'middling rentier', which is distinguished from the 'extreme rentiers' of Kuwait, Qatar, and the United Arab Emirates, who wield extremely high levels of resource abundance (Herb, 2014: 14). Indeed, as both Kropf (2010) and Herb (2014) illustrate, it is critical to understand the difference between resource abundance and dependence to better explore the impacts of resources on political economies. The comparison of Oman with the Norwegian case – a country which is resource abundant but is not usually considered as suffering from negative rentier effects – offers a useful example. While Norway exported resources amounting to 14 per cent of its GDP in 1970, Oman's resource exports were 89 per cent of its GDP (Kropf, 2010: 111). The crucial difference lay in the different levels of industrialisation between the two countries at the time, with Norway having a well-established manufacturing sector with the capacity to add value prior to export.

What we see in the remainder of this section is the degree to which Oman's resource abundance shaped the economy's dependence on oil. Although its manufacturing capacity has improved dramatically over the decades, the share of oil income in government revenue and in its export structure remains significant. In many ways, Oman's early oil-rentier economy follows the classic depiction where

> *the state becomes the main intermediary between the oil sector and the rest of the economy.* It receives revenues which are channelled to the economy through public expenditure, and since public expenditure generally represents a large proportion of national income, the allocation of these public funds among alternative uses has great significance for the future development pattern of the economy. (Abdel-Fadil, 1987: 83)

As shown in Figure 4.1, government expenditure in Oman has been the function of crude oil export income. This has been the case since the 1970s. The government adjusts its expenditure to how much it has in its coffer from oil income. This relationship between oil income and government expenditure is expected in the short run, but the fact that it has been a consistent story for almost five decades calls into question the ability of the government to sustain its income in the long run, hence its ability to maintain fiscal balance.

This is all the more evident with a view of past oil price drop episodes and their aftermath impact on public finance and the overall performance of the economy. For example, following the 1986 oil price drop of 37 per cent, and the dampened oil prices throughout the 1980s and 1990s, government expenditure increased by just 1 per cent annually in 1986–99. In the same period, GDP experienced an average real growth of 4 per cent annually, down from an annual real growth rate of 10 per cent from 1971 to 1985. Therefore, from the mid-1980s to the end of the 1990s, the Omani government undertook a variety of measures to increase income and restrict expenditure.

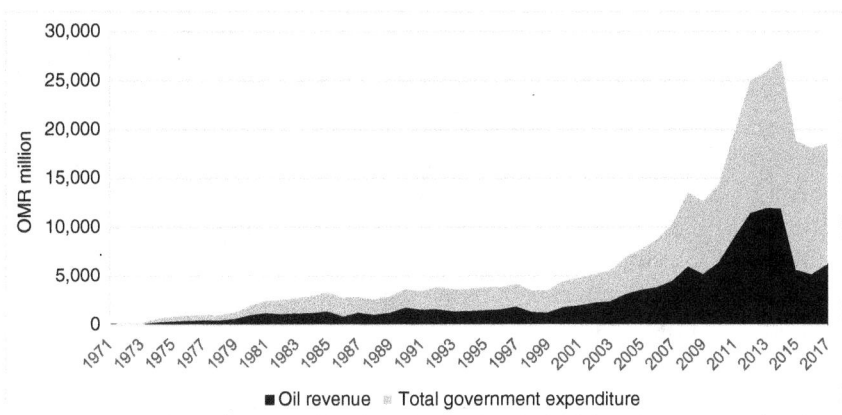

Figure 4.1 Oman's oil exports income and government expenditure, 1971–2017 (in OMR million). Source: NCSI (1981–2018).

In 1986, Oman even devalued its currency peg by about 10 per cent against the US dollar (Luciani, 2011: 27). The reforms implemented during this period allowed expenditure to remain contained, and afforded the government the opportunity to begin discussing a long-term development strategy known as Vision 2020 (World Bank, 1994; Ministry of National Economy, 2007). This Vision was released in 1995, and the five-year plans which followed it were to take the government in steps towards its ambitious diversification and development goals. The contraction of government revenue was the impetus behind the long-term strategy and economic plans.

Then the oil boom of the 2000s hit, allowing a renewed expenditure surge. The data show that from 2000 to 2014 government expenditure hiked by 14 per cent annually following oil income increases of about 17 per cent annually. This oil boom was exceptional in its incline and in the corresponding spending hike. The pace, growth, and duration of high expenditure patterns during this period make the spending even more difficult to reduce during times of austerity. This second wave of plenty saw the average real GDP grow by 13 per cent annually. Expenditure rates increased alongside growth. The data over the past five decades (see Figure 4.1) suggests difficulty in adjusting spending patterns downward.

The sustainability of such a structural pattern becomes even more questionable during oil downturns. Understanding the strength of the relationship between oil income and expenditure underpins the policy challenges that emerge during times of austerity. Public expectations of the state's redistribution of this largesse, not only in the form of infrastructural development and public goods delivery but also in the form of public-sector jobs and subsidised energy inputs and limited taxation, foster both social tensions and policy making quandaries. This section illustrates how central oil is to economic development in Oman's modern economic history. It establishes the background to our claim that the Omani policy environment is trapped between the competing interests of capital, the pressures of austerity prompted by fiscal realities, and socio-economic demands for the jobs and public goods the country and region has grown accustomed to. We agree with Hanieh's claim that, for ruling elites in all Gulf Cooperation Council (GCC) states under the current austerity pressures, 'navigating the tight line between cutting public sector employment and other forms of government provision, while attempting to maintain a social base of citizen support, is a complex and difficult task' (Hanieh, 2018: 225).

Oil exports and government expenditure

This pattern might not be surprising to those familiar with development paths in resource-dependent economies. GDP growth corresponds with oil

Table 4.1 Oman's oil dependence.

	Crude oil exports, % of GDP	Oil revenues, % of total revenues	Crude oil exports, % of total exports
1972	65	100	100
1982	58	92	92
1992	44	84	82
2002	42	76	67
2012	49	85	59
2013	46	86	57
2014	44	84	57
2015	31	75	50
2016	25	68	48
2017	28	73	46

Source: NCSI (1981–2018).

export earnings, which in turn correspond with government revenues. As Table 4.1 illustrates, Oman's oil dependence runs deep; oil dominates the economy and consistently comprises a high share of government revenue, exports, and GDP.

Although the share of government revenue from oil income dropped from 100 per cent in 1971 to 79 per cent in 1986, the overall share of non-oil income has not changed much since then. The non-oil income share increases during periods of international oil price slumps and decreases when oil prices recover (see Figure 4.2). In fact, the average share of non-oil income from 1986 to 2017 is 21 per cent, increasing to 29 per cent in 1986, when oil prices dropped by 32 per cent. Its share increased even more to 32 per cent in 2016, when oil prices dropped by 47 per cent. Overall, the government income structure has not changed much since 1986, and non-oil income plays a minor role in financing government expenditure.

Between 1970 and 1976, government expenditure was used to develop and build basic infrastructure, such as roads, airports, and ports, and provide basic services such as water, electricity, and basic health care. Hence, most of the government's expenses were invested in development projects, peaking to 43 per cent of total government expenditure in 1974 (see Figure 4.3). However, the share of development expenditure in total government expenditure has declined since then, and in 2017, the share was 23 per cent. In addition, recurrent government expenditure increased at substantial rates and peaked at 49 per cent of total government expenditure in 1998 and 1999. In 2017, government recurrent expenditure was 49 per cent of total

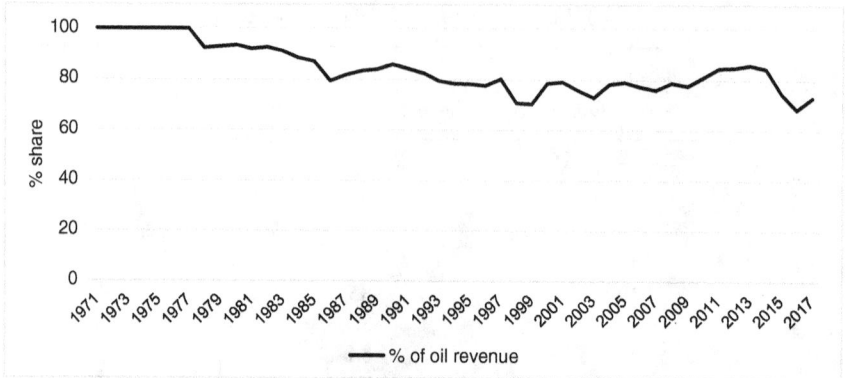

Figure 4.2 Oman's oil revenue, 1971–2017 (percentage share of total government revenue). Source: Ministry of Finance (1975–2018), NCSI (1981–2018).

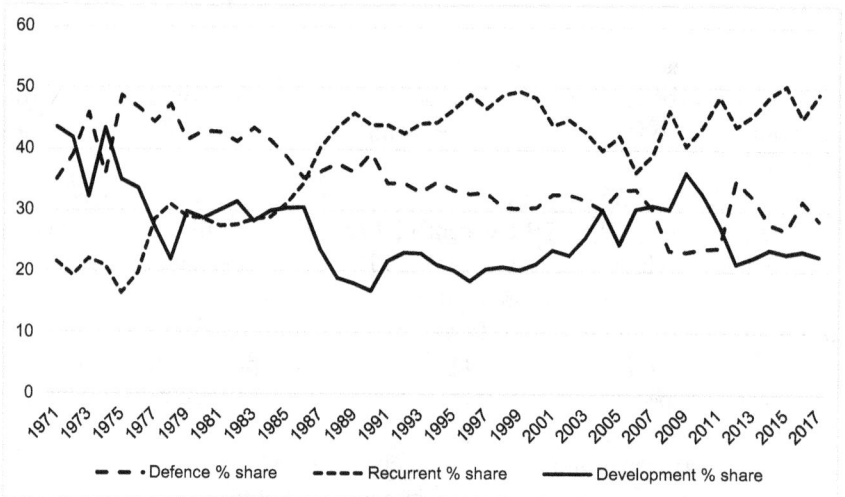

Figure 4.3 Oman's government expenditure structure, 1971–2017 (percentage share of total government spending). Source: NCSI (1981–2018).

government expenditure. Furthermore, defence-recurrent expenditure in Oman is significant, having reached 49 per cent of total government expenditure in 1975.[1] Defence expenditure, as a share of total government expenditure, has declined since 1975 but remains significant, with the average annual defence expenditure in 1976–2017 being 35 per cent. Recurrent and defence expenditure accounted for approximately 73 per cent of government expense in 1971–2017.

Once again, the period between 2014 and 2017 saw real average GDP growth down to 3 per cent, with a negative growth of 0.3 per cent in 2017. While state coffers certainly feel the strain, the government finds expenditure reduction difficult. In fact, changes in oil price levels have direct and immediate impact on revenue levels, revealing the high degree of fiscal dependence on oil income (see Figure 4.4). Meanwhile, government expenditure, while mirroring fluctuations in the oil price, oscillates less intensely, underlining the stickiness of government spending. There is a time lag, hesitancy to announce policy change, and difficulty in implementing unpopular government cutbacks.

Much rentier state literature suggests that governments utilise oil rents to support high government spending during oil windfalls as a means of securing political loyalty and repressing dissent. The case of Oman seems to suggest a less linear narrative. The data presented in this chapter illustrates the stickiness of government expenditure. It also suggests that periods of oil market stagnation seem to maintain or increase the impulse of governments to spend. In the Omani case, efforts to reduce spending and increase non-oil revenue have been amped up since 2014, but overall expenditure contraction proves difficult. As will be shown below, even minor austerity measures tend to be reversed in the face of social anxiety. On one level, this indicates the government's difficulty in balancing fiscal instability with social instability concerns in contexts of entrenched expectations. On another level, it also suggests that scholars may need to revisit some of the core causal mechanisms accepted in much of the literature on rentier states. Now that the global economy seems to be embedded in a 'low for longer' oil price scenario (Malek, 2018), more empirical cases emerge which can be compared with each other, and also with the policy reactions during the 1980s.

Indeed, similar expenditure patterns can be observed in other oil-dependent economies. This seems contingent on the fiscal robustness of the country case under examination. Moreover, the trend does not appear to be limited to authoritarian contexts but extends to democratic ones as well. For example, new research on Argentina suggests that government expenditure in the form of 'patronage spending' increases during oil-rent downturns, especially when this is reflected in 'job destruction in the oil sector' (González, 2018: 101). The Argentine case studies suggest that patronage is used as a means of containing social discontent. Similar to our observations in Oman and other Gulf rentiers, González shows evidence of countercyclical behaviour, with increased state spending during times of crisis to secure political loyalty and fight unemployment. During oil highs, the government is less tied to patronage expenditure and invests in infrastructure and other public goods

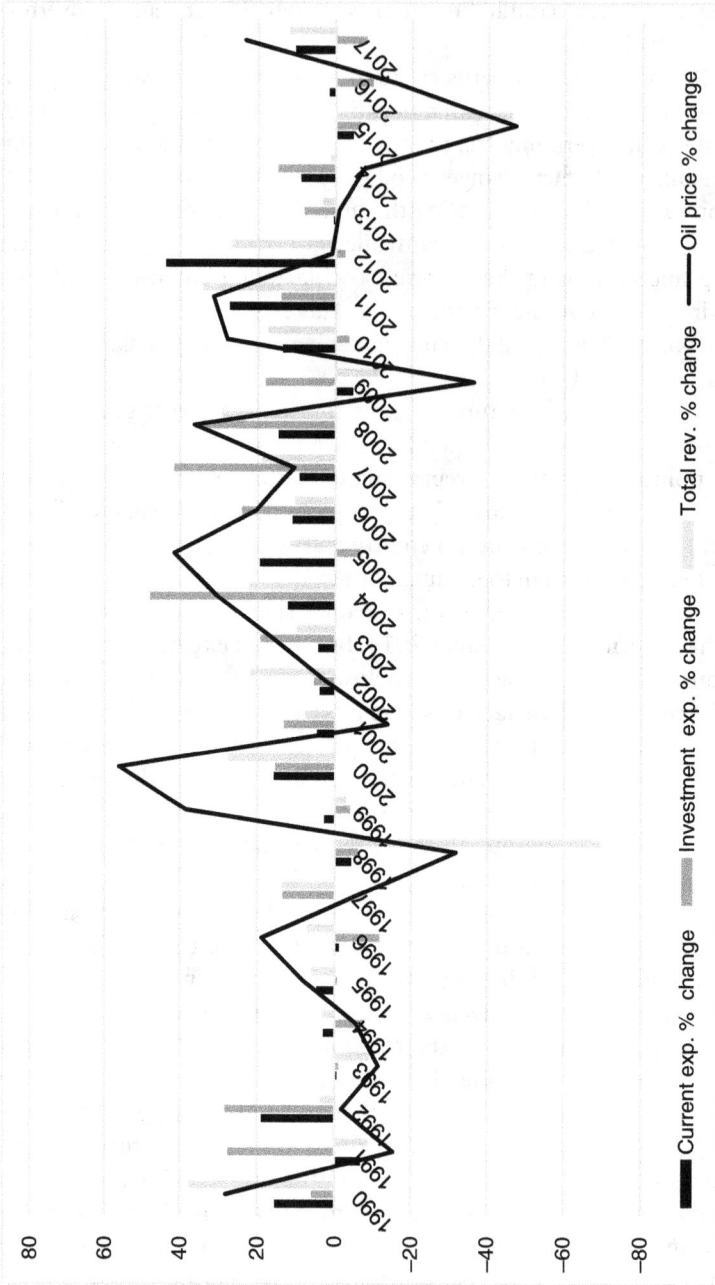

Figure 4.4 Oman's annual change in government expenditure and oil income (percentage). Source: NCSI (1981–2018).

(González, 2018). What is most interesting for the purpose of this chapter is that similar expenditure and patronage patterns are occurring in democratic states and Gulf rentiers alike. This suggests that a concern with stability both motivates and cripples government policy choices across political systems. A large study of multiple country cases with a cross section of political systems would be useful to examine the generalisability of such claims. However, it is important to note here because it helps us de-exceptionalise the narratives around economic choices in the Gulf economies. Our focus in the sections that follow on policy choices and initiatives of the Omani government in the most recent oil downturn illustrates our argument about the tensions between fiscal realities and sociopolitical anxieties.

Sustaining government expenditure during decline

After the oil price declined sharply in the last quarter of 2014, the government had to undertake a variety of actions. Importantly, they reviewed overall income and expenditure with the goal of lowering the break-even oil price. Resolving this meant increasing alternative revenue sources and cutting spending. Several adjustments had to be made, but such reform is always faced with tensions between the demands of fiscal sustainability and social expectations.

In the remainder of this chapter, we explore several reform categories to examine Oman's efforts to come to terms with recent economic constraints while at the same time balancing the need to mitigate the negative social repercussions of austerity. First, we examine the significant increase in government borrowing and the forms and debates around taxation. These debates have a regional dimension, as Gulf states convened to discuss ways of responding to the downturn and craft collective GCC policies – including the introduction of an excise duty and a value-added tax (VAT). Second, we turn to the labour market. As employment pressures mounted, the state pushed the private sector more forcefully to hire local labour and made it more difficult to recruit workers from abroad. Third, we examine subsidy reforms. Finally, we explore a newer trend of state–society engagement around economic development planning. This was made especially obvious through the ways the government engaged with society and the private sector over the shape of economic plans under the auspices of the *tanfidh* programme. *Tanfidh*, or the National Programme for Embracing Economic Diversification, was introduced in collaboration with Malaysia's Performance Management and Delivery Unit (PEMANDU), with the goal of accelerating the pace of diversification away from oil dependence. It was operationalised as a way for societal sectors and actors to have an input in the construction of economic strategies and plans (Tanfeedh, 2017). This marked the first time the government engaged with ideas 'from below' in the process of economic planning, and the format has continued through the visioning process for Vision 2040.

Debt and taxes

From 2014 to 2017, the Omani government's oil income dropped by an average of 10 per cent annually, with a drop of 52 per cent alone in 2015 (NCSI, 2018). To meet expenditure requirements, the government withdrew money from the State General Reserve Fund (SGRF) and borrowed from the domestic and international markets. Therefore, both the current account deficit and public debt increased (see Figure 4.5). In 2017, public debt reached 40 per cent of GDP and the current account deficit reached 15 per cent of GDP.

In June 2016, Oman issued its first sovereign bond since 1997. The bond issued was worth USD 2.5 billion. Again, the government issued an international sovereign *ṣukūk* (Islamic bond) in 2016 worth USD 0.5 billion. Since then, the government has been tapping the international market to meet its expenditure demand. In 2017, the government issued USD 5 billion worth of bonds, and in 2018 it issued USD 6.5 billion worth of bonds. By

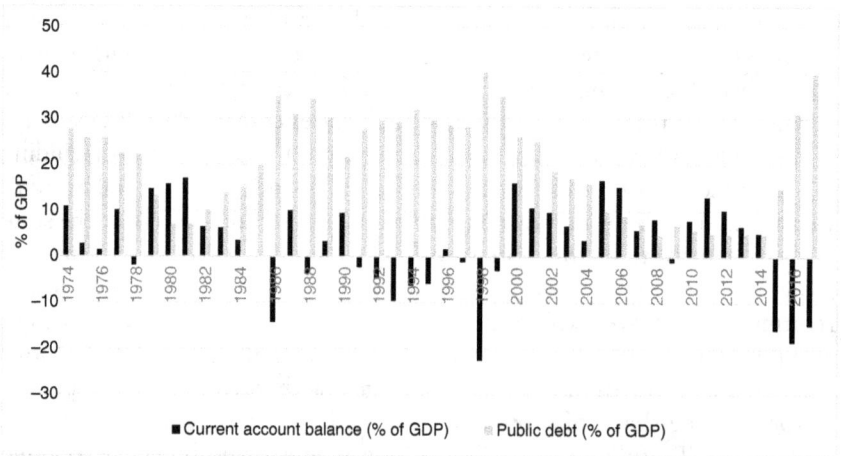

Figure 4.5 Oman's yearly current account and public debt, 1974–2017 (percentage of GDP). Source: World Bank (2019) for current account balance; public debt is calculated from Ministry of Finance (1975–2018).

the end of 2019, the debt-to-GDP ratio had reached 59.9 per cent, with the fiscal deficit at around 6.7 per cent (Bambridge, 2019). The government so far has never had real difficulty in financing its deficit by borrowing from domestic or international financial markets, hence the pressure to reduce government expenditure has been somewhat mediated.

However, the burden of debt has been increasing and the Omani government's sovereign bonds now have a sub-investment grade by all three major credit rating agencies, Standard & Poor, Fitch, and Moody's. In fact, Moody's downgraded the 2019 foreign currency credit rating of Oman on 5 March 2019, from 'Baa3' to 'Ba1' with a negative outlook (Moody's Investor Service, 2019). This policy measure illustrates the oscillating response to fiscal contraction. The state must choose between straining its financial position in global markets to sustain spending or enacting austerity measures and risking social outcry and instability. The Omani government has been walking a tightrope between the two since 2015.

The government not only used debt instruments to sustain spending, but also pursued measures to cut expenditure. Omani government expenditure decreased from Omani riyal (OMR) 15,172 million in 2014 to OMR 12,274 million in 2017 (OMR 1 is about USD 2.6). The average government expenditure reduction from 2015 to 2017 was 7 per cent yearly, in which both defence and recurrent expenditure decreased by an average of 6 per cent annually and development expenditure by an average of 8 per cent annually (Ministry of Finance, 1975–2018).

To increase the efficiency of government spending, state-owned enterprises have been directed to seek financing for their activities through local and international banks, and the government promised to privatise some of its state-owned enterprises in 2016. Until now, no state-owned enterprise has been privatised but various restructuring activities have been ongoing.

The other policy that is usually adopted in this case is to look for alternative sources of income. Fees and taxes are one such source. The government imposed new fees on telecom companies, raising royalties owed from 7 to 12 per cent (Al-Ansari, 2016). By Royal Decree 9/2017, a 35 per cent tax on liquefied natural gas (LNG) exports was issued, and fees for mineral exploitation (e.g. concerning quarries, crushers, and chrome) and for petrochemical companies were increased by 35 per cent (Qanoon.om, 1974–2019).

Moreover, under the amendments to the Income Tax Law, tax exemptions were removed and the corporate income tax rate was increased from 12 to 15 per cent. Changes also saw the removal of the tax exemption for the first OMR 30,000 profit bracket through Royal Decree 9/2017 (Qanoon.om, 1974–2019; KPMG, 2019). In addition, Royal Decree 23/2019 introduced a 100 per cent excise tax for tobacco products, energy drinks, and meat, while a 50 per cent tax was applied on carbonated drinks and alcohol in June 2019 (Qanoon.om, 1974–2019). The introduction of VAT was also announced to the public but not yet implemented by December 2020.

Interesting developments also occurred at the regional GCC level. The statement issued by the Financial and Economic Cooperation Committee of GCC Ministers of Finance on the unified VAT agreement and excise Taxation Agreement of October 2016 was clear about the need to impose taxes to cover government spending and as a tool for financial reform. The statement declared that

> Taxation (especially value-added tax) is one of the most important sources of public revenues to cover the public expenditure and an effective tool in implementing financial, economic, and social policies, because taxes contribute to the provision of financial resources so that governments can finance development projects and provide services. (GCC, 2016; authors' translation)

This means, among other things, that the Gulf states consider income from taxes to be an alternative, important source of income from oil. Moreover, the statement confirms for the first time that the time of dependence on oil revenues to implement fiscal, economic, and social policies is over, and that taxes will become a major source of funding for infrastructure projects and basic services. There is also a tacit reference in the statement that taxes, not just VAT, will be part of economic policies in which governments will work to finance public expenditure. This could also mean that governments of the GCC are looking into other types of taxes.

Oman has yet to introduce VAT, which appears to be an indeterminant delay. It did, however, implement the excise tax in June 2019. Initially, the introduction of VAT was delayed in 2017 with the expectation it would be implemented in September 2019. Yet when the annual state budget was published in January 2019, it made no mention of VAT, and by September 2019 VAT was delayed yet again; this time indefinitely. Despite this, businesses are being encouraged to prepare in advance. According to Sulaiman bin Salim Al-Ameri, General Manager at the General Secretariat of Taxation, the Ministry of Finance does aim to implement VAT, but this depends on the technical preparedness of the system (*Oman Daily*, 2018; Mushtaq and Shah, 2019).

Preparedness of the system is a valid concern, as can be observed in the expected growing pains of neighbour states which have already implemented VAT (Clarke, 2019). However, a cynical assessment could also interpret this as yet another way of delaying uncomfortable reforms. While some of the above government efforts were rewarded with an average increase in non-oil revenue of 3 per cent in 2015–17, the non-oil share of government revenue remains low. Government revenue increased from OMR 7,583 million in 2015 to OMR 8,514 million in 2017, but the non-oil share of this was only 27 per cent in 2017.

Labour market regulation

Alongside increasing indebtedness, some of the greatest challenges facing Oman's economy stem from the labour market. The unrest during 2011 threw the concern of rising unemployment among Oman's youth into the limelight. Since late 2017, employment demands from young graduates have resurfaced forcefully once again – with loud discontent on social media and in demonstrations in front of the Ministry of Manpower. Recent International Labour Organization (ILO) estimates suggest that the overall youth unemployment rate in Oman was 13.7 per cent in 2017, with female youth unemployment estimated at 33.9 per cent (Ennis, 2019a, 2019b; ILO, 2019).[2] Other reports suggest the overall unemployment rate was 17 per cent, with youth unemployment as high as 49 per cent in the same year (World Bank, 2018). At the same time as unemployment concerns sit at the forefront of public policy discourse, the demand for non-national labour persists (Ennis and Al-Jamali, 2014). Data from mid-2018 shows that the private-sector economy continues to rely on expatriates across skill levels (see Table 4.2).

As foreign nationals are not covered by Oman's minimum wage regulation, the cost of expatriate labour results in labour markets where unskilled and skilled-worker salary levels are determined by the market rate in labour-sending countries. This means wages for lower-skilled workers are much lower than the minimum salary an Omani national would be willing to work for and below the cost of living. In many sectors, then, the labour market is segmented, and citizens and non-citizens do not compete for the same jobs. At higher professional levels, the wage differential flattens, and there is more competition for jobs (Hertog, 2013: 179–80; Ennis and

Table 4.2 Oman's private-sector labour nationalisation rates.

Skill level	% of Omanis
Specialists	24.81
Technicians	25.89
Professional occupations	14.81
Skilled workers	9.57
Unskilled workers	8.51
Total private-sector labour	12.53

Source: Ministry of Manpower (2018).

Walton-Roberts, 2018: 173). Oman's history of economic development planning since the early 1970s reveals an acute awareness of the need to reduce its dependence on oil and expatriate workers through economic diversification initiatives. Yet the picture has changed little. The private sector maintains a preference for foreign recruitment, and citizens have an expectation of public-sector employment. This persists in the background even while it has grown increasingly obvious in the public imagination that such labour market bifurcation and expectations are unsustainable.

The government targets national job creation and labour segmentation through two types of interventions. On the one hand, the government uses a combination of human capital investment (education) and more direct labour market interventions like quotas for certain percentages of positions to be reserved for citizens. These policies are known as Omanisation. On the other hand, the government works on the supply side of labour and targets the inflow of foreigners through increasing visa fees for employers or blocking recruitment in certain occupations. The successful implementation of Omanisation policies and foreign visa restrictions has been varied and sector dependent. While the public sector, banking, and insurance achieved high Omanisation ratios, other sectors have stagnated. When oil prices are low, the country faces greater pressure to create jobs for citizens.

As outlined by Ennis and Al-Jamali (2014), successful Omanisation gains during the 2000s were quickly lost after the unrest of 2011, which prompted a series of government measures to quickly quell the discontent. Several of these reversed existing incentives to encourage citizens to seek positions in the private sector. In particular, the creation of 50,000 new government jobs actually saw nationals in the private sector resign to compete for these jobs. Data show that Omani participation in the private sector remained flat for a two-year period, with the overall private sector Omanisation level dropping from 16 per cent in 2010 to under 12 per cent by the end of 2012. This appears to be a direct result of government policy, as decreases occurred during the same months as policy announcements (for data visualisation and discussion, see Ennis and Al-Jamali, 2014: 12–13).

The urgency of job creation has intensified since renewed youth protests for jobs in the autumn of 2017, which have continued on- and offline. In late 2017, the government announced it would again create 25,000 new jobs in the early months of the new year. It soon became evident that these could not all be found in the public sector. Even with the announcement, the Council of Ministers 'urged all private sector establishments to take the initiative and shoulder their national responsibilities in the issue of employment of Omanis' (Ministry of Foreign Affairs, 2017). By January, pressure had grown on various sectors to increase their recruitment of Omani graduates. At the same time, companies were banned from foreign recruitment of eighty-seven different professions for six months in 2018 (*Arab News*, 2018). By May 2018, the Ministry of Manpower had penalised 161 companies for not hiring even one Omani citizen (*Times of Oman*, 2018c). This visa ban has since been extended several times, first until January 2019, then again to the end of June 2019 (*Muscat Daily*, 2019).

On the one hand, the government has pressure from growing numbers of job seekers. On the other, the private sector lobbies for greater labour market flexibility and ease of hiring from abroad. Interests diverge, and pressure for different forms of regulation and deregulation across the labour market try government policy and expand the tension between policy choices and informality. It is no surprise, then, that when oil prices rise, pressure is relieved, and the country easily slips back into a reliance on public-sector hiring and foreign labour recruitment. During times of higher regulation of migrant labour, businesses seek informal channels or personal bureaucratic connections to achieve exceptions or ways around stringent regulatory demands. Perceptions of unwilling private-sector employers among job seekers only exacerbate employment tensions and blend these with allegations of corruption and privilege.

Persistent challenges from the labour market carry the risk of social and political instability, which comes to the forefront of public concern during periods of financial constraint and fiscal uncertainty. The challenges of the labour market are not new. Both an acknowledgement of the problem and various Omanisation goals reappear in each iteration of the five-year development plans, in long-term visions, and in sectoral development strategies (from health care to tourism). The failure to resolve labour market problems rests in the labour market's embeddedness in economic structures and in the state's entanglement between competing directions and competing interests in how to resolve these pressures.

Closing the subsidy tap: Cutbacks and leaks

Similar to the story of the labour market, high fiscal pressure since 2014 has also trapped the government between competing desires for stability as it grapples with subsidy reforms. External pressure from international financial institutions comes in the forms of warnings concerning the necessity of austerity measures to reel in government expenditure (IMF, 2018). Internal pressure from below reacts to mentions of price increases. The government is caught between the threat of financial market instability and social instability. This is especially evident around fuel subsidy reforms.

The first price adjustments targeted industrial producers rather than prices that directly affected citizens. For instance, in January of 2015 the government doubled the price of natural gas for industrial producers, and in January of 2017 increased electricity prices for large industrial, commercial, and government electricity consumers (*Times of Oman*, 2016: 40; IMF, 2017: 40). It was not until the rise in fuel prices in January 2016 that households were directly affected by subsidy reductions. Petrol prices increased by 23 per cent, with monthly adjustments thereafter to reflect price changes in global oil markets and fuel-price changes in neighbouring economies.

Social reaction to these changes was immediate, and by February 2017 a public protest was staged outside of the Ministry of Oil and Gas against what was seen as a 75 per cent increase in the cost of fuel at the pump over the year (Al-Shaibany, 2017a). Moreover, *majlis al-shūrā*[3] members and trade unionists argued for more stable prices or ways of easing the burden on low-income earners (Gulf Business, 2017a, 2017b). The reaction of the government was equally swift, revealing the deep concern about spiralling social unrest. The same week as the protests, the Council of Ministers announced a price cap (Al-Shaibany, 2017b). By August, a pilot study of a fuel subsidy scheme for the poor was rolled out, and by mid-December the government had adopted a fuel subsidy card scheme to replace the cap (*Times of Oman*, 2017a, 2017b). All citizens over eighteen years of age earning less than OMR 600 a month were eligible for a subsidy card which guaranteed they would pay no more than OMR 0.18 per litre (approximately USD 0.43) of fuel for up to 200 litres per month. Within just ten days of launching the card, over 100,000 citizens had registered for the national fuel subsidy system (*Times of Oman*, 2017c). By June 2018, the subsidy scheme was expanded to include all citizens earning less than OMR 950 per month, increasing the eligibility to around 69 per cent of the working population (*Times of Oman*, 2018a; Oman Fuel Pricing Committee, 2019).

Government action and reaction to reducing fuel subsidies reveals the tension around such economic policy choices. The tension is reflected across all discussions of subsidy reductions that directly impact households. 2018 also broached the topic of electricity subsidy reform for households. This public discussion was accompanied by careful messaging about the level of subsidies and the necessity to reduce them with protections for low-income earners in mind. When electricity authorities conducted a study, they learned that a section of the Omani population was not aware that they were being provided energy subsidies (*Times of Oman*, 2018b). Thus, the discussion has entailed raising awareness of the subsidies in place, their cost, the opportunity cost involved, and demonstrating an awareness of the necessity of a social safety net being in place. The reaction to fuel subsidy changes revealed the necessity for incremental government messaging to prepare the public for any changes.

The tanfīdh *programme and state engagement with the private sector and society*

Amid the uncertainty stemming from the oil price shock, the government introduced the *tanfīdh* programme to implement the ninth five-year development plan (2016–20) with the aim of 'deepening' economic diversification, reducing dependency on oil sector activities, and diversifying the base of government income (Tanfeedh, 2017). *Tanfīdh*, introduced at the start of this section, is described as an 'action-oriented' programme:

> [It] aims to identify the responsibilities, resources, and timeframes needed for implementation of initiatives that drive economic diversification; set clear standards and Key Performance Indicators (KPI) for said initiatives; and provide periodic reports on the progress achieved in the implementation of the initiatives to ensure that the public is routinely informed of the Program's progress. (Tanfeedh, 2017: 8)

This programme is the short form and most popular name for the development programme called the National Programme for Enhancing Economic Diversification. The *tanfīdh* initiative was especially remarkable for efforts to engage society in decision-making. The Arabic word *tanfīdh* literally means implementation and inspires a sense of a proactive government interested in strategising and then executing the ideas curated from the societal constituents. The government for the very first time invited the private sector, civil society, and the broader public to participate in crafting and prioritising *tanfīdh* programmes. This follows earlier messaging from higher ranks of government all the way to Sultan Qaboos, who, in his Annual Council of Oman speech in November 2012, stated: 'As the necessary infrastructure is almost complete, we have directed the government to focus on future social development' (Al-Said, 2015: 549).

Within the framework of *tanfīdh* in October 2016, civil society, the private sector, and officials were invited to participate in the programme laboratories. The labs lasted for six weeks. The concurrent labs focused on manufacturing, transport and logistics, tourism, finance, and the labour market – all areas identified as key targets of economic planning. The programme designed 121 initiatives and projects with the aim of generating 67,000 jobs for the labour market by 2020, 30,000 of which would be technical and skilled labour positions reserved for Omanis (Tanfeedh, 2017: 110). The proposed initiatives were also intended to provide investment opportunities estimated to be worth OMR 14 billion, nearly 93 per cent of which was expected to be financed by the private sector (Tanfeedh, 2017: 30–1).

Further labs followed the first phase of *tanfīdh*, in energy and mining and the fisheries sectors. The energy and mining sector lab focused on renewable energy, gas, metal and non-ferrous metals, and the governance of the sector. The energy and mining programme came up with forty-three initiatives and projects. These initiatives are designed to generate 1,660 jobs for Omanis by 2023 and provide investment opportunities worth OMR 797 million, 99 per cent of which is expected to be financed by the private sector. The energy sector hopes to generate 11 per cent of energy requirements from renewables by 2023.

The objectives of the fisheries sector are to increase its contribution to GDP from OMR 225 million in 2016 to OMR 781 million in 2023, and to attract the private sector to finance 80 per cent of the sector projects and generate at least 8,600 new jobs by 2023. The ICT sector is currently being examined by the private sector and the government. On 1 July 2019, the sultan issued four royal decrees (50/51/52/53/2019) (Qanoon.om, 1974–2019), including laws of bankruptcy, public–private partnership, privatisation, and foreign capital investment, with the objective of creating a favourable regulatory environment for public and private partnership and investment.

It is still too early to evaluate the effectiveness of these initiatives. This is partly due to the fact that the Supreme Council for Planning (SCP) has not been releasing the promised regular reports on the programme. *Tanfīdh* is supposed to 'provide periodic reports on the progress achieved in the implementation of the initiatives to ensure that the public is routinely informed of the Program's progress' (Tanfeedh, 2017: 7). In news articles in the official dailies which announce job creation, there is usually one sentence that points to *tanfīdh*. However, official reporting and analysis has not yet appeared beyond policy discourse. As per the eight-step programme methodology, following the labs (step number four), KPI are supposed to be announced before implementation, and following implementation, an independent international audit firm is supposed to review and validate the results, after which the results are expected to be published. However, steps five, seven, and eight have not been forthcoming, and no explanation has been provided (Tanfeedh, 2017: 19). It thus remains to be seen how successful the *tanfīdh* programme will be in attracting private investment and generating jobs.

What the state has done is take efforts to continue to strengthen communication with the community on a continuous basis, including civil society. As one measure towards this end, the government established the Government Communication Centre in the General Secretariat of the Council of Ministers in October 2017 (*Muscat Daily*, 2017). Since its inception, the centre has taken upon itself the task of communicating with civil society on an ongoing basis to explain government plans and programmes, as well as the various policies adopted by the government, including policies that may not be popular, such as eliminating subsidies on fuel and utilities, VAT, and taxation. This reflects government learning on the importance of political messaging to prepare the public sphere for difficult policy decisions. This is intended to help ease negative social reaction. The centre adopted a direct approach of communication to civil society and seems to encourage key members of civil society organisations to speak to official and private media on its behalf. In this way, the state becomes more effective at framing the public discourse around economic policy.

Whether theatrical or genuine, the government appears to find this deliberative framing of public discourse and engagement with civil society to be an important activity. This practice has continued, and notably civil society was again invited to participate in Oman's Vision 2040 workshops. The workshops were intended to formulate the strategic directions and long-term objectives for Oman. There have been ongoing, meaningful attempts to engage with society in the construction of the vision. In many ways, the engagement with civil society appears to be a positive development. However, it may also indicate a desire to centralise discourse on alternative economic development programmes and, even more cynically, provide an additional means of identifying and surveilling civil society actors, concerns, and activities.

Conclusion

This chapter shows that economic adjustments in Oman reflect oil market fluctuations, but expenditure does not adjust as steeply during contractions in the oil market. We have argued that the Omani government is pulled in opposing reform directions as it tries to respond to the oil price downturns and is mired in what presents as an impossible pendulum between financial instability and social instability. If the state prioritises its declining global financial position and growing indebtedness in financial markets, it will have to take socially difficult austerity measures. The IMF Article IV consultations repeatedly underline its typical range of subsidy cutbacks and labour market reforms (e.g. IMF, 2018). At the same time, these types of measures prompt social discontent, and therefore the state regularly rolls back or revises reform announcements to calm social reaction, as illustrated above. Throughout this oscillation, we continue to observe high government expenditure. Even with government cutbacks, the revenue–expenditure gap widens. In illustration of this argument, this chapter has examined several interesting fiscal adjustment trends under oil price downturns in the Omani case. First, Oman has targeted the usual policy spaces to deal with contractions in its revenue stream. Increasing taxation regimes, and sources of taxation, combined with subsidy cutbacks have been key components of the policy discourse. Policy implementation hesitancy, reversals, and/or softening of the social impacts of these measures indicate not only a governmental concern with welfare, but also a concern about rising levels of discontent. The concern with discontent is also evident in the evolving policy measures around nationalising various occupations in the labour market and expanding available employment. This policy space also illustrates the broader tensions between private-sector interests and public expectations that pull the government in different directions. Second, the challenges of reforming the labour market demonstrate the stickiness of jobs among expenditure lines. Crucially, they underline the embedded structures of the economy that obstruct limited policy measures in the absence of deep structural transformation. Indeed, promises to increase the quantity of available jobs seem to be a sure measure to, at least temporarily, ease social pressure around employment. Third, the intensity of financial concerns and constraints is countered by the intensity of concerns around potential social instability. The government has navigated this tension by softening policy measures while delaying or rolling back others, as well as increasing its spending capacity by borrowing on the international market and drawing on sovereign funds.

At a minimum these findings show us the divergent concerns of the 'middling rentiers' versus the 'extreme rentiers' (Herb, 2014). More broadly, they suggest that an analysis of countercyclical spending across oil-dependent cases in the Middle East and also in other regions and across political spectrums will provide a stronger understanding of expenditure and patronage patterns in resource-abundant states. Observing patterns, or breaks from these, in contracting rentiers offers important insights to our theorising on extractive industries and on the economic and political behaviours of resource-dependent countries.

Notes

1 Defence expenditure was particularly high from 1970 to 1975 because of the Dhofar War. When Sultan Qaboos took power in 1970, he inherited a war from his father in the province of Dhofar in southern Oman.
2 To read more about the employment gender gap in Oman and other Gulf economies, see Ennis (2019a, 2019b).
3 *Majlis al-shūrā* refers to the Consultative Council. This is the elected, lower chamber of government.

References

Abdel-Fadil, M. (1987), 'The macro-behaviour of oil-rentier states in the Arab region', in H. Beblawi and G. Luciani (eds), *The rentier state*. London: Croom Helm, pp. 83–107.

Ahmed, A., and S. Al-Saqri (2012), 'Natural resource depletion, productivity and optimal fiscal strategy: Lessons from a small oil-exporting economy', *Journal of Applied Economics*, 11:1, 56–80.

Al-Ansari, O. b. A. (2016), 'Muʿashshir suq masqaṭ yartafaʿ 17 nuqṭa' [Muscat securities market index raises 17 points], *Oman Observer*, 29 December. www.omandaily.om/?p=421650 (accessed 4 March 2019).

Al-Said, S. Q. b. S. (2015), *The royal speeches of His Majesty Sultan Qaboos Bin Said*. Muscat: Ministry of Information.

Al-Shaibany, S. (2017a), 'Omanis call for government cap on fuel prices in first protest since 2011', *The National*, 2 February. https://bit.ly/2QFg7HT (accessed 4 March 2019).

Al-Shaibany, S. (2017b), 'Oman caps fuel price after protests', *The National*, 8 February. https://tinyurl.com/2vev97c8 (accessed 4 March 2019).

Al-Yousef, M. B. M. (1996), *Oil and the transformation of Oman: The socio-economic impact*. London: Stacey International.

Arab News (2018), 'Scores of companies fined for violating Oman's expat work ban', 15 May. www.arabnews.com/node/1302966/business-economy (accessed 16 May 2018).

Bambridge, J. (2019), 'Oman's economy at a glance', MEED Middle East Business Intelligence, 19 December. www.meed.com/omans-economy-at-glance/ (accessed 27 February 2020).

Beblawi, H., and G. Luciani (eds) (1987), *The rentier state*. London: Croom Helm.

Clarke, J. (2019), 'Challenges of VAT implementation in the UAE', MEED Middle East Business Intelligence, 20 February. www.meed.com/vat-regime-uae-pinsent-masons/ (accessed 17 July 2019).

Ennis, C. A. (2015), 'Between trend and necessity: Top-down entrepreneurship promotion in Oman and Qatar', *The Muslim World*, 105:1, 116–38.

Ennis, C. A. (2019a), 'The gendered complexities of promoting female entrepreneurship in the Gulf', *New Political Economy*, 24:3, 365–84.

Ennis, C. A. (2019b), 'Rentier-preneurship: Dependence and autonomy in women's entrepreneurship in the Gulf', *POMEPS Studies*, 33, 60–6. https://tinyurl.com/yavww8m5 (accessed 29 February 2020).

Ennis, C. A., and R. Z. Al-Jamali (2014), 'Elusive employment: Development planning and labour market trends in Oman', Chatham House, Research Paper, 16 September. https://bit.ly/3eOM1cF (accessed 29 February 2020).

Ennis, C. A., and M. Walton-Roberts (2018), 'Labour market regulation as global social policy: The case of nursing labour markets in Oman', *Global Social Policy*, 18:2, 169–88.

GCC (2016), 'al-Ittifaqiyya al-muwahhada li-daribat al-qima al-madafa li-duwal Majlis al-Taʿawun li-Duwal al-Khalij al-ʿArabiyya' [The common VAT agreement for the Gulf Cooperation Council states]. Abu Dhabi: Ministry of Finance. https://bit.ly/3sExIwJ (accessed 8 March 2019).

González, L. I. (2018), 'Oil rents and patronage: The fiscal effects of oil booms in the Argentine provinces', *Comparative Politics*, 51:1, 101–26.

Gray, M. (2011), 'A theory of "late rentierism" in the Arab states of the Gulf', Georgetown University Center for International and Regional Studies, Occasional Paper, 7. https://bit.ly/3qRlK1s (accessed 17 February 2020).

Gulf Business (2017a), 'Oman shura members call for fixed annual fuel price', 5 February. https://tinyurl.com/edzjsmsb (accessed 4 March 2019).

Gulf Business (2017b), 'Oman trade unionists urge government to introduce fuel stamps', 7 February. https://tinyurl.com/993s6b5a (accessed 4 March 2019).

Hanieh, A. (2018), *Money, markets, and monarchies: The Gulf Cooperation Council and the political economy of the contemporary Middle East.* Cambridge: Cambridge University Press.

Herb, M. (2014), *The wages of oil: Parliaments and economic development in Kuwait and the UAE.* Ithaca: Cornell University Press.

Hertog, S. (2013), 'State and private sector in the GCC after the Arab uprisings', *Journal of Arabian Studies*, 3:2, 174–95.

ILO (2019), 'Oman employment data', ILO Stat. https://bit.ly/3dUoSXH (accessed 14 August 2019).

IMF (2017), 'If not now, when? Energy price reform in Arab countries', Policy Papers. https://tinyurl.com/tb9hwn53 (accessed 3 March 2019).

IMF (2018), 'IMF staff concludes 2018 Article IV visit to Oman', 19 April. https://tinyurl.com/3h7uwnkz (accessed 7 March 2019).

KPMG (2019), 'Oman tax law amendments', February. https://tinyurl.com/88rrtk5j (accessed 8 March 2019).

Kropf, A. (2010), 'Resource abundance vs. resource dependence in cross-country growth regressions', *OPEC Energy Review*, 34:2, 107–30.

Luciani, G. (1990), *The Arab state.* Berkeley: University of California Press.

Luciani, G. (2011), 'The political economy of monetary integration and exchange rate regime in the GCC', in M. Legrenzi and B. Momani (eds), *Shifting geo-economic power of the Gulf: Oil, finance and institutions.* London: Routledge, pp. 23–38.

Malek, C. (2018), 'Oil's "lower for longer" reasserts itself', *Financial Times*, 22 November. https://on.ft.com/3eQRG1U (accessed 19 July 2019).

Ministry of Finance (1975–2018), *State final account, issues 1975–2018.* Muscat: Government of Oman.

Ministry of Foreign Affairs (2017), '25,000 jobs for Omanis in public, private sectors', 5 October. www.mofa.gov.om?p=10333&lang=en (accessed 3 March 2019).

Ministry of Manpower (2018), 'Open data portal'. www.manpower.gov.om/OpenData/home/home (accessed 1 March 2019).

Ministry of National Economy (2007), *Vision for Oman's economy – 2020: Long-term development strategy (1996–2020)*, 2nd edition. Muscat: Government of Oman.

Moody's Investor Service (2019), 'Moody's downgrades Oman's rating to Ba1, outlook negative', 5 March. https://tinyurl.com/vaxd5mn (accessed 7 March 2019).

Muscat Daily (2017), 'Government communication forum launched; focuses on transparency and accuracy of Information', 23 October. https://bit.ly/3eQKyT8 (accessed 2 December 2019).

Muscat Daily (2019), 'Visa ban on 87 jobs in private sector extended: MOM', 4 February. https://bit.ly/33esEEt (accessed 3 February 2019).

Mushtaq, I., and J. Shah (2019), 'VAT in Oman: Businesses should take practical steps now', *Times of Oman*, 6 February. https://bit.ly/3sENZSs (accessed 6 February 2019).

NCSI (1981–2018), *Oman: Statistical yearbook, 1981–2018.* Muscat: National Centre for Statistics and Information.

NCSI (2018), *Oman: Statistical yearbook, 1981–2018*. Muscat: National Centre for Statistics and Information. https://bit.ly/3aVWTUZ (accessed 20 July 2019).

Oman Daily (2018), 'Tawajuh li-l-tatbiq fi Sebtember 2019 wa iʿfaʾ tam li-94 salʿa wa qitaʿayy al-taʿlim wa-l-saha' [Effective from September 2019, with exemptions for 94 commodities and the health and education sectors], 25 September. www.omandaily.om/?p=631146 (accessed 8 March 2019).

Oman Fuel Pricing Committee (2019), 'National subsidy system'. https://nss.gov.om (accessed 3 March 2019).

Qanoon.om (1974–2019), 'al-Marasim al-sultaniyya' [Royal decrees electronic database]. https://qanoon.om/ (accessed 3 March 2019).

Tanfeedh (2017), 'Tanfeedh handbook', The National Program for Enhancing Economic Diversification (Tanfeedh), July. https://tinyurl.com/nwex8d6h (accessed 8 March 2019).

Times of Oman (2016), 'Higher tariff for large electricity consumers in Oman from January 2017', 12 October. https://bit.ly/3vsLX9o (accessed 3 March 2019).

Times of Oman (2017a), 'Fuel cards for poor in test phase in Oman, says Ministry of Oil and Gas', 15 August. https://bit.ly/3eMnyVF (accessed 4 March 2019).

Times of Oman (2017b), 'Oman government introduces fuel subsidy scheme', 13 December. https://bit.ly/3upewVf (accessed 4 March 2019).

Times of Oman (2017c), 'More than 100,000 Omanis register for fuel subsidy system', 23 December. https://bit.ly/3ubkgS8 (accessed 4 March 2019).

Times of Oman (2018a), 'Fuel subsidy net widened for Omanis', 11 June. https://bit.ly/3gUvvKY (accessed 4 March 2019).

Times of Oman (2018b), 'Power subsidies should protect the needy, not the greedy', 28 October. https://bit.ly/338xOCg (accessed 3 March 2019).

Times of Oman (2018c), '161 companies penalised for violating Omanisation law', 14 May. https://bit.ly/33bRCEs (accessed 19 May 2018).

Valeri, M. (2009), *Oman: Politics and society in the Qaboos state*. London: Hurst Publishers.

World Bank (1994), 'Sultanate of Oman: Sustainable growth and economic diversification', World Bank Report, 12199-OM.

World Bank (2018), 'Oman's economic outlook – April 2018'. https://tinyurl.com/azmks28p (accessed 4 March 2019).

World Bank (2019), 'World development indicators online database', World Bank Data. https://datacatalog.worldbank.org/dataset/world-development-indicators (accessed 4 March 2019).

5

Qatar: Leadership transition, regional crisis, and the imperatives for reform

Matthew Gray

Introduction

The plunge in global crude oil prices that began in mid-2014 had a range of sudden and sharp impacts on oil-exporting economies, including Qatar. This ushered in a remarkable set of challenges across the Gulf, but in the case of Qatar, matters were further complicated by the fact that power had only just been transferred to a young, new amir the previous year, after Amir Hamad (r. 1995–2013) abdicated the leadership to his son Tamim (r. 2013–). Formidable as the sudden economic challenges must have appeared to Amir Tamim, however, they were substantially offset by Qatar's unique features – such as its higher reliance on natural gas exports rather than oil, and its enormous hydrocarbon income per citizen – which helped it manage low oil prices far better than other Gulf states. Notwithstanding this, however, it was subsequently saddled with a set of further and unique complications in June 2017, when Saudi Arabia, the United Arab Emirates (UAE), Bahrain, and some other states cut diplomatic ties with Qatar and imposed sanctions and blockades, aiming to pressure Doha into dramatically altering foreign policy. While Qatar has seemingly outmanoeuvred and overcome what has become known as the 2017 Qatar crisis, this predicament nonetheless profoundly threatened its economy and security.

This chapter outlines and assesses how various actors and forces have shaped Qatar's political economy, especially since the decline in oil prices in 2014, examining the key dynamics at work while also providing a theoretical structure through which to best view these dynamics. In assessing the impacts of low oil prices since 2014, there is an emphasis on both the structural characteristics of Qatar's economy and the dynamics of the post-2017 crisis, with the leadership transition also considered. The overarching argument is that Qatar was uniquely positioned for the pressures it has faced since 2014: a dual combination of a unique economic structure and a highly effective set of regime strategies have allowed it to withstand economic shocks and sudden diplomatic pressures by underwriting a number of important reforms and limiting the threats to political stability. Its recent history has included examples of poor decisions, and some fundamental and needed reforms remain neglected – and these are not ignored in the pages that follow – but on balance, Qatar has shown considerable dexterity and efficacy in its handling of multiple challenges since 2014.

Qatar's economic trajectory from independence to Amir Hamad (1995–2013)

Prior to 1995, when Amir Hamad seized power from his father, Khalifa, and began a massive transformation of its political economy, Qatar was a sleepy, inconspicuous place. Oil production grew steadily and consistently in the 1950s and 1960s, giving the state greater capacity and a dominant role in the economy (Gray, 2013: 29–46), especially after the leap in oil prices in 1973–74 and in 1979–80. Oil revenues jumped from USD 122.4 million in 1970 to USD 5,387 million in 1980 (al-Naqeeb, 1990: 81), a stunning 4,400 per cent rise. This funded massive infrastructure development and the creation of an 'all-encompassing welfare state' (Ulrichsen, 2014: 24), expanding and deepening the rentier structure which, with only a few modifications, dominates the state–society relationship to the present day. The 1980s were a more challenging time after oil prices fell in the early 1980s, staying low by then-recent standards until the late 1990s. The 1980s were also marked by regional instability and anxiety created by the 1978–79 Iranian Revolution, the 1980–88 Iran–Iraq War, and then the 1990–91 Gulf War. The state continued to develop infrastructure and expand health care and education, with low oil prices offset somewhat by an expansion of natural gas and petrochemicals production. By the mid-1990s, Qatar's gross domestic product (GDP) per capita was USD 12,820, around the median point among the Gulf states and making it an upper income economy, if only just (Abdelkarim, 1999: 30), but its economy was dominated by oil and it had a passive foreign policy that yielded to Saudi Arabia on most key matters (Gray, 2013: 44–5).

Shaykh Hamad, heir apparent to his father, Khalifa, mounted the coup against the latter in June 1995 because he felt that the amir was too cautious and resistant to necessary reforms (Herb, 1999: 118–19). He quickly began transforming the political economy. He was energetic and ambitious, and through his inner circle actively drove and managed Qatar's economic, social, and diplomatic transformation. Hydrocarbon rents remained the dominant source of revenue, and the rentier arrangement that it funded remained as strong as previously, but Hamad's transformation was nonetheless

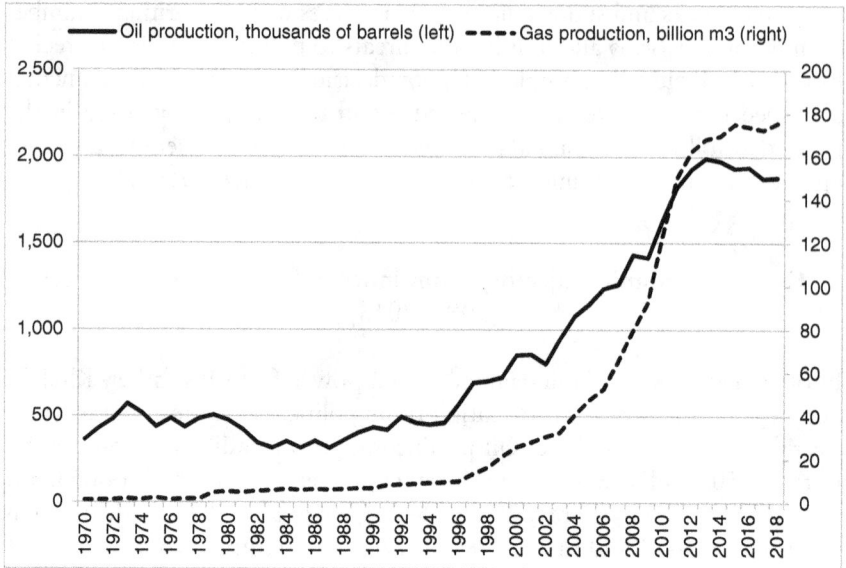

Figure 5.1 Qatari production of oil (in thousands of barrels) and gas (in billion m3), 1970–2015. Source: Derived from data provided in the Excel workbook in BP (2019).

dramatic. Oil production was further developed (Gray, 2013: 91–2; The Economist Intelligence Unit, 2017), but it was the natural gas industry that was most colossally expanded, driven by joint ventures in the giant North Dome field (see Figure 5.1). By the 2000s, gas had overtaken oil in economic importance. The North Dome is the largest non-associated gas field in the world, shared between Qatar and Iran (which calls it South Pars). It is thought to have about 33.5 tcm of recoverable gas, with up to 25.5 tcm in North Dome, about 95 per cent of total Qatari gas reserves (Dargin, 2011: 308). Gas extraction and production is managed by Qatar Petroleum (QP), through joint ventures with international firms under two main phased projects, Qatargas and RasGas (Gray, 2013: 93–7). These have used international expertise and capital effectively, while maintaining overall Qatari control of the projects. Other gas projects followed, such as the Dolphin Project to supply neighbouring Gulf states, as did initiatives such as in gas processing and experiments in new areas such as gas-to-liquid (GTL) technology (Coleman, 2000: 32–4; Gray, 2013: 98–100).

But it was in other sectors that Hamad most dramatically transformed Qatar. The revenue numbers illustrate this: total oil and gas revenues rose from about Qatari riyal (QAR) 39 billion (about USD 10.7 billion) in 2000 to over QAR 364 billion (over USD 100 billion) in 2011, yet their share of overall GDP remained roughly similar, at 60.4 per cent in 2000 and 57.7 per cent in 2011 (Gray, 2013: 87). Hamad focused on economic diversification, seeking, like other energy exporters, to lower the risks of energy price fluctuations and to create broader employment opportunities. The policy was driven by Qatar's main strategic document, the Qatar National Vision 2030 (QNV), which emphasised human development through a comprehensive reform of education; 'Qatarisation' of the workforce, seeking to expand the role of Qataris in the economy; and measures to increase commercial competitiveness (Gray, 2013: 119–22). Stemming from this were more business-friendly policies, including initiatives to attract investment, tax reforms, property law reforms, and institutional and governance reforms (Ibrahim and Harrigan, 2012: 12–18). A sovereign wealth fund (SWF), the Qatar Investment Authority (QIA), was created in 2005 to invest some of Qatar's rent long term, and quickly became an active international investor (Gray, 2013: 67–9, 105–9; Kamrava, 2013: 96–102).

This led to the emergence or expansion of several economic sectors. While they were and usually still are underwritten by energy rents, the ultimate aim was that they become independent of such support. The construction sector is an example, as it has seen a remarkable boom since the 1990s because rents have risen and it supports the development of other sectors such as transport, higher education, sports, and tourism. Major events, such as Qatar's hosting of the 2022 FIFA (Fédération Internationale de Football Association) World Cup, have been drivers, too. The rise in higher education has been an important component in Qatar's human development strategy, with branches of foreign campuses established at Education City and Qatar University substantially reformed. Banking and finance also grew in support of a more sophisticated economy: a stock market opened in 1997, state bonds became more common, and the state supported a growth in Islamic financial institutions. The creation of the Qatar Financial Centre (QFC) in 2005 offered concessionary tax and operational terms and sought to develop and (carefully) globalise the sector. Finally, transport and tourism were crucial to diversification. As one example, the state supported a massive modernisation by Qatar Airways after the 1990s and subsequently bought into the firm, using it to raise Qatar's profile and to encourage certain types of travel and tourism (Morakabati *et al.*, 2014: 416–18).

The airline is only one example of the national 'branding' strategy that began under Hamad, wherein the state has sought to cultivate a strong, positive global image of Qatar's society and economy, with the message that it is a modern, globalising society, open to the world and pro-business while also respecting its traditions and values (Peterson, 2006: 732–48; Roberts, 2017a: 93–101, 103–22). Branding has also been undertaken through major events – Doha's hosting of the 2006 Asian Games, then winning the bid to host the 2022 FIFA World Cup – and through cultural events, scholarships for foreign students, tourism, investments abroad, and even by the consistent use of its national colours. Most famously, perhaps, Al Jazeera television has been prominent in the branding strategy, painting Qatar as an open, reforming country. Branding has sought to build a positive view of Qatar, especially among foreign political and commercial elites with the power to influence their states' policies, but also as a means of nation-building at home. Linked to branding, Hamad also had an activist foreign policy, seeking an outsized prominent international role for Qatar in diplomatic, security, and at times military initiatives (Roberts, 2017a: 35–91). Thus, diplomacy, branding, and economics were linked and interwoven; this tripartite dynamic sought to make Qatar more attractive to the outside world, while building webs of external support as a form of security, thus reducing the risk of conflict, developing new commercial opportunities, and ultimately reinforcing the ruling family's position (Soubrier, 2017: 130–1).

By the end of his tenure, Hamad's record was solid, if not exclusively positive. As shown in Figure 5.2, Qatar remained highly reliant on rents throughout the 1990s and 2000s, even in periods of relatively low rents. The oil price boom of 2004–8 is evident, as is the short-run collapse of prices around the time of the 2008–9 global financial crisis, although by

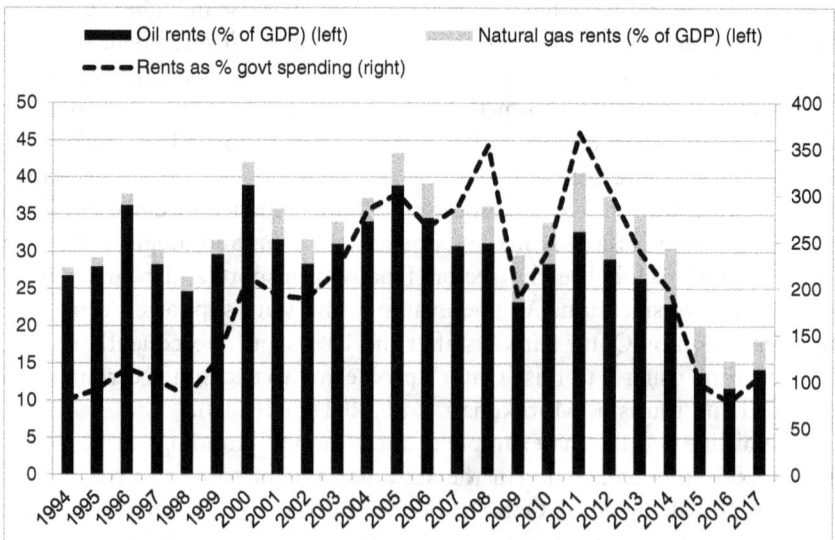

Figure 5.2 Qatar's oil and gas rents (percentage of GDP) and rents as percentage of state spending. Source: Derived from data downloaded from World Bank (n.d.).

then, Hamad's reform and diversification strategies were beginning to have an impact. Some reforms remained undone, however, such as labour market reforms, while others, such as democratic steps outlined in the 2004 Constitution, were promised but not implemented (Gray, 2013: 233–6). National branding seemed successful for a long time, although it laid the foundations for the grievances that prompted the diplomatic crises in 2014 and after June 2017.

Theorising Qatar's political economy

Three theoretical approaches are used here to explain Qatar which provide, in combination, a thorough explanatory framework for the political economy of Qatar under Hamad and Tamim (and even back into the Khalifa period to some extent). Hamad's era is important because it was the time when this strategy for consolidation, elite maintenance, and popular legitimisation was solidified. This period, and Tamim's, are best explained by a conglomeration of three theoretical arguments: rentier theory, and particularly the idea of 'late' or 'late-stage' rentierism; 'new' or 'entrepreneurial' state capitalism; and economic statecraft. At least two of these (rentierism and state capitalism) describe and elucidate dynamics that pre-date Hamad's reign, but these dynamics were strengthened and refined by him and continue after 2013 under Tamim.

Rentierism refers to systems dominated by rents, or income that is externally derived and largely unproductively earned such as oil and gas royalties which come to the state and are (partly) distributed to society, usually in lieu of taxation (Luciani, 1990: 65–84). This lets rulers buy support or acquiescence from society, purchase a repressive capability, and avoid democratic concessions. The original rentier state theory (RST) arguments were too simplistic and too ambitious in claiming that rents defined the very nature of the state. The idea of 'late' rentierism (Gray, 2011) sought to address this by refining the concept without denying its basic applicability. It accepts that rentier rulers are able to use rents to buy support and avoid democratic reforms but argues that leaders nonetheless have to be responsive to society to some extent and maintain a basic level of legitimacy. Hence, they have a clear and long-term development strategy, good economic policies focused on growth and confidence, and at least a partial openness to globalisation and its effects (Gray, 2011: 23–36) – none of which earlier RST explained effectively. Rents do not define the very nature of a state; they only create a political dynamic, albeit an absolutely central one.

A second feature of the Gulf states, including Qatar, is state capitalism. Specifically, Qatar is a case of a 'new', or 'entrepreneurial', type of state capitalism. As with older forms, under entrepreneurial state capitalism the state is a disproportionate owner in an economy and regulates it closely, but in contrast to earlier state capitalisms, is not doing so to achieve social change, such as addressing social class differences. In Qatar, the state is a major owner in the economy, and regulates some parts of it, but it also lets market forces set prices, seeks out foreign investment (in some areas, at least), and usually encourages state-owned firms to be efficient and profitable. This led to observers such as Bremmer (2010) describing such systems as 'new' state capitalism; many such cases, including Qatar under Hamad and now Tamim, are models of this system, which could also be called 'entrepreneurial' state capitalism (Gray, 2011: 34–5, 2013: 10–11), given that the state remains dominant in the economy but is also business-friendly, invites foreign investment (even if some areas of the economy remain off limits or tightly controlled), is tolerant of commercial risk, supports new industries, and in most cases encourages state-owned firms to focus on efficiency and profitability (Gray, 2013: 64–74; Ulrichsen, 2014: 86–90). While there was a form of state capitalism in place in Qatar long before 1995, Hamad's reform and refinement of it transformed it into a much more dynamic, entrepreneurial form.

Finally, Hamad's foreign and security policies demonstrate the utility of the third approach here, that of dynamic economic statecraft. In Qatar, this has encompassed the use of aid programmes, hosting of major sporting and cultural events, tourism, mass media outreach, and even foreign investment. Reinforcing this are strong and consistent messages that create an image, feel, or association with the country at home and abroad, thereby 'branding' it to prospective trading partners, investors, and even ordinary Qataris. Qatar's foreign policy has also been part of its economic statecraft. This includes close security, defence, and trade ties with the USA but also a functional relationship with Iran, which seeks to give Iran economic reasons to support the political status quo in Qatar, and links to groups such as Hamas, where recognition and aid were used to try to boost Qatar's influence and image in the region. Under this strategy, economic and diplomatic goals were closely linked to each other, ultimately seeking to enhance domestic political support for the state and regime. Arguably, elements of this policy backfired on Qatar after Tamim took power, but under Hamad it was a key component of the transformation of Qatar and largely remains so under Tamim.

The three theoretical approaches used here are laid out in Figure 5.3. There may be other arguments with some explanatory utility as well – arguments about societal agency do point to real influence wielded by social groups and forces, even in strong, effective rentier structures, for example – but the conglomeration of these three are overarching. Oil and gas wealth is central

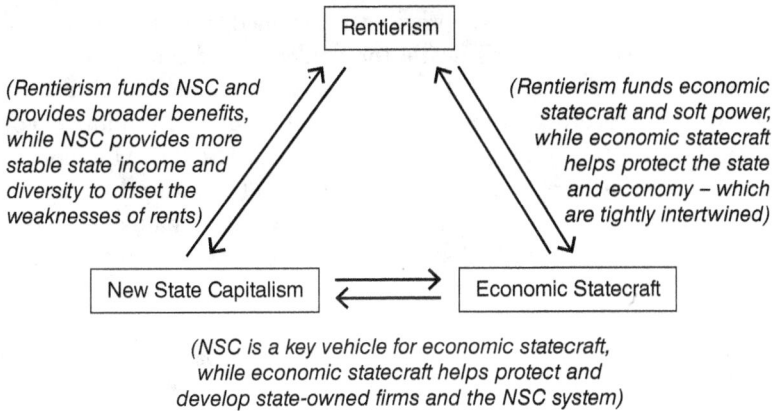

Figure 5.3 An explanatory framework for Qatar's political economy. Source: The author.

to all three: it provides the incomes needed by the state to fund the rentier arrangement with society, as well as the investment needed for entrepreneurial state capitalism to develop and function successfully. (New) state capitalism helps compensate for weaknesses in rentier systems, especially by providing a means for elite co-optation, and levels out economic performance as rent income fluctuates. Economic statecraft, done well, buys greater security for the political-economic system by developing new commercial links, attracting investment, and creating a favourable view of the country abroad, but relies on rentierism for some of the funds to operate and to ensure popular acquiescence. These three approaches also explain the state's ability to smoothly manage the economy, because they combine to give the state enormous resources and a deep but low-profile reach into the political economy and across societal segments such as classes, tribes, and generations.

Leadership transition and economic stasis and change, 2013–14

On 24 June 2013, Amir Hamad announced that he was handing power to his son and heir apparent, Tamim, who took over the following day. Tamim had been groomed for the role of amir for some years, having been put in charge of implementing the QNV strategy in 2008, also heading up preparations for the 2022 World Cup, and later taking on a higher diplomatic profile (Ulrichsen, 2014: 174–6). The transfer of power was carefully planned and executed, with some later rumours that Hamad maintained a strong but informal adviser role to his son after June 2013. This is noteworthy given that Qatari leaders had previously assumed power after removing their fathers or on the latter's death; for a ruler to abdicate as Hamad did, and orchestrate a smooth transfer of power, was unique (Ulrichsen, 2014: 175). At thirty-three years of age at the time, Tamim was a uniquely young ruler among his Arab Gulf peers (*The Economist*, 2013a). His reputation at the time was for being a little more religious and cautious than his father, but not excessively so; a balance between piety and dynamism seemed to be something that Hamad had sought in an heir (*The Economist*, 2013a). That the change in ruler occurred so quickly but smoothly is a reminder that there is sufficient elite consensus to ensure such stability, and another reminder, after the 2011 Arab Spring completely bypassed Qatar, that the regime's strategy for legitimisation and acquiescence had been broadly successful.

Tamim quickly consolidated power. By far the most powerful figure beyond the amir was the prime minister and foreign minister, Hamad bin Jassim, who was quickly removed on 25 June. Many of Hamad bin Jassim's allies and clients were subsequently purged as well. Tamim lost a considerable depth of experience from his inner circle, but he gained a loyal elite, whose roles were spread more widely than under Hamad (Ulrichsen, 2014: 175). Tamim's early focus was on a careful and discreet softening and adjustment of some of his father's policies. In his inaugural speech, he implied that he was less interested in regional influence – a pillar of his father's foreign policy – and avoided discussion of support for Islamist groups such as Hamas and regional uprisings such as in Syria (*The Economist*, 2013b). This probably reflected a reassessment of Qatari priorities, as well as Tamim's relative caution compared to his father. Tamim's priority was to be domestic politics. Hamad bin Jassim was widely considered to have put too much effort into foreign policy at the expense of domestic matters and allowed a haphazard expansion of government bodies and poor communication and often rivalry between competing bodies (*The Economist*, 2013c). Reforms were announced to reduce duplication, improve the effectiveness of the public sector, improve public infrastructure (such as expanding roads in and around the capital and improving water and sewage), all suggesting that Tamim wanted to be seen as addressing ordinary Qataris' concerns quickly (Hammond, 2014: 8). Beyond these announcements, though, there was a broad continuity in Qatari policy, especially economic policy.

In fairly short order, however, two substantial problems challenged the new amir. The first was a diplomatic crisis between Qatar and some other Arab Gulf states, which began on 5 March 2014 when Saudi Arabia, the UAE, and Bahrain withdrew their ambassadors from Doha and demanded that Qatar change some of its regional policies. They claimed that Tamim had promised the previous November not to interfere in its neighbours' affairs – code for Al Jazeera's sometimes-critical commentary on Gulf politics – and informally were also upset about Qatar's supportive stance towards the Egyptian Muslim Brothers (Dazi-Héni, 2014; *The Economist*, 2014). The two sides also had disagreed on policy towards Egypt after the Arab Spring, when the Muslim Brothers briefly dominated its politics. The crisis only ended in November, when diplomatic relations returned to normal, reportedly after Qatar committed to an agreement on non-interference in regional states' affairs and to cease support for the Muslim Brothers and other groups (Katzman, 2019: 7–8). These same issues would later emerge, however, in the far more serious crisis that began in June 2017.

The other, more important and deeper crisis that Tamim faced was the dramatic fall in oil prices that began in mid-2014. Table 5.1 shows various economic indicators for Qatar over 2012–18. The impacts of the drop in oil prices are very evident, although these impacts are milder than might have been predicted. Qatar proved able to ride out the drop in prices better than many other states, partly because of its low population and high per capita rents, and also because its sources of rents were broader and its economy a little more diversified than most. It also helped that the government responded quite steadily and promptly to the fall in rents, focusing on fiscal adjustment while also maintaining state investment in major projects. Non-hydrocarbon capital spending of USD 182 billion was allocated for 2014–18 (QNB, 2015: 7), which supported demand in the non-energy sector (QNB, 2015: 3) and helped sustain a high population growth rate (as demand for foreign workers continued to grow), in turn maintaining consumer demand. Nonetheless, it was an acute challenge for the government and the still-new amir.

Qatari policy responses after 2014

Qatar's response to the drop in oil prices after June 2014 has been based on the implementation of a consistent and strongly state-led set of policies. The guiding aim has been to maintain confidence and to buoy domestic demand, while cautiously implementing selected reforms. Substantial reforms in areas where dramatic change would entail serious political risks for the state have not been undertaken, however, and neoliberal macroeconomic reform has mostly been eschewed. This is consistent with the regime's survival and legitimisation strategies: the basic rentier structure of the political economy has been maintained, as has the entrepreneurial state capitalism structure of economic ownership, even if there has been some attempt at expanding the private sector.

Table 5.1 Qatar: Key economic indicators, 2013–22.

	2013	2014	2015	2016	2017	2018*	2019**	2020**	2021**	2022**
Real GDP growth (%) (2013 prices)	4.4	4.0	3.7	2.1	1.6	2.2	2.6	3.2	3.3	2.7
Hydrocarbon sector	0.1	-0.6	-0.6	-0.9	-0.7	-1.1	0.4	1.8	2.3	0.5
Non-hydrocarbon sector	10.4	9.8	8.5	5.3	3.8	5.3	4.6	4.3	4.2	4.4
Inflation (%)†	3.2	3.4	1.8	2.7	0.4	0.2	0.1	3.7	2.3	2.0
Budget balance (% GDP)	19.3	12.3	1.3	-9.2	-6.6	2.3	3.0	3.4	3.2	3.4
Revenue	47.6	45.7	43.6	30.9	26.9	31.7	32.8	31.7	30.8	29.9
Expenditure	28.3	33.4	42.3	40.1	33.5	29.4	29.8	28.3	27.7	26.5
Central government debt (% GDP)	30.9	24.9	35.5	46.7	49.8	48.4	52.7	45.9	40.6	37.1
Hydrocarbon exports (USD billion)	120.0	113.9	65.6	473.8	57.1	72.1	67.7	67.7	67.7	68.8
Crude oil production ('000bbls/day)	697.8	673.1	636.4	646.0	607.4	608.0	622.5	633.4	642.5	639.2
Natural gas exports (m tons/year)	91.8	90.9	91.3	91.1	90.3	92.4	92.3	91.8	92.5	92.4
Average oil price (USD/bbl)(Brent)	108.8	99.0	52.4	44.0	54.4	71.1	61.8	61.5	60.8	60.4
Current acct balance (% GDP)	60.5	49.4	13.8	-8.3	6.4	18.0	8.9	8.4	6.9	8.8
External debt (% GDP)	49.3	56.2	88.4	127.2	99.6	101.1	106.7	98.3	92.0	87.2
Exchange rate (USD/QR)	3.64	3.64	3.64	3.64	3.64	3.64	3.64	3.64	3.64	3.64
Population growth rate (%)	9.3	10.6	10.0	7.4	1.5	2.3	1.3	0.2	0.2	0.3

*Figures for 2018 are IMF estimates.
**Figures for 2019–22 inclusive are IMF forecasts.
†Consumer price inflation rate.
Source: Derived from data in IMF (2018: 25, 2019: 27).

As already noted, the Qatari economy was in solid shape overall in 2014. Rents and state revenues per capita were very high – giving the state a financial cushion and considerable room for policy manoeuvring – the budget was in surplus, and the banking and finance sector was relatively healthy (IMF, 2015a: 4–9, 11–15, 27–32). Debt may appear high and seem to have risen very sharply after 2014 (Table 5.1); however, it is important to keep in mind the decline in nominal total GDP after 2014, which impacts debt figures when represented as a percentage of GDP, and the most marked rise in debt was in privately held, not public, debt. As a generalisation, Qatar's debt is less worrying than its sheer size might otherwise suggest, because so much of it is long-term, productive debt, incurred in the development of its gas sector, for new public infrastructure, and other reasons more worthy than, say, speculative or consumption-driven debt. The state's immediate response when revenues fell has already been noted. Beyond this, other reforms were fairly narrow or selective: the introduction of competition in some industries (e.g. the taxi industry), anti-monopolistic legislation policies, and reforms to boost trade, increase commercial transparency, and clarify some capital market regulations (Bertelsmann Stiftung, 2018: 18–22).

Once it became obvious that oil prices were in a sustained slump, however, and facing a budget deficit in 2015, the state began to focus more on fiscal management. In spending, a dramatic reduction in welfare, subsidies, or state services would be felt quickly by the population and risked being seen by them as the state not living up to its rentier obligations. Cost cutting was therefore targeted at trimming fat in some state institutions. State subsidies to citizens are sensitive, too, as Krane has observed: '[C]entralized states are at a disadvantage when it comes to reforming subsidies or other benefits. Everyone knows who is behind any painful policy decision' (Krane, 2019: 138–9). While Tamim had made the case publicly and early in his tenure that reform to the rentier arrangement was necessary and that the government could not provide everything for its citizens (Bertelsmann Stiftung, 2018: 22–3), in practice subsidy cuts disproportionately targeted the foreign population in Qatar – who outnumber citizens by a ratio of almost 9:1 – rather than citizens. The most notable cuts were in utility rates, petrol prices, and postage costs. These were the first price rises faced by consumers of these goods and services in eight years – that is, since the last time there was a drop in rents and the government felt budget pressure (Bertelsmann Stiftung, 2018: 23). However, many basics, such as water and electricity, remained free for citizens, and plans for reforms in this regard are long-term ones, with utility companies having ten years to reach market price commercialisation (IMF, 2015a: 13). Fuel subsidy cuts were the most substantial: in January 2016, a cut in subsidies pushed prices up by 30 per cent, and the retail price, previously set by the state, was allowed in future to fluctuate,

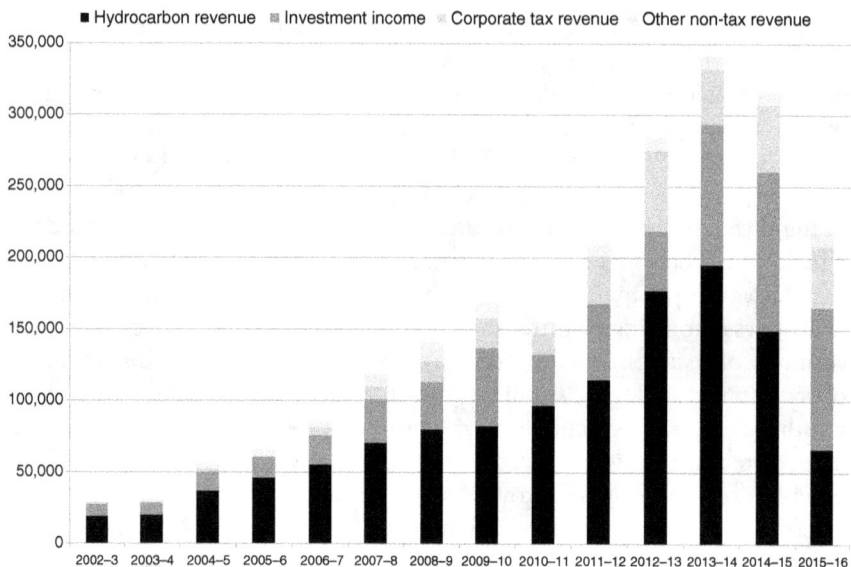

Figure 5.4 Qatari state revenue and the growth of tax revenue, 2002/3–2015/16 (in QAR billion). Source: Derived from data in Abdellatif *et al.* (2017: 676 (Table 1)).

leading to a total price increase over 2014–18 of 136 per cent for petrol and 106 per cent for diesel (Krane, 2019: 142).

The state has been equally cautious with revenue reforms. As Figure 5.4 shows, tax revenues are low – and the rise is mostly from taxes on state firms, above all QP, which also provides nearly all of the state's investment income as well. Otherwise, there is a 35 per cent tax on oil and gas company profits, a 10 per cent tax on foreign companies, a 5–7 per cent withholding tax for some payments to non-residents, a 5 per cent customs duty, and a range of fees for some government services (IMF, 2015b: 5). However, there is no personal income tax, no corporate income tax on wholly Qatari-owned private-sector companies, and indeed no other major taxes. There was no broad-based consumption tax in place in 2018 either, although a 5 per cent consumption tax was planned to have been introduced in 2020 (IMF, 2019: 16, 25). The only other new tax is the recent so-called sin tax, introduced at the start of 2019, which raised the excise on alcoholic drinks (reversed in mid-2019, in anticipation of the 2022 World Cup), energy drinks, pork products, and tobacco by 100 per cent and on soft drinks by 50 per cent (Karasz, 2019). Just as on the expenditure side, imposing too many or expensive taxes was and is probably viewed by the state as being too politically risky. The budget slipped into deficit in 2016 and 2017, but this was manage-able given Qatar's small population and large rent incomes.

Several other economic steps are also notable. In 2017, the state decided to increase long-term gas production by lifting a twelve-year moratorium on development of the North Field. This is expected to increase total gas output by 30 per cent by 2024 and will dominate investment in the hydro-carbon sector over 2019–24 (QNB, 2018: 3, 8). The decision was likely prompted more by the rise of other gas powers such as Australia, the USA, and Russia (Munro, 2017) than by economic urgency, but nonetheless, it will increase rent income and have multiplier effects on the broader economy, too (QNB, 2018: 8). This is a mixed sign for the long-term future of Qatar: it will provide the state with additional revenue after 2024 to help fund economic diversification and development, but at a political level, it may reinforce the already-deep rentier bargain and undermine the prospects for political reform.

Also importantly, foreign investment laws have been relaxed. In 2014, the 2010 Foreign Investment Law was amended, which had capped foreign ownership of businesses at 49 per cent (Bertelsmann Stiftung, 2018: 21). The government announced plans to allow foreign investors to own 100 per cent of a business in most sectors (but not hydrocarbons, banking, or insurance), placing them on an equal legal footing with local companies for the first time (OBG, 2019). The government also focused on expanding industrial zones. A Qatar Free Zone Authority was set up in 2018 to manage the regulation of two new free zones that would open in 2019 (OBG, 2018a): Umm al-Hul, near Qatar's main seaport, would focus on maritime, construction, building, metals, machinery, downstream petrochemicals, and logistics; and Ras Bufuntas, next to the airport, would focus on health care, medical services, automotive and aircraft industries, and business services. The use of free zones has been quite extensive in the Gulf and has had mixed results, but Qatar's experience with the QFC has been relatively successful. The free zone policy was provided with USD 3 billion to support foreign investors who come to the zones, with concessions such as 100 per cent foreign ownership of assets, customs duties exemptions, flexibility in the use of (foreign) labour, and corporate tax exemptions for up to 20 years (OBG, 2018a).

The legal and regulatory environment for commercial dispute resolution was amended and improved. The Civil and Commercial Arbitration Law No. 2 of 2017 introduced some substantial changes in this respect (OBG, n.d./a), bringing Qatar into line with international standards, especially those set out in the UN Commission on International Trade Law (OBG, n.d./a). These reforms clarified dispute resolution authority and the conditions for agreeing to a dispute resolution mechanism, and importantly, set new standards for the resolution of international civil and commercial disputes. Beyond the Civil and Commercial Arbitration Law, other commercial reforms around this time were more specific, but not inconsequential, aiming to support economic diversification and increase competitiveness and innovation in the private sector. As examples: in 2014, private taxi companies were allowed to begin competing with the state-owned traffic company; in 2016, a 2002 law was amended to improve the competitiveness of commercial agents; and again in 2016, reforms were made to improve the transparency of land ownership registration (Bertelsmann Stiftung, 2018: 19–20). To support the private sector, the banking and finance industry has been strengthened, and regulation of it is strong, while banks have also been encouraged to lend more flexibly to the private sector and to broaden their loans exposure (IMF, 2018: 14–16, 30–1). Capital market reforms are likely to come soon as well, given the government's emphasis on attracting international investors and its encouragement of large, privately owned firms to list on the stock market.

While doing the above, the state also maintained most of the major projects and initiatives already planned or underway, even once it was clear that oil prices were falling sharply and probably for a sustained period. A couple of petrochemical joint ventures were affected, with the Al Karaana one cancelled and the Al Sejeel one postponed, although other new projects arose later, such as an ethane cracker joint venture between QP and US firm CPChem announced in June 2019 (Argus, 2019). The government's first budget after the oil price fall allocated an enormous sum, some USD 182 billion, for capital spending until 2018, plus over USD 10 billion for hydrocarbon projects over 2015–17 (QNB, 2015: 7). The 2022 FIFA World Cup requires investment in a minimum of eight stadiums, at a cost of USD 10 billion (Al Heialy, 2016), plus 'soft' infrastructure which will also serve the economy longer term, such as roads, bridges, highways, and other public works (OBG, n.d./b). Public transport infrastructure projects include an enlargement of the international airport terminal, plus spending of USD 35–45 billion to construct the Doha Metro, a light rail project at Lusail City, and a long-distance railway. Given that this investment will assist the tourism industry, Qatar has also focused there, targeting high value visitors, abolishing visas for many nationalities or streamlining the process for others, and increasing promotion efforts in emerging markets such as China and India (OBG, 2018b).

The World Cup has prompted other reforms. The one that arguably received most attention was the 2016 reform to the labour and visa laws for foreign workers. Qatar came under pressure in 2015 and 2016 because of its *kafāla* ('sponsorship') visa system for foreign workers. This required workers to obtain a visa through a Qatari sponsor, who could also control their ability to leave the country or move jobs, and workers who completed their contract and left Qatar had to wait two years before they could return and change sponsor. In December 2016, the government announced that it was abolishing the *kafāla* system and replacing it with a contract-based system that would be fairer to workers and give them greater security and predictability (OBG, 2017a). Workers could change jobs, fines were increased for employers who breached workers' rights, and a grievance process was set up for disputes between workers and employers over permission to exit the country. Debate remains, however, over whether these reforms were sufficient: Qatari employers retain enormous power (Human Rights Watch, 2019), and the government appears reluctant to reform the system more than is necessary, perhaps because its close, oftentimes protective relationship with business is a component of Qatar's late rentierism.

The June 2017 crisis and the Qatari political economy

The Qatar crisis that began on 5 June 2017 is the most serious diplomatic crisis among the Arab Gulf monarchies in modern times. It began with Saudi Arabia, the UAE, Bahrain, Egypt, and Yemen accusing Qatar of supporting Islamic extremist groups, aiding in Iranian designs in and destabilisation of the region, and meddling in the internal affairs of other Arab states (Barnard and Kirkpatrick, 2017). These claims were very similar to those made in 2014, but this time the crisis was far more dramatic. The Saudis closed their land border with Qatar – Qatar's only land border – and all of the states accusing Qatar, except Egypt, evicted Qataris from their countries and ordered their own nationals home. They also cut off their air and sea space to Qatar. On 22 June, they presented Doha with a dramatic list of demands, including that it close Al Jazeera and pay reparations for their past actions. These were demands to which Tamim could simply not acquiesce (Katzman, 2019: 7–9), and although later reframed as a shorter set of 'principles', by then both sides were entrenched. The power dynamic underlying the crisis was complex, and less imbalanced than might be assumed. The number and power of the states lined up against Qatar were substantial – Saudi Arabia is the region's largest economy, Egypt its most populous state, and the UAE a regional trade entrepôt and highly diversified economy – but Qatar had its own sources of power and influence, including a very wealthy economy and ties with several key states such as Turkey, Iran, Iraq, the USA, and Russia.

Qatar had clearly overplayed its hand and become overconfident, perhaps due to its leadership's relative lack of internal dissent and the boost in regional popularity it gained by its support of the 2011 Arab uprisings

(Roberts, 2017b: 5–7). It forgot, however, that Qatar's soft power has limits, and was even a source of the crisis, and hard diplomatic power still matters. Still, the government responded to the crisis emphatically, and the population supported the government and the amir strongly; in fact, the crisis had a nation-building effect and was a boon for Tamim. Given the fiscal strength of its economy and high rents per capita, it withstood the economic pressures imposed by the crisis quite well, and some initial economic impacts were short-lived (IMF, 2018: 4–7). There was a sudden drop in trade, including the need for urgent new import sources for essentials such as foodstuffs, some 40 per cent of which had previously come through Saudi Arabia (Katzman, 2019: 21), but Iran, Turkey, India, and others quickly began supplying most key goods, and new suppliers and trade routes were found for most other products. The banking system was badly affected at first, as some international financing could not proceed and foreigners leaving Qatar withdrew funds, but the Qatar Central Bank stepped in to support liquidity, which limited financial impacts to the short term. Stock market prices, spreads, and other indicators all returned to normal quickly, with only long-term sovereign bond rates proving stubborn (IMF, 2018: 5).

Qatar cleverly ensured that the crisis did not have any significant impact on its international energy trade or its reputation as a supplier. It assured importing states that gas commitments would be honoured, exporting through its own waters and via friendly or sympathetic states such as Oman and Iran. There was really no impact on international financial markets from the crisis, suggesting that this strategy was successful. Longer term, it also increases the chances that importing states will renew expiring contracts, thus helping maintain Qatar's gas competitiveness. This was a wise strategy: although Doha could have leveraged its energy exports to encourage importing states into pressuring the states arrayed against Qatar, as Zafirov (2017: 196) has noted, such tactics would have been self-defeating in the longer term.

In all likelihood, Qatar's response to the crisis was not driven by the low oil price environment of 2017, and it would probably have reacted in a similar way had oil and gas rents been a lot higher, or for that matter a lot lower, at the time. At most, the amir may have felt a little more confident than otherwise, since arguably he had more room for policy manoeuvring than did regional actors with less diverse economies or lower rents per capita – most notably in this context, Saudi Arabia. In fact, in the long term, the crisis may actually be a net positive for the Qatari economy. It prompted the government to accelerate its economic diversification plans and focus more on self-sufficiency, something that Tamim had indicated a desire to do as early as his first speech as amir. To these ends, investments were made in industries such as dairy – which will make Qatar completely self-sufficient in milk – meat and poultry production, food processing, and light manufacturing (OBG, 2017b). This investment will diversify its economy, while reinforcing its entrepreneurial state capitalism, both of which are useful outcomes for a young amir looking to consolidate his position and an elite concerned to ensure systemic durability.

The other policy shift that Qatar has made was the decision, effective from 1 January 2019, to leave the Organization of the Petroleum Exporting Countries (OPEC). This liberates the government from OPEC agreements on output and pricing, but is a problem for OPEC, as it undermines the body's strength and increases the risk that other states might follow Qatar out the door. Qatar claimed that its motivation in leaving was that it wanted to focus on the gas industry instead. Certainly, it is expanding gas production, while oil output has been declining, but the decision was also driven, at least partly, by Saudi domination of OPEC. Saad al-Kaabi, the Qatari minister of state for energy affairs, acknowledged this: 'We are not saying we are going to get out of the oil business, but it is controlled by an organization managed by a country.' Moreover, 'we are a very small player in [OPEC] and I don't have a say in what happens [in it]' (quoted in Knecht, 2018). The decision likely will have a net benefit on Qatar's economy, especially since Qatar does appear to be shifting its focus to the gas sector.

The 2017 crisis was perhaps likely in hindsight, given the high stakes of Qatar's ambitious and activist foreign policy since the 1990s. Recalling Qatar's combination of specific forms of rentierism, state capitalism, and economic statecraft, this high-stakes approach makes sense, and may even be unavoidable to some extent. However, Doha overplayed its hand more than once. It underestimated the anger that would be created among its Arab Gulf neighbours by its relatively open approach to Iran and some Sunni Islamist groups, which required it to go against Riyadh directly. Its branding strategy was broadly successful, from it hosting major events to its airline to its foreign aid profile, but the prominence and freedom it gave Al Jazeera, while an important symbol of Hamad's reformist proclivities, upset several important actors, from the USA to Saudi Arabia. Despite all this, Doha has had functional, at times effective, relationships with the other Arab Gulf monarchies, working well in certain respects within the Gulf Cooperation Council (GCC), such as on trade liberalisation, labour market reforms, and fostering commercial innovation (Gray, 2013: 186–91). But even if Qatar emerges from the crisis relatively undamaged, regional tensions will persist, and Doha must accept at least some responsibility for this.

A cautious conclusion

Qatar's economy has proven quite resilient and, overall, successful. It has made errors and missed opportunities, but for this small country to rise so quickly and now be among the wealthiest per capita in the world is quite remarkable. It has weathered sustained low oil prices since 2014 quite well, even after the blockade began in June 2017. The amir confronted all this as a young ruler soon after coming to power, before he could fully consolidate his position.

The prospects for Qatar in the coming years seem good, barring a dramatic turn of events. More likely, the crisis simply marked the changing power dynamics in the Gulf, as various actors' economic and foreign policy interests increasingly diverge. Regardless, there remains the need for further reforms if Qatar is to maximise its economic potential. To achieve genuine economic diversification away from hydrocarbons – say, to reach a point where rents are around one quarter of GDP rather than around half, as now – would mean a transformation far more profound than seems possible at present. It would require a dramatic expansion of the private sector, and a two- or threefold population increase for domestic demand to sustain such a private sector. Although it is building new industries, whether these businesses can stand on their own, as globally competitive operations and with no support from the state, is questionable. It is certainly unlikely for the time being. Moreover, being globally competitive includes having a labour market – for Qatari nationals, not just foreign workers – that is much more flexible and dynamic than at present, and where the best young people have the incentives to choose a career in the private sector over the public sector (Gray, 2013: 221–6). For now, that is not the case, with the public- to private-sector wages gap standing at 171 per cent (IMF, 2018: 12), among the highest in the world. The Qatari government is aware of this and may turn to addressing it (IMF, 2018: 12–13), but a profound policy shift is, for now, politically unfeasible.

Several other reforms are also required, above all in taxation. Higher tax revenues would smooth out the effects of rent fluctuations, reallocate wealth better than through normal rentier structures, and improve intergenerational equity (Abdellatif *et al.*, 2017: 680). The planned consumption tax will broaden the sources of state revenue and help the population adjust to having a tax obligation, but further tax reforms are needed, including an alignment of corporate income tax between Qatari and foreign businesses (Abdellatif *et al.*, 2017: 674, 678–80). An income tax on individuals is highly desirable, but virtually impossible, even if oil prices fall still further, because in effect it would end the late rentier bargain that is the foundation of the regime's durability and support. On the expenditure side, further subsidy reforms, improvements in public-sector management and accountability, and initiatives to promote innovation and entrepreneurialism are all desirable (IMF, 2018: 32). Further improvements to the education system, and especially to training, would assist economic diversification.

When the 2017 Qatar crisis is finally over, assuming it ends peacefully and without worsening any further, Qatar's prospects are strong. The state has deep reach into society and the economy and is able to steer the economy while maintaining strong popular support, a key reason for its smooth handling of the economy since 2014. Qatar also has events such as the FIFA World Cup to look forward to, which will boost the economy and then leave behind new infrastructure and expertise. How thoroughly it realises its potential, however, comes down to several factors, including the state's ability to maintain and develop its support and legitimacy while implementing economic policies that will probably prove highly contentious, as well as its ability to operate more cautiously on the world stage. These are not minor challenges and would require an even deeper transformation than the reforms of the past few decades, which were central to making Qatar the wealthy, dynamic state that it is today.

References

Abdelkarim, A. (ed.) (1999), *Change and development in the Gulf*. Basingstoke: Macmillan.

Abdellatif, M. M., A. G. Eid, and B. Tran-Nam (2017), 'Qatar – Oil price fluctuations and the need for tax policy reform in Qatar', *Bulletin for International Taxation*, 71:12, 674–80.

Al Heialy, Y. (2016), 'Qatar's World Cup stadiums to cost $10bn, official says', Arabian Business, 22 May. https://tinyurl.com/dachxh43 (accessed 27 November 2018).

Argus (2019), 'Analysis: Qatar petchem sector gets a much-needed boost', 25 June. https://tinyurl.com/hj7nx7yz (accessed 12 July 2019).

Barnard, A., and D. D. Kirkpatrick (2017), '5 Arab nations move to isolate Qatar, putting the U.S. in a bind', *The New York Times*, 5 June. https://tinyurl.com/zt38ytnr (accessed 14 July 2017).

Bertelsmann Stiftung (2018), *BTI 2018 country report – Qatar*. Gütersloh: Bertelsmann Stiftung.

BP (2019), 'Statistical review of world energy June 2019'. https://tinyurl.com/y4kp6vaw (accessed 29 October 2019).

Bremmer, I. (2010), *The end of the free market: Who wins the war between states and corporations?* New York: Portfolio.

Coleman, G. (2000), 'Qatar encourages foreign investment and joint ventures', *The Middle East*, 297, 32–4.

Dargin, J. (2011), 'Qatar's gas revolution', in B. Fattouh and J. P. Stern (eds), *Natural gas markets in the Middle East and North Africa*. Oxford: Oxford University Press, pp. 306–42.

Dazi-Héni, F. (2014), *Qatar's regional ambitions and the new emir*. Washington, DC: Middle East Institute. www.mei.edu/publications/qatars-regional-ambitions-and-the-new-emir (accessed 19 February 2019).

Gray, M. (2011), 'A theory of "late rentierism" in the Arab states of the Gulf', Center for International and Regional Studies, Occasional Paper, 7. https://bit.ly/3qRlK1s (accessed 17 February 2020).

Gray, M. (2013), *Qatar: Politics and the challenges of development*. Boulder: Lynne Rienner.

Hammond, A. (2014), 'Qatar's leadership transition: Like father, like son', European Council on Foreign Relations, Policy Brief, 95. https://bit.ly/3nHTcr7 (accessed 5 May 2021).

Herb, M. (1999), *All in the family: Absolutism, revolution, and democracy in the Middle Eastern monarchies*. Albany: State University of New York Press.

Human Rights Watch (2019), 'Qatar: Partial reforms risk undermining progress', 17 January. https://tinyurl.com/jc98h9u8 (accessed 11 June 2019).

Ibrahim, I., and F. Harrigan (2012), 'Qatar's economy: Past, present and future', *QScienceConnect*, 9, 12–18.

IMF (2015a), 'Qatar: 2015 article IV consultation – Staff report; and press release'. Washington, DC: International Monetary Fund.

IMF (2015b), 'Qatar – selected issues', IMF Country Report, 15/87. Washington, DC: International Monetary Fund.

IMF (2018), 'Qatar: 2018 article IV consultation – Press release; staff report; and statement by the executive director for Qatar'. Washington, DC: International Monetary Fund.

IMF (2019), 'Qatar: 2019 article IV consultation – Press release; staff report'. Washington, DC: International Monetary Fund.

Kamrava, M. (2013), *Qatar: Small state, big politics*. Ithaca: Cornell University Press.

Karasz, P. (2019), 'A 6-pack of beer for $26? Qatar doubles the price of alcohol', *The New York Times*, 1 January. https://tinyurl.com/brbkfntx (accessed 26 February 2019).

Katzman, K. (2019), 'Qatar: Governance, security, and U.S. policy', Congressional Research Service Report, R44533. Washington, DC: Congressional Research Service.

Knecht, E. (2018), 'Gas-focused Qatar to exit OPEC in swipe at Saudi influence', Reuters, 3 December. https://tinyurl.com/fnbwvm3n (accessed 26 February 2019).

Krane, J. (2019), *Energy kingdoms: Oil and political survival in the Persian Gulf*. New York: Columbia University Press.

Luciani, G. (1990), 'Allocation vs. production states: A theoretical framework', in G. Luciani (ed.), *The Arab state*. London: Routledge, pp. 65–84.

Morakabati, Y., J. Beavis, and J. Fletcher (2014), 'Planning for a Qatar without oil: Tourism and economic diversification, a battle of perceptions', *Tourism Planning & Development*, 11:4, 415–34.

Munro, D. (2017), 'Qatar moves to ensure LNG dominance', The Arab Gulf States Institute in Washington, 17 April. https://agsiw.org/qatar-moves-ensure-lng-dominance/ (accessed 24 February 2019).

al-Naqeeb, K. H. (1990), *Society and state in the Gulf and Arab Peninsula: A different perspective*. London: Routledge.

OBG (2017a), 'Qatar year in review 2016'. https://oxfordbusinessgroup.com/news/qatar-year-review-2016 (accessed 20 February 2019).

OBG (2017b), 'Qatar year in review 2017'. https://oxfordbusinessgroup.com/news/qatar-year-review-2017 (accessed 20 February 2019).

OBG (2018a), 'Launch of new free zones to boost Qatar's foreign investment and diversification plans'. https://bit.ly/3e3r8vG (accessed 25 February 2019).

OBG (2018b), 'Qatar looks to new high-potential tourism markets'. https://tinyurl.com/4wfwvyd2 (accessed 19 February 2019).

OBG (2019), 'Qatar: Year in review 2018'. https://tinyurl.com/3n75h7uy (accessed 25 February 2019).

OBG (n.d./a), 'New legal rules and regulations in Qatar'. https://tinyurl.com/4bbdy847 (accessed 19 February 2019).

OBG (n.d./b), 'Spending in all modes of transport gathering pace in Qatar'. https://tinyurl.com/w9mdnhh2 (accessed 24 February 2019).

Peterson, J. E. (2006), 'Qatar and the world: Branding for a micro-state', *Middle East Journal*, 60:4, 732–48.

QNB (2015), 'Qatar economic insight 2015'. Doha: Qatar National Bank.

QNB (2018), 'Qatar economic insight September 2018'. Doha: Qatar National Bank.

Roberts, D. B. (2017a), *Qatar: Securing the global ambitions of a city-state*. London: Hurst.

Roberts, D. B. (2017b), 'Securing the Qatari state', The Arab Gulf States Institute in Washington, Issues Paper, 7. https://tinyurl.com/23ym4j54 (accessed 28 July 2017).

Soubrier, E. (2017), 'Evolving foreign and security policies: A comparative study of Qatar and the United Arab Emirates', in K. S. Almezaini and J. Rickli (eds), *The small Gulf states: Foreign and security policies before and after the Arab Spring*. London: Routledge, pp. 123–43.

The Economist (2013a), 'A hard act to follow', 29 June. https://econ.st/3sNSFoS (accessed 24 February 2019).

The Economist (2013b), 'Change of tack', 15 July. https://econ.st/2OiVOOy (accessed 25 February 2019).

The Economist (2013c), 'No more own goals', 2 October. https://econ.st/3rdLIgK (accessed 25 February 2019).

The Economist (2014), 'No one is happy', 7 March. https://econ.st/2OiWaoE (accessed 25 February 2019).

The Economist Intelligence Unit (2017), 'Total, Qatar Petroleum to jointly run oilfield for 25 years', 13 July. https://bit.ly/3sO1hvJ (accessed 23 February 2019).

Ulrichsen, K. C. (2014), *Qatar and the Arab Spring*. London: Hurst.

World Bank (n.d.), 'Qatar', World Bank Data. https://data.worldbank.org/country/qatar?view=chart (accessed 30 October 2019).

Zafirov, M. (2017), 'The Qatar crisis – Why the blockade failed', *Israel Journal of Foreign Affairs*, 11:2, 191–201.

6

The nexus between state-led economic reform programmes, security, and reputation damage in the Kingdom of Saudi Arabia

Robert Mason

Introduction

While rentierism is still a relevant theory in explaining state–society relations of energy exporters, this chapter argues that the breakneck changes taking place in the Saudi economy cannot be disconnected from equally profound changes taking place in Saudi Arabia's domestic and international politics. Rentier state theory (RST) is much poorer at modelling the impact of these wider variables, and yet they are already affecting the Kingdom's oil revenues and its immediate ability to retain capital and attract foreign direct investment (FDI) to sustain economic change, especially job creation.

State-led capitalism, especially in rapid expansionist mode concerning mega-projects, provides some income, as does hedging in the non-oil sector, but it is not adept at providing jobs, which is a key aspect for economic development and social cohesion. The Saudi Deputy Minister for Economic Affairs Abdulaziz Al-Rasheed noted in 2018 that job creation is important since 100,000 Saudis joined the workforce in the final quarter of 2017 alone (Kane, 2018).

In the Saudi adjustment strategy to lower oil prices post-2014, we see that Saudi Arabia has experienced a range of short- and medium-term risks and opportunities. In the lead-up to 2014, they included the fallout from the global financial crisis in 2008–9 and the Arab uprisings from late 2010. The Arab uprisings in particular sparked major Saudi and other Gulf Cooperation Council (GCC) member state efforts at targeted financial aid and military assistance to states such as Egypt and Bahrain.

Short-term impacts post-2014 include the Joint Comprehensive Plan of Action (JCPOA) with Iran in July 2015, which could have significantly lowered international oil prices. The main post-2014 adjustment programme, Saudi Vision 2030, was launched by Crown Prince Muhammad bin Salman (MBS) in April 2016. This was quickly followed by the detention and arrests of prominent Saudi princes, government ministers, and businesspeople at the Ritz Carlton, Riyadh, from 2017 to 2019, based on an anti-corruption premise that netted the government more than USD 100 billion in income. The fallout from the murder of Jamal Khashoggi from 2 October 2018 also falls into this bracket, since the international response has been muted so far. From the election of US President Donald Trump in November 2016, opportunities have presented themselves through Saudi economic diversification efforts within Vision 2030 and by the Trump administration's focus on economic engagement.

Trump's Middle East policy effectively amounted to a 'pay and play' transactional policy. So long as Saudi Arabia paid 'cash' for US military support to shore up its oil defences, the Trump administration continued to support it with US troop deployments (Martin, 2019). This extended to vetoing a congressional bill passed in April 2019 which was designed to end US military involvement in the Yemen conflict. As Mason (2014a) noted, investment, arms, and civil nuclear energy cooperation remain high on Saudi and US political agendas, particularly in the context of economic opportunity and addressing regional security challenges (i.e. Iran).

In the medium to long term, security and economic risks and constraints are to be associated with Iranian threats to Saudi security and oil revenues, costs and risks associated with the spillover effects, and insecurities from Saudi-led intervention in Yemen, the US shale and gas boom, and the shift towards renewable energy. The chapter follows up on these issues in the sections below.

Literature review: Saudi political economy and international relations

As Diwan (2019) and Hertog (2019) point out in separate publications, RST remains relevant in explaining the economic stagnation of Saudi Arabia, due to the rising population and incomes, which makes diversification in the short term so much more difficult. Unlike authoritarian neoliberal models such as China, Saudi economic growth is not explicitly export led. Although oil exports play a crucial role, it is also about state-directed investments across a diverse portfolio. As Kurlantzick (2016) notes, state-led capitalism has also been a function of wider challenges in the region.

A constantly fluctuating oil price certainly doesn't help matters. As Hertog (2019) and Diwan (2019) found, the Saudi focus should shift to a more modest rentier adaptation whereby the Kingdom doesn't seek to replace oil exports with new industries so long as it can cover its import bill, but at least provide gainful employment to the educated youth in high-productivity jobs while reducing reliance on foreign expatriate labour.

The response to lower oil prices in Saudi Arabia and elsewhere in the Gulf has been to refocus attention on what Baldwin called 'economic statecraft', which lacks a narrow definition but generally comprises a policy instrument in an attempt to influence other international actors. It also takes into account dimensions of the target's behaviour such as attitudes and expectations (Baldwin, 1985: 32). Mason (2014b) asserts that Saudi Arabia engages its economic resources in the foreign policy realm, often referred to as 'Riyal Politik', in particular through alliance construction with security guarantors such as the USA and through alliance deconstruction of adversaries such as Iran.

Kamrava's (2018) work linked to RST suggests that, whereas oil has stunted institutional development as discussed in Beck and Richter (2021), it can be a blessing in certain circumstances, in particular where a political opening presents itself, such as a further opportunity for state power maximisation, measures aimed at ensuring popular legitimacy, or being able to action an exceptional political vision (Kamrava, 2018). That is borne out in a number of examples, especially in implementing Vision 2030.

The campaign on climate change, tackling pollution, and the Kingdom's reputation for joyriding and car crashes involving youth, partly linked to cheap petrol prices, have also been enabling factors. This is especially the case since Saudi Arabia tied its Paris Accord commitments to reducing domestic energy demand by raising prices, at least to a more sustainable level (Krane, 2019: 140).

Miller asserts that socio-economic challenges that lead to inequality and unrest pose more of a threat to the Kingdom than Iran or any other external security threat (Miller, 2016: 305). As Springborg (2018) notes, the 'deep state' in Saudi Arabia consists of around 20,000 princes, many in the military and intelligence services, which arguably have greater legitimacy than in other countries of the Middle East and North Africa (MENA). But given concerns over inequality and unrest, their role becomes a paramount concern for the continued survival of the Saudi monarchy.

The economy

The precipitous drop in the international oil price in 2014 required a swift response to balance a budget already heavily invested in providing economic largesse and maintaining social stability and the political status quo. Saudi Arabia has a large foreign reserve cushion, but the Kingdom started to burn through it by 16 per cent between 2014 and 2015, dropping it down to USD 555 billion in 2016 (Daiss, 2016). This was in addition to running a

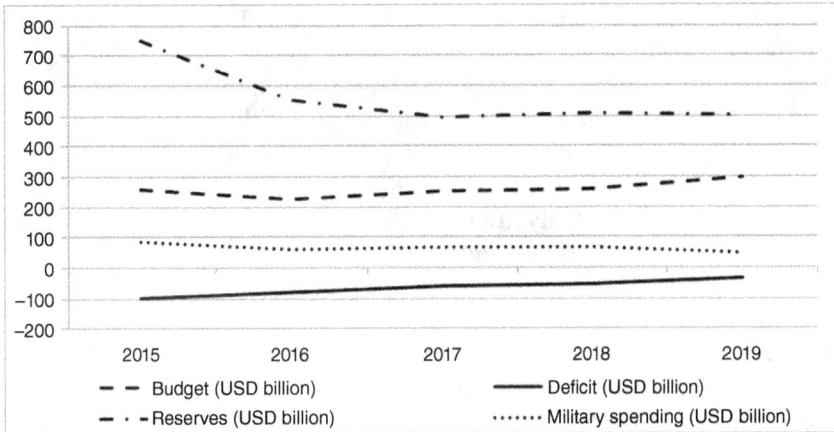

Figure 6.1 Saudi budget, deficit, foreign reserves, and military spending, 2015–19 (in USD billion). Source: Al-Khateeb (2015), Daiss (2016), Wald (2017), Arabian Business (2018), France 24 (2018), Trading Economics (2018a), CEIC (2019), Reuters (2019a). Note: 2019 figures are projected.

USD 98 billion deficit in 2015 (Daiss, 2016). Figure 6.1 highlights these trends.

Even though Saudi Arabia maintains one of the lowest costs of production in the GCC, the Kingdom can only afford falling foreign reserves and higher deficits for a few years. Indeed, the International Monetary Fund (IMF) warned Saudi Arabia about soaring state spending on social welfare programmes that cost the Kingdom USD 265.2 billion in 2013 on top of big infrastructure projects (Dokoupil, 2014). The IMF also said that eroding foreign reserves were causing a repeated annual budget deficit, with revenues estimated to drop by about 55 per cent between 2013 and 2019 at a time of lower oil revenues and to only be able to fund six and a half months of spending in 2019. What actually occurred is that economic reforms quickened in 2019, which generated more revenue, but still the IMF anticipated that the budget deficit would reach 7 per cent of gross domestic product (GDP) in 2019 as opposed to the 4.2 per cent projected by the Saudi government (Reuters, 2019a).

Despite a strong economic argument for maintaining high oil prices through so-called OPEC+ – an oil production quota system established between OPEC and major producers beyond OPEC, mainly Russia, in November 2016 – there are strong domestic economic and geopolitical reasons not to cut production and maintain a lower price. First and foremost, if a higher oil price is supported, higher cost producers such as Iran and Russia could gain market share at Saudi Arabia's expense (Freeman Jr, 2015). Market share was also at stake after President Obama signed the JCPOA with Iran, which included sanctions relief on oil exports, although restrictions on financial transactions with Iran made for a cautious business climate. Loss of market share would have cut Saudi Arabia's long-term revenues and geopolitical influence. Only after the USA offered a significant new arms package were Saudi anxieties addressed.

Relatively low oil prices also enhance Saudi relations with key customers such as China, India, and Japan (which is switching back to fossil fuels after the Fukushima nuclear disaster). However, Saudi Arabia is likely to remain dominant in energy markets for many years to come, as restructuring high-energy industry and civilisations to rely on renewable energy will be more challenging than switching from one fossil fuel to another (Smil, 2017). Cutting oil production would do little to affect Saudi–USA energy relations, since fracking in the USA has become more efficient through new technology and through cutting labour costs. The US fracking industry can therefore more easily withstand lower oil prices.

Since the accession of King Salman to the throne in 2015, economic expansionism has been evident. The momentum became even more noticeable after his son, Muhammad bin Salman (MBS), became Crown prince and responsible for the economy and defence portfolios from June 2017. Vision 2030 is a plan to reduce Saudi Arabia's dependence on oil, diversify its economy, and develop public sectors such as health, education, infrastructure, and tourism. Programmes cover economy and society, industrial development and logistics, national transformation, financial sector development, quality of life, privatisation, fiscal balance, a public investment fund (PIF), and realisation. The National Transformation Plan (NTP) was designed to support Vision 2030 with more details cascaded down, including highlighting the main challenges to implementation, outlining strategic objectives, and referencing indicators and targets. While the mainstay of the Saudi economy for the foreseeable future will be oil, Vision 2030 outlines a new economic path including: tackling energy subsidies, creating new revenue streams through fees and a 5 per cent value-added tax (VAT), attracting more FDI, and using the PIF as an engine for economic growth (Young, 2018a).

Vision 2030 proposes developing or adding new productive sectors such as entertainment and family-friendly tourism, including a Qiddiya theme park near Riyadh, slated to open in 2022. Cinemas, football events, natural attractions such as al-'Ula in Medina, and luxury tourism on the Red Sea such as the Amaala project are all expected to generate revenues. Less certain are the Nium and Red Sea reef projects. Tourism is expected to generate 224,000 jobs per year by 2030 and add USD 5 billion to the GDP (Reuters,

2019b), but that would make the Kingdom comparable to Thailand or Germany in tourism terms, which appears rather ambitious.

Another area of job creation is defence manufacturing. Saudi Arabian Military Industries was founded in May 2017 as part of Vision 2030 to serve national defence requirements, especially border control and protecting critical infrastructure. Rahman (2019) reports that 40,000 jobs are expected to be created over twelve years with some USD 3.7 billion in GDP contributions. As part of the plan, a joint venture has been signed with Boeing to maintain and repair fixed and rotary-wing military aircraft in the Kingdom, with expected revenues of USD 22 billion by 2030 and 6,000 jobs (Boeing, 2018).

To avoid social unrest, much hinges on new cash stipends to offset some of these measures, as well as jobs for youth and women's economic inclusion. The NTP was revised in 2017 and again in 2018, accompanied by high state expenditure on social benefits and public wages (Young, 2018b). The expectation of the plan that twenty-four government agencies will work together to deliver is perhaps ambitious given Saudi Arabia General Investment Authority's (SAGIA) previous experience in the 2000s. Much hinges on the political will to convert policy rhetoric into concrete and measurable actions.

There are other issues in the Saudi economy, such as structural reform, which have had a bearing in the past and in the adjustment policies we see now. For example, most FDI into the Kingdom took place between 1981 and 1984, the years of state-driven industrial expansion (Hertog, 2010: 145). From then until 2005, most FDI was limited and went mainly into joint ventures with companies that were more insulated from state bureaucracy, such as Saudi Basic Industries Corporation (SABIC). FDI peaked again in 2008 at around USD 39.5 billion (World Bank, 2019), mainly related to the work of SAGIA. Along with World Trade Organization influence, it worked towards deregulation of markets, up to Saudi accession in 2005. Utilities, education, health services, parts of the telecom sector, retail, and electricity transmission were opened up to foreign investors.

There is some speculation as to SAGIA's effectiveness given its limited interactions with other government agencies and their willingness to engage. This is also true of the business community, which has for the large part remained reactive and focused on maintaining clientelist privileges vis-à-vis the state (Hertog, 2010). With different government interests and bureaucratic brokers charged with chasing and finalising deals on behalf of local and foreign firms, it is easy to see that relatively low and uneven FDI could be expected. FDI suffered again from 2009 onwards due to a combination of the effects of the global financial crisis, negative intra-country loans, and divestments by foreign multinationals such as UK/Dutch Shell Group and the sale of its 50 per cent (USD 820 million) stake in Sadaf petrochemicals to SABIC in August 2017 (Dudley, 2018).

At the individual level, some senior executives at PIF have left recently because they were not happy with the corporate culture and micromanagement of MBS (Jones, 2018). There is clearly an investment climate in Saudi Arabia, but it is becoming clear that attracting the most innovative and high-tech businesses to the Kingdom, to fill business opportunities in cities such as Nium, is a challenge. Due to Saudisation and *nitāqāt* policies, expatriate levies and fees have increased, but these are short-term measures to encourage the employment of Saudi labour rather than as part of a fiscal adjustment strategy. In some cases, they have been temporarily rolled back to stimulate economic growth (Shahine, 2019).

The importance of job creation in Saudi political economy

Like many rentier states, Saudi Arabia has managed a bloated public sector in order to provide gainful employment to those who choose this path. In 2017, about two-thirds of Saudi citizens worked in the public sector. But only about 38 per cent of Saudi citizens (or 20 per cent of female Saudis) choose to work, compared to up to 80 per cent of citizens in advanced countries, due to the generous social welfare programme (Hertog, 2018). Most foreigners are employed in the private sector in low-skilled sectors such as retail, services, or construction. They command private wages of as little as Saudi riyal (SAR) 3,855 a month (or approximately USD 1,000 a month) compared to the Saudi national equivalent of SAR 7,717. However, the better paid public-sector jobs are the ones which most Saudis still hold out for because the salaries rise to SAR 10,589 a month (Hertog, 2018).

Given the pressures associated with the Arab uprisings, Saudisation (the requirement of Saudi companies to employ Saudi nationals up to certain levels) is perhaps more consequent to Saudi political economy and immediate concerns about political and social stability. Saudi Arabia needs 1.2 million jobs by 2022 to reduce its unemployment figure down to 9 per cent from 12.8 per cent, with many jobs expected to be created in the retail sector (Rashad and Kalin, 2018). Although there is movement on the job front and associated social change, for example women being allowed to drive for the first time in 2017, there are still patriarchal barriers to finding work, and the kinds of jobs that recent graduates might prefer to take are not available. Approximately 230,000 graduates enter the workforce each year and will double the size of the national workforce by 2030 (England and Al Omran, 2019a). The popular demands from the Arab uprisings – improved living standards, civil and political freedoms, and combating corruption – still represent a threat to the Saudi monarchy.

In 2011, it could have become an existential threat. For example, in April 2011 rare protests were witnessed in Riyadh and Jeddah (Alsharif and Benham, 2011). King Abdullah moved quickly to commit USD 37 billion for pay rises to offset inflation, as well as to more unemployment benefits and affordable family housing for 18 million low- to middle-income Saudis (Laessing, 2011). The measures did not immediately address broader requests by some Saudis for constitutional reform, national dialogue, or greater female political participation. However, by 2013, King Abdullah did appoint women to the previously all-male consultative *majlis al-shūrā* (Salih, 2014).

Economic and fiscal policy changes

The Kingdom is a classic rentier state both in terms of distribution and dependence on hydrocarbons for most of the national budget. Wald (2017) notes that the Saudi break-even oil price may be nearer to USD 95 a barrel, but current prices are about half that. To fund the shortfall, the Saudi government initially responded by implementing austerity measures, with spending cuts in ministries at 5 per cent in 2016 (Wald, 2017). By 2018, it looked like recovery was set to accelerate again. In recognition that a high deficit would maintain public spending, the government announced its largest budget in 2019 to facilitate economic transformation and diversification.

The Saudi break-even price for oil is still higher because spending commitments are higher, and that's largely due to population growth, as Table 6.1 shows.

While there was a growth in population up to 2018, which is likely to be even more pronounced out to 2050, there is no accompanying growth in foreign reserves. The reserves accumulated during the last era of high oil prices were probably a one-time opportunity. GDP could double, and energy exports could rise by a third to 2030, but there is still a declining long-term demand curve for energy. The Saudi economy is likely to suffer from low growth even if it diversifies into other petrochemical businesses such as fertilisers, synthetic fibres, and rubber. In the worst case, if there is no significant economic growth from Vision 2030, Saudi Arabia could experience a currency crisis from limited foreign currency reserves. It could, in a

Table 6.1 Saudi population in millions, 2009–18.

2009	2010	2011	2012	2013	2014	2015	2016	2017	2018
26.66	27.56	28.37	29.2	29.99	30.77	31.52	31.74	32.61	33.41

Source: Trading Economics (2018b).

worst-case scenario, transition from a high-income country with gross national income (GNI) per capita at around USD 54,000 now, to an upper-middle-income GNI per capita of USD 3,896 to USD 12,055 (World Bank, 2018).

Non-oil GDP amounted to just 1 per cent in 2017. This was expected to rise to 2.8 per cent in 2018 (Saadi, 2018). Much of the increase in non-oil revenue, including a 63 per cent rise in non-oil income in the first quarter of 2018, can be attributed to improved and higher tax collection (Khaiche and Noueihed, 2018). Tax revenues were estimated to be USD 44.2 billion in 2018, a year-on-year increase of 89.4 per cent (Abusaaq, 2019).

Saudi Arabia has attempted to generate more revenue by issuing bonds. The Kingdom issued a USD 17.5 billion sovereign bond in 2016, and another smaller bond for USD 7.5 billion in January 2019 costing slightly more than before given the heightened political and economic risks after the Khashoggi affair (Pronina *et al.*, 2019). The Saudi government was about to list a 5 per cent stake in Saudi Aramco for USD 100 billion before it was pulled in August 2018. The reason for this remains unclear but could be linked to concerns over achieving the target investment. Instead, the Saudi government arranged for Aramco to buy a USD 70 billion stake in SABIC (Azhar *et al.*, 2018). The Aramco purchase was made possible by using an USD 11 billion loan, which effectively brings in funding to feed the PIF, but this kind of solution is not sustainable. It does not provide jobs, whereas more infrastructure spending could, but due to a limited budget, limited interest from Saudis in such jobs, and due to competition from Chinese firms, this option is also problematic. By December 2019, the Saudi Aramco initial public offering (IPO) of 1.5 per cent of the company generated USD 25.6 billion, valuing the company at USD 1.7 trillion, and was the biggest IPO in history (*The Economist*, 2019).

Saudi Arabia is also using subsidy removal and energy pricing implemented in 2018 to curb domestic energy use by 1.5 million barrels of oil down to 2 million barrels by 2030 (Gamal and Carvalho, 2019). The measures should free up energy supplies to export and boost government coffers in a period of falling or relatively low oil prices compared to the period 2011–14. The measures below illustrate that the more recent annual budgets have been supported by a range of tax, fiscal, and other economic reforms:

- Increase visa fees except for hajj and *'umrah* from 2 October 2016
- Introduction of white land tax (clear land) from March 2017
- Expats to pay 'family tax' upfront when leaving Saudi Arabia from June 2017
- Monthly employment tax on foreign labourers and dependents from July 2017
- Reduction of Aramco tax rate from 85 per cent to 50 per cent to attract investors for IPO from March 2017

- Increase in transport fuel prices from December 2017
- Increase in electricity and water tariffs from December 2017
- VAT of 5 per cent from January 2018
- Introduction of real estate taxes: 1 per cent sales fee, plus transaction fee, and mortgage registration fee from February 2018
- Increase in excise tax on tobacco, energy drinks, and soft drinks from 50 per cent in June 2017 to 100 per cent in May 2019 (Young, 2018a).

These have generally gone through the political decision-making process without a debate with the business community. Prior to the Khashoggi incident, the onus was very much on foreign investors and relatively easy wins by raising taxes and tariffs. However, with the political and business environment having changed after the Ritz Carlton detentions and arrests post-2017, the onus has moved back to local private sector-led reform through public–private partnership projects and privatisations, of which there were few (England and Al Omran, 2019b). But now business–state trust appears to be lacking until there is a prolonged period without further public relations (PR) upsets.

Major construction projects

The new fiscal environment in which the Kingdom found itself after 2014 meant some delays and rethinking of construction projects which were commissioned in the boom years before King Abdullah's death. These projects included the Riyadh financial district (to be named King Abdullah Financial District), King Abdulaziz Airport in Jeddah, and an economic city on the Red Sea coast (Paul *et al.*, 2018). All would be reborn under MBS leadership. However, the rationale for these developments is unclear. Hertog (2017: 276) notes that such mega-projects have been 'tools of diversification, vehicles of self-enrichment for ruling elites, attempts to deepen strategic alliances with the world's leading economic powers, or simply princely follies driven by surpluses too large for any government to handle rationally.' The lack of formal political influence by most Saudis means that there are few political constraints over the rollout of such projects.

The Kingdom has some 4,700 construction projects worth USD 852 billion, and FDI has again become a central pillar of Saudi economic planning and, to some extent, political legitimacy. These mega-projects include the USD 26.6 billion restoration of the Grand Mosque in Mecca. The Saudi Binladin Group was responsible for this. Yet, MBS's economic reforms and attempts to tackle corruption may have been motivated to some extent by the billions that the government allegedly owed to such companies – possibly up to USD 40 billion to the Saudi Binladin Group alone (Paul *et al.*, 2018). If proven, this factor of indebtedness could be just as consequential to the trajectory of reforms and Saudi political economy more broadly as any other factor, such as the cyclical nature of the construction industry. It could also be categorised as a feature of authoritarian upgrading, since it involves 'reconfiguring authoritarian governance to accommodate and manage changing political, economic and social conditions' (Heydemann, 2007: 1).

One of the government's first moves in 2015 was to review state spending and contracts, with a focus on big projects such as those carried out by the Saudi Binladin Group, where negligence or corruption may have been evident (Paul *et al.*, 2018). When a Binladin Group crane collapsed during the expansion of the Mecca Grand Mosque project in 2015, killing 107 people, mostly pilgrims, the government moved to suspend further state contracts. This in turn led to financial pressure, non-payment of workers, and riots, and finally caused the government to review all its other projects (Paul *et al.*, 2018). It is difficult to know if this incident led to a necessary re-evaluation of state–business relations or simply provided a useful justification at the right time to make changes. It appears that the Saudi Binladin Group in a public–private form will continue to construct Nium, which is also expected to include robots, drone taxis, and a bridge to Africa. At least five board members have left this project since the Khashoggi killing, though, including former US energy secretary Ernest Moniz (Bendix, 2018).

The Saudi Binladin Group is also finishing Abraj Kudai, which will be the world's largest hotel (with 10,000 hotel rooms) and is located in Mecca. It was set to open in 2017 but suffered in 2015 due to the financial review and should now open in 2021. It clearly shows Saudi government interest in supporting Islamic tourism as it seeks to diversify the economy. Since most of the other projects are yet to be completed, there is no guarantee that they will all receive the domestic and/or external funding commitments necessary to make them a reality. However, they are proposed as follows:

- Al Faisaliah Smart City is a USD 25 billion high-tech city on the Red Sea that relies on renewable energy. It is said that it will create 1 million housing units by 2050 and 1 million jobs in technology, health, and education.
- Riyadh metro will connect eighty-five stations on six new lines across the city. It is worth USD 22.5 billion and expected to open in 2021.
- Jeddah economic city is another project, slated to cost USD 20 billion, being developed by Kingdom Holdings, including the world's tallest tower, Jeddah Tower (at an additional cost of USD 2.2 billion), and is set to open in 2026.
- The construction of King Abdullah Financial District at a cost of USD 10 billion is at the centre of efforts aimed at diversification in support of Vision 2030. It will soon house banks, law firms, and the stock exchange, including a monorail and network of skywalks.

- King Abdulaziz International Airport in Jeddah cost USD 9.6 billion and will soon facilitate pilgrims travelling between Mecca and Medina.
- USD 7.3 billion high-speed Haramain Railway was inaugurated in September 2018 and was to link Mecca and Medina with Jeddah. It caught fire at about the same time Saudi Arabia is alleged to have suffered major battlefield losses in Yemen in late September 2019.
- Phase five of King Abdullah Security Compound will cost USD 8 billion. It is expected to house investigation and prosecution buildings, as well as government security and immigration departments.
- Downtown Jeddah is a USD 5 billion waterfront development, comprising, among other things, 12,000 new housing units.
- Mall of Saudi and City Centre Ishbiliyya will cost USD 3.7 billion.

Reputation damage and its effect on undermining the Saudi business and investment climate

MBS has attempted to shift the centre of political legitimacy from the Wahhabi establishment to his reform agenda focused on social change, including the younger generation and women, and the economy. But a combination of repressive tendencies, including methods used to consolidate political power, tackle corruption, and deal with social and political activists, have served to undermine Western opinion, FDI appetite, and arguably, the future viability of some mega-projects. There are also questions as to why the PIF takes such a central role in the conclusion of major deals, such as that with AMC Cinemas, when the Saudi private sector could have stepped in under the right conditions and regulatory framework (Sakoui, 2018).

The murder of Jamal Khashoggi in the Saudi consulate in Istanbul has confirmed that this Saudi government – a CIA assessment has confirmed MBS culpability in the Khashoggi case (Harris *et al.*, 2018) – is pursuing every measure to weaken internal dissent. Riedel (2018) points out that this represents a quick and fairly cheap solution to complex and difficult challenges abroad.

Saudi Arabia has for a decade or more also been pursuing activists in the social media space. It started as a response to the Arab uprisings and the development of a growing cyber arsenal based on the Emirati model (Bing and Schectman, 2019). It then shifted to counter threats posed by Iran and Qatar, Shiʻi militias, Islamic State, and al-Qaʻida, but has increasingly turned towards internal dissenters (Haykel and Bunzel, 2019). The Saudi online operation was led by Saud al-Qahtani, who was later fired in 2018 for his role in the killing of Jamal Khashoggi (Ignatius, 2018), but has not so far faced prosecution.

Repression has also been coercive. For example, there are a series of incidents of Saudi princes disappearing, including Prince Sultan bin Turki, who had for years lived in exile before his Paris-bound plane was diverted to Riyadh in 2017 (Watson, 2018). Prince Turki bin Bandar Al Saud, who posted critical tweets and videos about the Kingdom, disappeared in 2015. Saudi officials discussed the possibility of assassinating the leader of the Quds Force, Qassim Suleimani (Mazzetti *et al.*, 2018). Indeed, the transition of power to MBS was also coercive. For example, Muhammad bin Nayef as Crown prince was allegedly forced to swear allegiance to MBS (Chulov, 2017).

Had the reform agenda been dominated by changes such as rolling back some of the powers of the *muṭawwi ʿīn* (religious police), allowing women to drive, and opening cinemas and other entertainment venues, the domestic social sphere might also be more positive, and was for a time (Ignatius, 2016). The detention of some of the same people as those calling for these changes, often women, has illustrated the precarious position of those calling for moderate reforms in the Kingdom and complicated the international response in turn.

Costly Saudi foreign policies and 'Riyal Politik' undermine post-2014 adjustment policies

Since MBS became prominent in the conduct of Saudi external affairs from 2016, he has been responsible for a series of missteps, miscalculations, and misadventures on the regional and international stage. The impetuousness can be seen in the cessation of military aid to Lebanon, more than USD 4 billion, which appears to go against the spirit of the Ta'if agreement. This 1989 Saudi-brokered agreement aimed at returning Lebanon to political normalcy. Instead, the perceived threat from Iranian Revolutionary Guards forces and Hizbullah in Syria has pushed Saudi Arabia into unprecedented tactics. These included briefly detaining Lebanese Prime Minister Saad Hariri in Riyadh, days after the forced detention of the Saudi business elite at the Ritz Carlton in Riyadh in November 2017. The crisis was only resolved after some intense diplomacy by French President Emmanuel Macron. As of November 2018, there is a new Saudi ambassador in Lebanon, illustrative of a new consensus in Riyadh (Rose, 2018). Saudi Arabia is also engaging in Iraq, where it reopened its embassy in 2016, and is attempting to claw back influence lost to Iran since the 2003 US-led intervention.

Since the United Arab Emirates (UAE) reopened its embassy in Damascus in December 2018, there are also prospects that Saudi Arabia will follow suit and recognise that its Syria policy has not been as fruitful as it had hoped. The Saudi minister for Gulf affairs visited Raqqa with Bret McGurk, when he was still Special Envoy to the Coalition against the Islamic State, and paid USD 100 million in October 2018 for stabilisation projects in areas held by the People's Protection Units (YPG) (Baskan, 2019). But this same group was under threat from a Turkish offensive and US withdrawal from northern Syria in October 2019. Relations with Turkey have continued to sour due to Turkey's support of Qatar during the Saudi-led boycott against Qatar, President Recep Tayyip Erdoğan's pro-Islamist leanings regionally, and over the Khashoggi murder.

Saudi Arabia paid USD 8 billion to Egypt from May 2012 to mid-2017, but Egypt has so far only repaid USD 600 million. In return for agreeing to 3 per cent interest in January 2019, Egypt will delay repaying USD 2.6 billion for another year and, according to MEMO (2019), has agreed terms to extend repaying the first three Saudi deposits totalling USD 6.6 billion for another year. Although there is strong Egyptian support for the Saudi boycott of Qatar, it has resisted invitations to join in the intervention in Yemen. More telling of the investment climate in Saudi Arabia is the Egyptian billionaire Naguib Sawiris saying he is still not convinced about law and order in Saudi Arabia and not, therefore, comfortable in investing there (Tan, 2019).

The escalating conflict in Yemen has allegedly cost Saudi Arabia at least USD 5–6 billion a month (Riedel, 2017), has also caused an ongoing humanitarian crisis, and has led to Saudi battlefield casualties, notably in September 2019 (Wintour, 2019). The Yemen conflict initially put emphasis on Saudi and UAE relations with those regional states in the Horn of Africa which supported their foreign policy aims. For example, in August 2015, the Saudis deposited over USD 2 billion in Sudan's central bank. Some 8,000 (and by some estimates up to 14,000) Sudanese fighters have been sent to Yemen (Kirkpatrick, 2018).

Saudi relations with the UAE have become strained after the UAE started to draw down its military forces in Yemen from August 2019. The UAE now aims to enhance its mediation efforts in the region and sent a peace delegation to Iran in July 2019 amid ongoing tensions in the Gulf.

To illustrate the splits between Saudi Arabia and the international community over Khashoggi and human rights in general, even in the West, Canada and Saudi Arabia have a growing list of contentions. *The Globe and Mail* (2019) points specifically to Canadian support for civil and women's activists such as Samar Badawi, who is currently in Saudi detention. After the Khashoggi killing, Germany halted its arms exports to the Kingdom and asked other EU states to do the same (Chazen and Pitel, 2018). It was highly unlikely that states such as the UK would do the same given the size of arms exports, worth GBP 4.7 billion since 2015 (Dehghan, 2018). But a UK High Court challenge in June 2019 did prompt the UK government to suspend arms sales to Saudi Arabia (Human Rights Watch, 2019).

The USA, at the executive level at least, remains wedded to MBS and the status quo in Saudi Arabia, especially given its importance in isolating and pressuring Iran. Saudi Arabia continues to engage in extensive lobbying activities in Washington, DC, spending USD 76.9 million on issues ranging from nuclear power to 9/11 claims, although five firms quit after the Khashoggi murder (Brody *et al.*, 2018). Even though US Treasury Secretary Steven Mnuchin skipped the Future Investment Initiative (FII) in 2017 and 2018, Halliburton and Baker Hughes GE both signed memos of understanding with Aramco at the FII in 2018.

Saudi Arabia is Silicon Valley's biggest investor, with stakes in Uber, Magic Leap, View, and Zume. Some are channelled by the Vision Fund, a collaboration between Japan's Softbank and Saudi Arabia's PIF (Greenwald, 2018). Tesla was also part of Saudi plans to expand investments to support Vision 2030. But Saudi Arabia has since hedged against this position after Elon Musk overstated Saudi support for his plans to delist Tesla shares from the Nasdaq stock exchange (DiChristopher, 2019).

Saudi Arabia maintains a centrality in US calculations concerning expected losses of Iranian oil output following further sanctions (Torchia, 2018). Still, US sanctions have been imposed against those the USA considers responsible for the death of Khashoggi, through the Global Magnitsky Act. However, some in the Saudi camp, such as the general manager of the state-owned Al Arabiya channel, Turki Aldakhil, have highlighted that Saudi Arabia could theoretically retaliate against further US sanctions by increasing oil prices, pricing oil in Yuan instead of US dollars, putting an end to intelligence sharing, and entering into a military alliance with Russia (Torchia and Mohammed, 2018). Whether or not any or all of these are realistic options for a kingdom under pressure remain to be seen; however, they do illustrate that Saudi Arabia retains value to the international community on multiple levels.

A Saudi–Russia alliance is particularly concerning given the latter's willingness to involve itself in the construction of civil nuclear reactors in a state which has uranium deposits and has threatened to develop nuclear weapons if Iran does. There is also the prospect that civil nuclear power will address the growing domestic energy demand in the Kingdom, although there are many questions over whether Saudi Arabia will sign up to the same gold standard (also known as a 123 Agreement) that the UAE has with the USA already. Saudi Arabia has an unacknowledged nuclear partnership with Pakistan, which will provide weapons at short notice if needed (Riedel, 2008). This gives current Saudi policy an additional international security dimension.

Riyadh continues to develop ever closer energy relations with Russia, China, and India, partly triggered by a diminishing market share in the Chinese oil market compared to Russia. In China, Saudi Arabia is also targeting start-ups such as ZhongAn (insurtech), Bytedance (social media), and Sensetime (facial recognition/AI) (Greenwald, 2018). Given the role of PIF in Vision 2030, most of these investments should be seen as dual use: technology transfer and job creation. Saudi Arabia and Russia look set to try to harmonise oil and gas supply and demand, illustrated by a temporary agreement in December 2018 to cut oil production. Over the longer term, Russian interests will be in securing a higher oil price, and relations with Qatar may prove to undermine relations with the Kingdom. Beyond economics, China and Russia have good relations with Iran, which Saudi Arabia will attempt to influence through equal or greater economic engagements. It already tried to usurp the Russia–Iran S-300 deal through the purchase of the S-400 (Mason, 2017).

In 2018, King Salman attempted to draw a line under the Khashoggi case with unconditional support for his son and a PR tour around the Kingdom. Meanwhile MBS undertook a PR tour of Western capitals and went on 60 Minutes to discuss the role of Saudi women. MBS has avoided further criticism by launching new tourism projects. For example, in February 2019, he travelled to al-'Ula between Medina and Tabuk, which is said to rival Petra for tourism potential. In October 2019, Saudi Arabia announced a new tourist visa for American, European, and Asian tourists to visit the Kingdom to boost this nascent industry.

Still, Khashoggi remains in the international public sphere due to Turkey's slow release of information surrounding the murder. The USA has also been critical at the congressional level, with strong words from Nancy Pelosi, the House Speaker, about military support for Saudi Arabia against Iran in the context of the Khashoggi murder and Iranian threats (DeBonis, 2019). Furthermore, a Public Broadcasting Service (PBS) Frontline documentary about Khashoggi was aired in October 2019.

Conclusion

In other times, the 2014 fall in oil prices might have been the number one issue facing Saudi Arabia. Given Saudi objectives in the Organization of the Petroleum Exporting Countries (OPEC) concerning long-term alliance patterns and the maintenance or development of relatively new Asian markets, it actually supports a range of short-term Saudi economic and geostrategic objectives. However, turbulence over the last few years has constrained the elite's room for manoeuvre. The oil price has put pressure on the national budget and increased the need for borrowing from international markets, putting the spotlight on the Kingdom's reputation and any associated political risk. The youth bulge and demographic trajectory are adding pressure to the current Saudi job strategy, and if it fails for whatever reason, the Saudi government will have to foot the bill. Social liberalisation and government targets are also driving women's expectations of labour participation.

With most mega-cites serving multiple purposes (housing, innovation, cooperation on borders) but not scheduled for completion until 2025 at the earliest, they will be slow to contribute to economic growth. It is also hard to see how possible vertical diversification into fertilisers, synthetic fibres, and rubber would significantly change the economic picture should oil prices remain depressed. A state-led investment programme coupled with reputation damage is unlikely to secure growth prospects for the Kingdom at a critical juncture, and we could therefore see more authoritarian upgrades in future. These might include economic patronage reductions in favour of further repressive measures and rolling back women's rights and other social reforms. Ultimately the Kingdom could become a 'bunker state' which abandons any attempt to manage social forces and legitimacy based on economic credentials.

The question is whether existing or new Saudi missteps will raise the prospect of civil unrest, economic decline, or the loss of critical international allies, and push Saudi Arabia into open conflict with adversaries such as Iran. The 2019 attacks on Abqaiq and al-Khurais, which temporary knocked out 50 per cent of Saudi export capacity, serves as a timely reminder that there is potential for this eventuality. Over the longer term, if Saudi oil policy is a successful part of maintaining constructive alliance patterns, including growing oil-based relations with China, India, and Japan, it could sustain the relevance of RST over the medium term if reform measures falter or fail.

References

Abusaaq, H. (2019), 'VAT increases Saudi Arabia's tax revenue by over 89%', *Forbes*, 1 October. https://tinyurl.com/prftwh2u (accessed 15 October 2019).

Al-Khateeb, L. (2015), 'Saudi Arabia's economic time bomb', Brookings, 30 December. https://tinyurl.com/htu6su8 (accessed 6 June 2019).

Alsharif, A., and J. Benham (2011), 'Saudi unemployed graduates protest to demand jobs', Reuters, 10 April. https://reut.rs/3bVWYI5 (accessed 6 June 2019).

Arabian Business (2018), 'Saudi Arabia to slash defence spending', 19 December. https://tinyurl.com/chyhk7n4 (accessed 6 June 2019).

Azhar, S., H. A. Sayegh, and C. Denina (2018), 'Bruised bankers seek consolation prizes after shelved Aramco IPO', Reuters, 31 August. https://reut.rs/2MLXqjI (accessed 6 June 2019).

Baldwin, D. A. (1985), *Economic statecraft*. Princeton: Princeton University Press.

Baskan, B. (2019). 'A new Turkey–Saudi crisis is brewing', Middle East Institute, 8 January. www.mei.edu/publications/new-turkey-saudi-crisis-brewing (accessed 6 June 2019).

Beck, M., and T. Richter (2021), 'Pressured by the decreased price of oil: Post-2014 adjustment policies in the Arab Gulf and beyond', in M. Beck and T. Richter (eds), *Oil and the political economy in the Middle East: Post-2014 adjustment policies of the Arab Gulf and beyond*. Manchester: Manchester University Press, pp. 1–35.

Bendix, A. (2018), 'Saudi Arabia's $500 billion megacity is in jeopardy after a Saudi journalist's death: Here's the status of the Kingdom's other multi-billion-dollar infrastructure projects', Business Insider, 21 November. https://bit.ly/3nF0ADP (accessed 6 June 2019).

Bing, C., and J. Schectman (2019), 'Inside the UAE's secret hacking team of American mercenaries', Reuters, 30 January. https://tinyurl.com/548e9ebs (accessed 6 June 2019).

Boeing (2018), 'Saudi Arabian military industries and Boeing form joint venture partnership targeting 55% localization', 30 March. https://bit.ly/3e7x3zr (accessed 23 June 2019).

Brody, B., N. Nix, and B. Allison (2018), 'Saudi lobbyists in D.C. hunker down after Khashoggi killing', Bloomberg, 26 October. https://bloom.bg/3vCdEMU (accessed 6 June 2019).

CEIC (2019), 'Saudi Arabia foreign exchange reserves', CEIC data. https://tinyurl.com/hv7mbkhf (accessed 6 June 2019).

Chazen, G., and L. Pitel (2018), 'Germany halts arms exports to Saudi Arabia after Khashoggi's death', *Financial Times*, 22 October. https://on.ft.com/3nDbzO8 (accessed 6 June 2019).

Chulov, M. (2017). 'Deposed Saudi Crown prince confined to palace', *The Guardian*, 29 June. https://bit.ly/3qdYV7Z (accessed 6 June 2019).

Daiss, T. (2016), 'Saudi Arabia burns through foreign reserves as oil prices slide', *Forbes*, 30 August. https://bit.ly/30axLEd (accessed 6 June 2019).

DeBonis, M. (2019), 'Citing Khashoggi's killing, Pelosi says U.S. should not bomb Iran at Saudi's behest', *The Washington Post*, 20 September. https://wapo.st/33eY5OZ (accessed 14 October 2019).

Dehghan, S. K. (2018), 'Dozens dead in Yemen as bus carrying children hit by airstrike', *The Guardian*, 9 August. https://bit.ly/3be2Jlr (accessed 6 June 2019).

DiChristopher, T. (2019), 'Saudi Arabia – which Elon Musk claimed would back a buyout – cut its Tesla exposure: FT', CNBC, 28 January. https://cnb.cx/3be6eIB (accessed 6 June 2019).

Diwan, I. (2019), 'A landing strategy for Saudi Arabia', *POMEPS Studies*, 33, 25–8. https://bit.ly/2PxphVN (accessed 2 June 2019).

Dokoupil, M. (2014), 'Saudi could see budget deficit next year, risks draining reserves – IMF', Reuters, 24 September. https://reut.rs/30auqFs (accessed 15 October 2019).

Dudley, D. (2018), 'Saudi Arabia suffers shock collapse of inward investment', *Forbes*, 7 June. https://bit.ly/2O1fuqt (accessed 6 June 2019).

England, A., and A. Al Omran (2019a), 'Saudi Arabia: Why jobs overhaul could define MBS's rule', *Financial Times*, 27 February. https://on.ft.com/3vyrWhC (accessed 15 October 2019).

England, A., and A. Al Omran (2019b), 'Saudi Business feels "pain" of the Crown prince's reforms', *Financial Times*, 8 June. https://on.ft.com/3xK8jFs (accessed 19 October 2019).

France 24 (2018), 'Saudi announces budget deficit for sixth straight year', 18 December. https://tinyurl.com/u5bv8sax (accessed 6 June 2019).

Freeman Jr, C. (2015), 'Saudi Arabia and the oil price collapse: Remarks to a panel at the Center for the National Interest', Washington, DC, 27 January. https://bit.ly/3h1OdAB (accessed 6 June 2019).

Gamal, R. E., and S. Carvalho (2019), 'Saudi Arabia sees domestic energy use falling, plans renewables push', Reuters, 15 January. https://reut.rs/2NVuN3W (accessed 6 June 2019).

Greenwald, M. B. (2018), 'The Saudi economy moves closer to Russia and China', *The National Interest*, 5 December. https://tinyurl.com/h7x3sdbz (accessed 15 October 2019).

Harris, C., G. Miller, and J. Dawsey (2018), 'CIA concludes Saudi Crown prince ordered Jamal Khashoggi's assassination', *The Washington Post*, 17 November. https://wapo.st/3aVDZxQ (accessed 6 June 2019).

Haykel, B., and C. Bunzel (2019), 'Messages to Arabia: Al Qaeda attacks MBS and the Saudi monarchy', Jihadica, 24 January. www.jihadica.com/messages-to-arabia/ (accessed 6 June 2019).

Hertog, S. (2010), *Princes, brokers, and bureaucrats: Oil and the state in Saudi Arabia*. Ithaca: Cornell University Press.

Hertog, S. (2017), 'A quest for significance: Gulf oil monarchies' international "soft power" strategies and their local urban dimensions', London School of Economics and Political Science, Kuwait Programme on Development, Governance and Globalisation in the Gulf States, Research Papers, 42. http://eprints.lse.ac.uk/69883/1/Hertog_42_2017.pdf (accessed 2 June 2019).

Hertog, S. (2018), 'Can we Saudize the labor market without damaging the private sector?', King Faisal Center for Research and Islamic Studies, Special Report. https://bit.ly/3kHhYGy (accessed 2 June 2019).

Hertog, S. (2019), 'What would the Saudi economy have to look like to be "post-rentier"?', *POMEPS Studies*, 33, 29–34. https://bit.ly/2PxphVN (accessed 2 June 2019).

Heydemann, S. (2007), 'Upgrading authoritarianism in the Arab world', The Saban Center for Middle East Policy at the Brookings Institution, Analysis Paper, 1. https://tinyurl.com/euaxvywv (accessed 2 June 2019).

Human Rights Watch (2019), 'UK arms sales to Saudis suspended after landmark ruling', 20 June. https://bit.ly/3bW5WVV (accessed 23 June 2019).

Ignatius, D. (2016), 'A 30-year-old Saudi prince could jump start the Kingdom – or drive it off a cliff', *The Washington Post*, 28 June. https://wapo.st/3nGUrqF (accessed 6 June 2019).

Ignatius, D. (2018), 'How a chilling Saudi cyberwar ensnared Jamal Khashoggi', *The Washington Post*, 7 December. https://wapo.st/3uecI18 (accessed 6 June 2019).

Jones, R. (2018), 'Expats flee Saudi fund, bemoan Crown prince control', *The Wall Street Journal*. 31 December. https://tinyurl.com/7pecayf8 (accessed 6 June 2019).

Kamrava, M. (2018), 'Oil and institutional stasis in the Persian Gulf', *Journal of Arabian Studies*, 8:1, 1–12.

Kane, F. (2018), 'Job creation as important as growth says Saudi minister', *Arab News*, 3 May. www.arabnews.com/node/1296041/business-economy (accessed 14 October 2019).

Khaiche, D., and L. Noueihed (2018), 'Saudi Arabia's non-oil income rises 63 percent on new taxes', Bloomberg, 7 May. https://bloom.bg/3eIX5bo (accessed 6 June 2019).

Kirkpatrick, D. (2018) 'On the front line of the Saudi War in Yemen: Child soldiers from Darfur', *The New York Times*, 28 December. https://tinyurl.com/479zcfnn (accessed 6 June 2019).

Krane, J. (2019), *Energy kingdoms: Oil and political survival in the Persian Gulf.* New York: Columbia University Press.

Kurlantzick, J. (2016), *State capitalism: How the return of statism is transforming the world.* New York: Oxford University Press.

Laessing, U. (2011), 'Saudi King back home, orders $37 billion in handouts', *Financial Times.* 23 February. https://reut.rs/3qc1w2b (accessed 6 June 2019).

Martin, J. (2019), 'Donald Trump once said Saudi Arabia should fight its own wars. Now he's deploying 3,000 U.S. troops there', *Newsweek*, 11 October. https://bit.ly/3kGi2Gs (accessed 14 October 2019).

Mason, R. (2014a), 'Back to realism for an enduring U.S.-Saudi relationship', *Middle East Policy*, 21:4, 32–44.

Mason, R. (2014b), *Foreign policy in Iran and Saudi Arabia: Economics and diplomacy in the Middle East.* London: I.B. Tauris.

Mason, R. (2017), 'Saudi visit to Moscow: Driving a wedge between Russia and Iran?', The Globe Post, 10 October. https://theglobepost.com/2017/10/10/saudi-salman-russia-iran/ (accessed 6 June 2019).

Mazzetti, M., R. Bergman, and D. Kirkpatrick (2018), 'Saudis close to Crown prince discussed killing other enemies a year before Khashoggi's death', *The New York Times*, 11 November. https://nyti.ms/3qcPzt4 (accessed 6 June 2019).

MEMO (2019), 'Egypt postpones $6.6 billion repayment to Saudi Arabia', 23 January. https://bit.ly/3ubBsHa (accessed 6 June 2019).

Miller, R. (2016), *Desert kingdoms to global powers.* New Haven: Yale University Press.

Paul, K., T. Arnold, M. Rashad, and S. Kalin (2018), 'As a Saudi prince rose, the Bin Laden business empire crumbled', Reuters, 27 September. https://reut.rs/3xGRgny (accessed 6 June 2019).

Pronina, L., N. Ismail, and A. Narayanan (2019), 'Saudis lure investors to $7.5 billion debt after Khashoggi', Bloomberg, 10 January. https://bloom.bg/3eMbCmO (accessed 6 June 2019).

Rahman, F. (2019), 'UAE, Saudi Arabia firms to boost defence cooperation', *Gulf News*, 20 February. https://bit.ly/3sEKY4q (accessed 23 June 2019).

Rashad, M., and S. Kalin (2018), 'Saudi Arabia needs 1.2 million jobs by 2022 to hit unemployment target: Official', Reuters, 25 April. https://reut.rs/3kFchsy (accessed 6 June 2019).

Reuters (2019a), 'IMF expects Saudi budget deficit to hit 7% of GDP this year', 15 May. https://reut.rs/3bYx2vj (accessed 14 October 2019).

Reuters (2019b), 'Saudi Arabia eyes billions of dollars in entertainment investments', 22 January. https://reut.rs/3uNBOEF (accessed 23 June 2019).

Riedel, B. (2008), 'Saudi Arabia: Nervously watching Pakistan', Brookings, 28 January. https://tinyurl.com/n8prjrms (accessed 6 June 2019).

Riedel, B. (2017), 'In Yemen, Iran outsmarts Saudi Arabia again', Brookings, 6 December. https://brook.gs/3rpXrsN (accessed 6 June 2019).

Riedel, B. (2018), 'Ricochet: When a covert operation goes bad', Brookings, December. https://tinyurl.com/3c9p5av8 (accessed 6 June 2019).

Rose, S. (2018), 'New Saudi ambassador arrives in Lebanon', *The National*, 29 November. https://tinyurl.com/5y7upw7s (accessed 6 June 2019).

Saadi, D. (2018), 'Saudi economy to grow 1.8% in 2018 on non-oil GDP expansion, NCB says', *The National*, 31 May. https://bit.ly/3bSKPDM (accessed 6 June 2019).

Sakoui, A. (2018), 'AMC cinemas tiptoes into Saudi Arabia as theater ban lifted', Bloomberg, 4 April. https://bloom.bg/3nFBfcQ (accessed 6 June 2019).

Salih, M. (2014), *Economic development and political action in the Arab world*. Abingdon: Routledge.

Shahine, A. (2019), 'Saudi King approves $3.1 billion plan to ease expat fee costs', Bloomberg, 10 February. https://bloom.bg/3xEi5st (accessed 6 June 2019).

Smil, V. (2017), *Energy transitions: Global and national perspective*. Denver: Praeger.

Springborg, R. (2018), 'Deep states in MENA', *Middle East Policy Journal*, 25:1, 136–57.

Tan, W. (2019), 'I would not invest in Saudi Arabia, says Egyptian billionaire Naguib Sawiris', CNBC, 12 February. https://cnb.cx/3e1EgS9 (accessed 6 June 2019).

The Economist (2019), 'Saudi Aramco's IPO the biggest ever', 7 December. https://tinyurl.com/2pe73u3r (accessed 15 January 2020).

The Globe and Mail (2019), 'Saudi–Canadian relations and the arms deal: A guide to the story so far', 2 January. https://tgam.ca/2OjRYEU (accessed 6 June 2019).

Torchia, A. (2018), 'Saudi economy starts to recover, set to accelerate as oil output rises', Reuters, 1 July. https://reut.rs/3e1SHFK (accessed 6 June 2019).

Torchia, A., and A. Mohammed (2018), 'Saudi Arabia says will retaliate against any sanctions over Khashoggi case', Reuters, 14 October. https://reut.rs/3bbpshS (accessed 6 June 2019).

Trading Economics (2018a), 'Saudi Arabia military expenditure'. https://tinyurl.com/ynbs7pxz (accessed 6 June 2019).

Trading Economics (2018b), 'Saudi population growth (annual %)'. https://bit.ly/2On1YgD (accessed 6 June 2019).

Wald, E. (2017), 'Saudi Arabia's 2018 budget is the country's largest ever', *Forbes*, 19 December. https://bit.ly/3bbpM04 (accessed 6 June 2019).

Watson, J. (2018), 'Jamal Khashoggi murder the latest in a string of suspected Saudi abductions and assassinations', ABC News, 9 November. https://ab.co/3sPxQd1 (accessed 6 June 2019).

Wintour, P. (2019), 'Houthis claim to have killed 500 Saudi soldiers in major attack', *The Guardian*, 29 September. https://bit.ly/382ogLB (accessed 15 October 2019).

World Bank (2018), 'New country classifications by income level: 2018–2019'. https://bit.ly/3vuV9tM (accessed 15 October 2019).

World Bank (2019), 'Foreign direct investment, new inflows (BoP, current US$)', World Bank Data. https://tinyurl.com/5zb5eyw9 (accessed 6 June 2019).

Young, K. (2018a), 'The difficult promise of economic reform in the Gulf', James A. Baker III Institute for Public Policy of Rice University. https://bit.ly/3874cI3 (accessed 2 June 2019).

Young, K. (2018b), 'Saudi Arabia's crisis is economic and demographic', Remarks from the Hoover Institution's Panel, 'Saudi Arabia in Crisis', 8 November. https://bit.ly/3kIcZ8l (accessed 15 October 2019).

7

Federal benefits: How federalism encourages economic diversification in the United Arab Emirates

Karen E. Young

Introduction

Since the sharp decline in oil prices in late 2014, the Gulf Cooperation Council (GCC) members have become more competitive with each other in altering their policy landscapes to streamline fiscal expenditure, to create new sources of government revenue, and to attract foreign investment and resident investors. We can see diversity in approaches to economic diversification in many policy choices, but labour market regulation and rights of foreign investors stand out to distinguish how the GCC states view their relationship with outsiders. However, this competitive landscape is not limited to the state level. In fact, as the most successful case of economic diversification of the GCC, the United Arab Emirates (UAE) benefits from a competitive policy environment within its federal system. The federation allows policy experimentation, particularly as some emirates face stronger pressures without equal distribution of oil revenues, as most oil reserves are in the emirate of Abu Dhabi. This policy learning and differentiation between emirates pre-dates the 2014 oil price collapse, as Dubai has been a diversification policy accelerator, historically willing to try new policies for government revenue generation to supplement its lack of resource wealth. The dynamic between emirates in the federal system has set the stage for a pattern of policy flexibility towards diversification now embraced and implemented by the central government between 2015 and 2019.

While other members of the GCC struggle to protect labour markets for nationals, the UAE has created new categories of long-term visas to encourage investors and skilled experts to live and work in the country. While other GCC states rely on debt to finance budget deficits, the UAE has yet to issue bonds at the federal level. The UAE has been a first mover among GCC states to implement new taxes and fees and decrease fuel and energy subsidies at the emirate and federal level over the course of several years, such that the public (both expatriate residents and citizens) is familiar with the imposition. Moreover, the UAE has been extremely effective in attracting foreign direct investment (FDI). In short, the UAE has shown some policy flexibility and success in meeting the challenge of economic diversification by creating alternative revenue streams for the government, reducing expenditure on subsidies, and encouraging new sectors of growth.

The goal of economic diversification away from oil revenue dependence is not new, but there has been an urgency to new policy responses since late 2014, when oil prices sharply declined from a ten-year boom cycle. The UAE has been successful in part because of the federal structure of the government, which has encouraged policy differentiation, learning, and even competition between the seven emirates. Because resource endowment is not equal across the federation and proceeds of oil wealth not evenly distributed, policy experimentation and change has flourished, furthering the pace of economic diversification. This policy learning and differentiation between emirates pre-dates the 2014 oil price collapse, but it set the stage for a pattern of policy flexibility now embraced and accelerated by the central government between 2015 and 2019. Even as the central government under Abu Dhabi leadership has increased its leverage over the federation, the structure and differences between emirates continue to create room for policy difference and experimentation.

This chapter focuses on a series of policy choices and reforms, from labour market regulation to fiscal reform, to policies affecting the investor and business climate. One broad trend in the post-2014 era has been a strengthening of the role of the state in the economy across the Gulf states. In the UAE, while we have seen centralisation of state authority, especially in foreign policy, there remains a benefit to new economic policy formulation, experimentation, and competition between emirates. Breaking down the policy agenda post-2014 in the UAE, we can identify three core economic response policy areas: fiscal policy, social development policy, and diversification policies. Each of these areas is a response to the demands of a lower oil price environment and the mounting challenges of forty years of rentier policy. This chapter dissects these policy shifts in detail and analyses how these policy choices fit together in a reform and diversification agenda for the UAE.

New international dynamics: Debt capital markets, geopolitics, and climate change

Across the GCC, governments face enormous strains on public finances and challenging economic outlooks due to fluctuating oil prices, demographic pressures, high unemployment rates, and a lack of economic diversification.

Debt has become a tool of choice, but the capacity to repay and the capacity to grow are both beginning to differentiate the GCC states (Young, 2019a). Wealth stored as reserve assets and capacity to attract new investment and forces for job creation and overall economic growth are not the same. Those states that borrow heavily now, without the capacity to generate new wealth in the future, will be at a disadvantage. Diversified economies will create growth that generates new revenue streams (arguably by taxes and fees) for the state, or they will rely on state-owned investment vehicles, including externally focused sovereign wealth funds, to generate it for them. The Gulf states are now major issuers of emerging market sovereign debt globally, flooding the bond market with new issues since 2015.

As Figure 7.1 demonstrates, the Gulf states were largely absent as major sovereign issuers of emerging market debt before 2009, when they were only minor issuers in international debt capital markets compared to other regions. This trend changed considerably after 2015, as Gulf governments sought to remedy their fiscal imbalances with debt. The largest issuer since 2015 has been Saudi Arabia. And while there has been consistent interest from investors to buy Gulf government bonds, a growing reliance on debt financing will become a burden for more states to service and repay that debt. In the stronger economies of the UAE, Qatar, and Kuwait, with their smaller populations and higher oil revenues per capita, debt issuance is an easy way to access credit that is not difficult to repay. For weaker economies

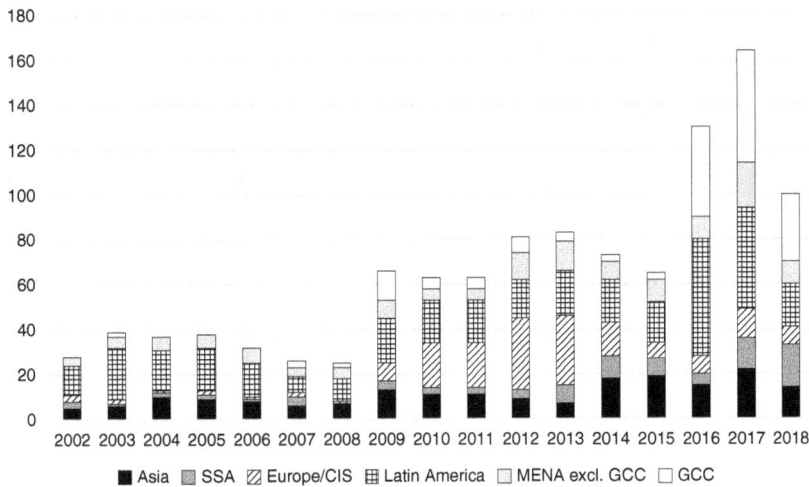

Figure 7.1 Emerging markets sovereign debt issuance (in USD billion). Source: Goldman Sachs Global Investment Research (2018).

like Bahrain and Oman, a mounting debt to gross domestic product (GDP) ratio means servicing that debt becomes more of a burden on fiscal policy decision-making in the years ahead, starting in 2022 for Oman. Some social services may not be part of future budgets if debt repayment has to be a government priority in spending at the risk of a credit rating downgrade and access to continued financing.

For the UAE, the sovereign debt burden is negligible. Individual emirates have their own access to debt capital markets, their own credit ratings, and their own responsibilities for repayment. This allows the federal government, and the emirate of Abu Dhabi, to act as a source of financial support to other emirates, as it did for Dubai in 2009–10. More recently, the federal government has moved to establish its own credit rating and ability to issue debt on behalf of the UAE as a whole. This opens up an enormous potential line of credit of easy-to-access cash for federal development and fiscal planning, lessening the pressure on the federal government. Even still, the UAE has been less reliant than some of its peers on fiscal outlays in social benefits to relieve or prevent criticism of new taxes and fees and reduced energy subsidies.

While debt repayment will constrict future fiscal policy space for some Gulf states, there are also environmental and geopolitical challenges to confront. The rising impact of climate change further complicates socio-economic challenges. The Gulf will be a more difficult place to sustain, and the world will find new ways beyond hydrocarbons to meet its energy needs. The region faces a future in which it will be more difficult to sustain a population both in comfort and in economic means, yet the region remains central in global geopolitical and security concerns. There is a concerted 'eastward turn' in the Gulf states' orientation towards its major energy consumers in Asia (Young, 2019b). The international order is undergoing a fundamental transition, with significant implications for the Middle East and the Gulf states. Inward and nationalist political shifts in the West, China's increased engagement with the Middle East through its Belt and Road Initiative, the ongoing conflicts in Syria and Yemen, and the potential breakdown of the Joint Comprehensive Plan of Action (JCPOA) necessitate a role for the Gulf states in international institutions and negotiated frameworks.

The diversification agenda in the UAE in the domestic regional and global contexts

There is a delicate balance in how the UAE accommodates the demands of its citizen population along with demands for growth, whether growth targets are in the non-oil sectors or some combination of value-added energy products. The state must position itself to meet not just domestic challenges but also global forces. How will global megatrends affect the UAE? How can the UAE adjust and diversify its economy in order to meet the dual challenges of fluctuating oil prices and climate change? How can the state adjust labour markets to absorb and support women and youth? How will the GCC crisis affect the region moving forwards? How will foreign relations with international actors shift in the coming years? These are all questions requiring flexible and decisive policy responses. The policy environment post-2014 is fraught with high stakes – how should the demands of citizens be met without losing power or going bankrupt, while also sustaining or amplifying the state's position in the global political economy? The UAE is frequently commended, and at times derided, for its ambition and the way it uses oil wealth to achieve both domestic and foreign policy goals (*The Economist*, 2017).

This chapter focuses on how the UAE seeks to position itself, given these domestic, regional, and global challenges. The chapter looks closely at a series of policy choices and reforms, from labour market regulation to fiscal reform, to policies affecting the investor and business climate. There is an important distinction between the idea of 'reform' and 'policy choice', as the shifts often entail some backtracking in diversification efforts and protectionism of labour markets for national employment and continuation of special provisions for citizens in social services. Reform denotes a move towards liberalisation, or a dismantling of the rentier model when, in many cases, policy changes are an invigoration of those enduring rentier paradigms (Gray, 2019).[1] In the case of the UAE, there has been both policy reform and flexibility in policy choice. There is evidence of significant efforts at subsidy reduction in energy and water, as well as examples of new revenue generation through taxes and fees, but there is also a consistent effort by the state to provide citizen benefits, especially in housing policy. Because the demographics of the UAE are tilted towards a majority expatriate population, foreigners pay more to the state in taxes and fees than citizens.

One broad trend in the post-2014 era has been a strengthening of the role of the state in the economy across the Gulf states. The state is in charge of the development agenda, especially in its choices of reforms to labour markets and foreign ownership restrictions and its enticements to foreign investors. In the UAE, we find an emboldened central state, in the power of Abu Dhabi under Crown Prince Muhammad bin Zayed driving the diversification agenda, but at the same time relying on subnational policy to try out reform and development policy approaches, most notably in Dubai.

UAE diversification efforts ahead of the competition

For the UAE, its position as a small state in a difficult neighbourhood has meant a foreign policy of increasing assertiveness since 2011 (Young, 2013). The 2010s have challenged the state's capacity to execute a demanding foreign policy, including military action, the distribution of foreign aid, and foreign investment to support key regional allies. The UAE's entry into the Yemen Civil War in 2015 has further stretched its defence expenditure and operational bandwidth, widening the small country's sphere of influence into the Horn of Africa with new basing infrastructure (Vertin, 2019). While recent port and military installations have dominated analysis of a security presence, there is a long-standing effort by the UAE to use its investment capacity abroad as a lever to its diversification efforts. Some of those investments have been geographically concentrated in Africa, but most have been placed in major equity markets in New York or London, real estate investments in Europe and the USA, and more recent partnerships with energy consumers and competitors from Russia to China to Pakistan (Young, 2019b). The UAE, like its GCC peers, is a global investor, and it intends its outward allocation of capital to use oil wealth to diversify holdings outside of the energy sector and to amplify that wealth for future generations. However, we are seeing a return to investment focused on energy assets via sovereign wealth funds, especially with partnerships or hedging strategies with the Russian Direct Investment Fund (Mubadala, 2018).

With diminished oil revenues after 2014, these foreign policy objectives and economic statecraft efforts increasingly compete with domestic demands for job creation and continued economic expansion. In many ways, the UAE is a leader among its peers in the GCC for its adaptation of policies friendly to foreign resident investors and its willingness to adopt taxes and measures to supplement government revenue. But the UAE is distinct from most of the GCC (Qatar is similar) in its dependence on foreign workers as an overwhelming majority of the population. Of its roughly ten million people, nine million are not citizens. The UAE is a distant second in population size to Saudi Arabia among the six GCC states.

According to the consultancy Fitch Solutions (2019), the UAE is the most diversified economy of the GCC and scores highest in their measures of its business environment and rapid pace of economic diversification reforms (Figure 7.2). In its rankings, Fitch Solutions (2019) credits the UAE's targeting of fiscal stimulus towards business-friendly measures, rather than strictly capital expenditure on less productive government development projects. Foreign actors may now own 100 per cent of local entities, and the government has begun issuing its first ten-year visas for investors and skilled

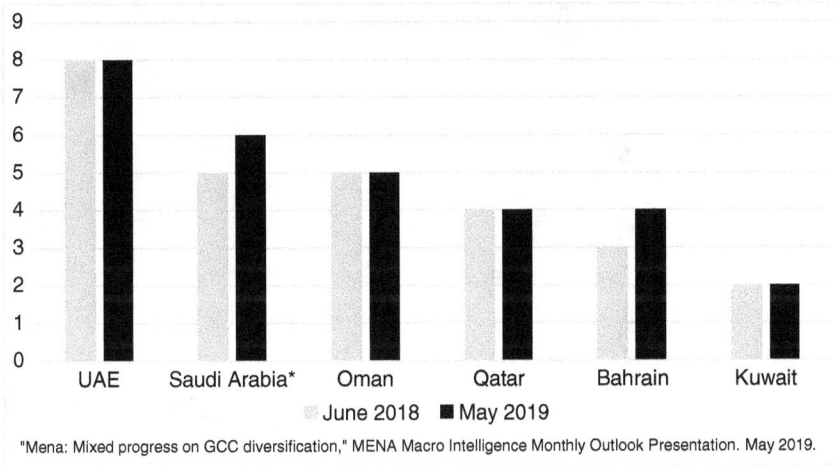

"Mena: Mixed progress on GCC diversification," MENA Macro Intelligence Monthly Outlook Presentation. May 2019.

Figure 7.2 Fitch Solutions GCC diversification scorecard. Note: Scale of 0–10. Higher scores represent a higher pace of reform and better prospects of diversification. *Note: Saudi Arabia's 2018 score was revised down to 5 from 6 in October, to reflect flagging foreign investor confidence stemming from perceptions of increased political risk. Source: Fitch Solutions (2019).

workers, though there remain significant barriers in the number of qualified awardees of the visas and the quantity and quality of sectors exempt to allow foreign ownership. Non-oil economic activity in the UAE has traditionally been a factor of Dubai's economy. Dubai's economic growth, however, has relied on property markets and tourism, which have proved sensitive to regional political tensions. According to research by Standard Chartered Bank (2019), both hotel occupancy rates and passenger arrivals at Dubai Airport weakened between 2016 and 2019. As Standard Chartered analysts also note, the UAE Central Bank lowered its 2019 growth forecast to 2 per cent based on non-oil sector growth of 1.8 per cent in the first quarter of 2019 (Standard Chartered Bank, 2019).

Between 2014 and 2018, the UAE, like all of the GCC states except Oman, experienced a decline in non-hydrocarbon GDP growth (Figure 7.3). The low points in this period have varied, signally some variation in overall economic sensitivity to a decline in oil revenues, and various points of intervention or injection of investment that have had an effect on an overall decline in growth. Oman, for example, had an interesting spike in growth, likely due to the investment and construction in its port in al-Duqm (funded by Chinese loans) and its port in Suhar in this period. Kuwait, meanwhile, rebounded somewhat between 2015 and 2016. Lower oil for longer is a

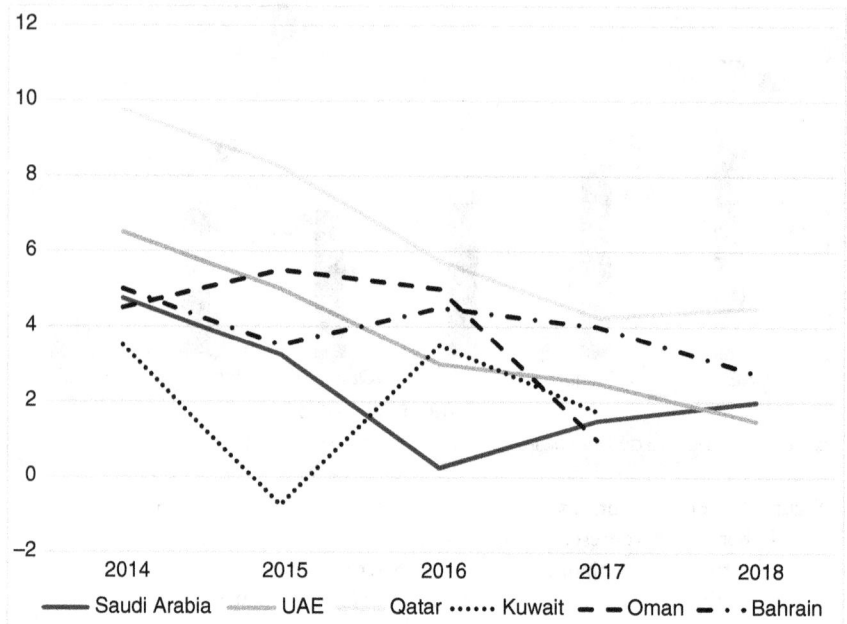

Figure 7.3 Non-hydrocarbon real GDP growth (per cent). Note: UAE 2018 figure based on statement made by the minister of economy. Source: Fitch Solutions (2019).

new reality for the Gulf states, which has introduced some volatility into their growth patterns and necessitated a flurry of government efforts to create new ways to generate economic growth.

Even non-oil growth links back to oil, and threats to oil production

The UAE is a leader in economic diversification across the GCC, but even the UAE is struggling to grow in the post-2014 oil price environment. Where we do see economic growth, we can often trace it back to investment in the oil sector or government spending efforts fuelled by oil revenue. The post-2014 lower oil price environment continues to weigh down economic activity across the Gulf states. And political tensions in the Gulf, which were heightened in the spring through the winter of 2019, did not have the effect of spiking oil prices. Instead, they tend to further weigh down private-sector business sentiment. Tanker threats and tensions in the Persian Gulf may have exacerbated already sluggish non-oil and private-sector growth, making economic interests a motivating factor in outreach and possible Emirati de-escalation talks with Iran in late 2019 (Vahdat and Batrawy,

2019). The UAE economy is heavily dependent on its expatriate population and tourists to drive consumption locally. A security threat, even without attack, is a major economic risk with long-term consequences.

Oil is still central in the economies of the GCC states, and even their most diversified members take heed of a threat to their primary source of economic activity. There has been little silver lining to the increasing political risk in the Persian Gulf, as oil prices remained stubbornly low in the USD 60 per barrel range (Meredith, 2019). The softer side of Gulf economies, non-oil GDP, is dependent on foreign investment, tourism, and expatriate labour and is especially sensitive to political risk. The result is a stalemate in which diversification efforts lag and oil dependency deepens, but oil is even less reliable as a long-term economic growth strategy.

While oil prices have somewhat rebounded from a recent low point in December 2018 at USD 50 per barrel, 'lower for longer' is the mantra, with prices hovering in the USD 60 range in 2019, despite significant regional political tension. Budget deficits continue to be the new normal across the GCC, with an average deficit of about 5 per cent for 2019 and the forecast being similar for 2020 according to CEIC Global Database and HSBC Investment Banking research (Nasser, 2019).

However, Abu Dhabi's real GDP grew 5.7 per cent (year-on-year) in the first quarter of 2019, all due to the oil sector, according to research by EFG Hermes (Abu Basha, 2019). The non-oil economy in Abu Dhabi continues to contract. Weak job growth, lower foreign tourist arrivals, and a slow hospitality sector have hit the emirate hard. While the slowdown in the property market in Dubai has attracted media and investor attention, the capital economy to the south is also weak and struggling to calibrate its foreign policy objectives with the realities of global oil markets and its stated goals of becoming a diversified economy (Kerr, 2019; *The National*, 2019).

The economic policy solution for the last four years for most of the Gulf states has been to draw upon international capital debt markets and enact fiscal stimulus – to borrow and keep spending. Abu Dhabi has been more cautious in borrowing than its neighbours Qatar and Saudi Arabia, issuing just one jumbo USD 10 billion bond in 2017 (Carvalho and Barbuscia, 2018). More recently, Abu Dhabi has relied on its state-owned entities to borrow independently of the government.

Gulf governments and their state-related entities, including sovereign wealth funds, are now major sources of emerging market debt. With its new federal debt law, the UAE is poised to be able to issue at the federal level, but has not yet (Smith *et al.*, 2018). There have been few debt issuances at the emirate level, but more recently from government-related entities, as Mubadala issued a ten-year USD 800 million bond in late 2018, and in May this year refinanced and expanded a 2016 USD 1.75 billion loan (Barbuscia, 2018a, 2019).

Economic diversification has increasingly meant an expansion of national oil companies into downstream products and partnerships with key energy-product consumers. For the UAE and the Abu Dhabi National Oil Company (ADNOC), in particular, this has meant partnerships and co-investments with China (Nehme, 2019a). There is no Chinese miracle, however, as new political ties and co-investments do not necessarily translate into immediate benefits in tourism or private-sector growth in Abu Dhabi. While tourist arrivals in Abu Dhabi from China peaked in the third quarter of 2017, with a near 90 per cent year-on-year increase from 2016, there has since been a declining rate of growth, with negative growth in the last quarter of 2018 and the first quarter of 2019, according to the Abu Dhabi Department of Culture and Tourism and EFG Hermes (Abu Basha, 2019).

Fiscal stimulus from Abu Dhabi has also been a bit slow, or at least more conservative, than outlays by neighbouring Saudi Arabia (Young, 2019c). Abu Dhabi's Ghadan 21 initiative is different from other Gulf stimulus measures in that it focuses more on shifting the business environment and creating incentives for investment rather than providing capital directly from the state (Sanderson and Khan, 2019). While this is a commendable longer-term approach to changing economic behaviour rather than enduring a cycle of government outlays, the short-term horizon looks difficult. Abu Dhabi investment spending is a fraction of what it was in 2012, when project awards neared USD 45 billion, while the 2019 investment spend forecast by MEED (2019) is closer to USD 23 billion.

Abu Dhabi also has the weakest labour market across the UAE. A series of mergers in state-owned entities in the finance sector has led to thousands of job losses (Sharif and Fattah, 2019). According to data by the UAE Central Bank and EFG Hermes, Abu Dhabi's labour market is the weakest of the UAE, with negative quarter-on-quarter changes in employment rates throughout 2018 (Abu Basha, 2019).

Economic diversification is beginning to look like a game of survival of the fittest in the GCC, as the strongest non-oil economies look to weather the current political tensions and a less reliable source of increasing government revenue in oil. For Abu Dhabi, tightened purse strings, doubling down on oil and energy investments, and delayed diversification may be a survival strategy, but for the federation as a whole, the costs will be much higher.

Tracking the reform agenda, and the shifting policy landscape

Breaking down the policy agenda post-2014 in the UAE, we can identify three core policy areas: fiscal policy, social development policy, and diversification policies. Each of these areas is a response to the demands of a lower oil price environment and the mounting challenges of forty years of rentier policy. Not all are reforms, and many are measures to calibrate citizen demands for employment and social services with the dependency on foreigners as investors and workers. The following subsections dissect these policy shifts in detail and analyse how these policy choices fit together in a reform and diversification agenda for the UAE.

Fiscal policy

In fiscal policy, the UAE has some structural characteristics that are distinct from its GCC peers. As a federation, revenue creation and revenue sharing emanate from the centre, from Abu Dhabi. The seven emirates of the federation are not equal in their endowments of capital or human resources. Oil resources are concentrated in the emirate of Abu Dhabi, and the emirate, as the centre of government, is the largest contributor to the federal budget (Young, 2017). The five smaller emirates (Fujairah, Ajman, Sharjah, Ras al-Khaimah, Umm al-Quwain) all depend on fiscal transfers for basic provision of electricity and social services to their populations, of which there are generally fewer foreigners than in Dubai or Abu Dhabi. Dubai is the national leader in economic diversification, as it has very little oil production, with its strong port logistics facility in Jabal ʿAli (soon to be rivalled by Port Khalifa in Abu Dhabi), its tourism sector, some financial services, and its real estate market. Dubai is the second largest contributor to the federal budget, though it has also been the recipient of Abu Dhabi financial support in the wake of the 2009 financial crisis and its own debt restructuring. Therefore, shifts in federal fiscal policy have uneven ramifications across the emirates and there is a constant political balancing more frequently weighted towards the growing influence and centralisation of authority in Abu Dhabi.

Tracking shifts in fiscal policy since 2014, we can identify several trends. The first is the implementation of taxes, which have been disproportionately borne by the foreign population. The GCC, as a customs union, hoped to have a cohesive approach to the implementation of a value-added tax (VAT) in 2018, but only the UAE and Saudi Arabia followed through as planned. Since 2018, only Bahrain has joined the two in the VAT group, in early 2019 (Cape, 2018). In 2017, the UAE announced excise taxes on tobacco, energy, and soft drinks. For the UAE, excise taxes and the value-added consumption tax weigh heavily on the expatriate population, which now serves as a convenient tax base. The implementation of fees for government services and visas have long been part of Dubai government revenue generation, as Dubai was the first in the Gulf region to implement a road-toll system in 2007, which took advantage of its commuter, tourist, and expatriate traffic along its main highway (Elsheshtawy and Al Bastaki, 2011). In 2019, Abu Dhabi followed suit (Odeh and Abu Omar, 2019). We can see some learning transfer in the implementation of tax and fees at the emirate level, which then becomes a basis for federal policy. In terms of revenue sharing, VAT provided an opportunity for some discussion (among leadership, not the citizenry) on equity between emirates. Even though VAT implementation began in early 2018, it was not until May 2019 that the UAE government announced how VAT revenue would be collected and distributed between federal units. Thirty per cent of VAT revenue is to go to the federal government, with 70 per cent to go to the local emirate government, collected at the point of sale (Emirates News Agency, 2019a).

The second most notable shift in Emirati fiscal policy since 2014 has been on debt. Though it has received little analytical attention, the 2018 legislation to allow the federal government to issue debt on behalf of the federation has further cemented central authority, but also now presents a dilemma in debt repayment, which would draw from shared resources. Prior to 2018, only individual emirates established international credit ratings and their own loan and bond issues. Because debt has become such a central policy mechanism to deal with the competing demands of government expenditure and reduced oil revenue since 2014, we can expect the federal government, under the leadership of Abu Dhabi, to establish itself on international debt markets as well. As at the start of 2021, the federal government has not yet raised capital through international debt markets, though a federally owned bank has issued a bond after the passage of the new debt law (Abu Omar, 2020).

However, the emirate of Abu Dhabi continues to raise capital on its own, and via its related entities. For example, Abu Dhabi raised USD 10 billion in bonds in late 2017, while Sharjah went to market just five months later in March 2018 to raise USD 1 billion in an Islamic bond or *ṣukūk* (Lohade, 2017; Barbuscia, 2018b). Abu Dhabi entities ADNOC and Mubadala, the emirate's oil company and one of its sovereign wealth funds, both issued bonds over USD 1 billion between 2016 and 2018 (Barbuscia, 2019). Dubai's government-related entities are also frequent borrowers on international capital markets, with Dubai Aerospace and DP World both issuing bonds worth more than USD 1 billion in 2018. The borrowing power between emirates and among state-related entities is not equal. Moreover, it remains to be seen how investors evaluate the difference in credibility and risk among emirates, the federal unit, and state-related entities.

Third, subsidies are a tremendous drag on the fiscal policies of the Middle East and especially the Gulf states. In 2014, the International Monetary Fund (IMF) estimated that pre-tax subsidies cost the Middle East and North Africa region USD 237 billion annually, equivalent to almost a quarter of all government revenues at the time (Sdralevich *et al.*, 2014). In a measure to reduce fiscal outlays in subsidies of fuel, electricity, and water, the UAE began an ambitious effort to rationalise prices beginning in 2015. Dubai again was a trailblazer and model for the federal government in this policy mechanism, similar to the experience with the Salik road toll. Dubai began to eliminate electricity and water subsidies beginning in 2011 as a model for subsidy reduction across the region (Krane and Monaldi, 2017). Because Dubai's diversification efforts predate the 2014 oil price decline, many of its public policy initiatives have provided a learning curve for other emirates and the federation as a whole. After 2014, the policy shift to reduce fuel, water, and electricity subsidies normalised across the GCC. In August 2015, the UAE took the step of price-setting fuel on a monthly basis at the federal level, led by a central committee, when a barrel of Brent crude was trading at USD 54. Since then, a fuel committee has set petrol prices based on a monthly review of global averages and operating costs (Hickin, 2018).

Fourth, in addition to cutting costs, the UAE has tried to generate new sources of government revenue beyond VAT and excise taxes. In this policy shift, we do see some balancing between efforts, as the government seeks to calibrate a business-friendly environment with demands for new revenue. Fees on licensing businesses and new fees for visas and tourist arrivals at airports have all been popular measures at both the federal and emirate levels, with varying degrees of permanence or success (*The National*, 2016). Rental fees on property rentals have targeted expatriates as a source of emirate revenue (Duncan, 2018). Abu Dhabi introduced a 3 per cent municipal fee on expat property rentals in April 2016, while Sharjah doubled its city administration fee for attesting rental contracts (Zriqat, 2016). But by 2019, a reversal on government administration fees took hold, as the federal government and UAE cabinet announced the cancellation of fees for 1,500 government services relating to business licensing (Warrier, 2019). The business-friendly policies are a diversification agenda in themselves, part of Ghadan 21, an economic development initiative of Abu Dhabi centred on new business and innovation, which is further addressed below (UAE Government, 2019).

Social development policy

In the area of social development policy, the UAE has tried to address core concerns of citizens across the Gulf rentier economies: housing and jobs. But in balancing the interests of residents and citizens, the UAE has a divided constituency. Dependent on expatriate labour and foreign investment in its real estate markets and local businesses, foreigners will not remain if they are seen solely as a revenue source to the state.

Housing is a traditional lever of patronage and social service provision in the Gulf (Young, 2018). The UAE has used two policy mechanisms to help provide access to affordable housing for its citizens, with some benefit as well to foreign residents. Developing a mortgage industry has been essential to encourage lending but also to develop the financial service sector, in itself a major employer of nationals. The UAE Central Bank lifted a cap on real estate lending as a percentage of bank deposits to encourage banks to extend more housing loans in 2018. In February 2019, the federal government also created a USD 8.7 billion plan for 34,000 housing units for citizens, many free or in no-interest loan packages (Nehme, 2019b). At the emirate level there are also regular disbursements of housing benefits, which encourage patronage networks to individual emirate rulers. For example, Sharjah provided USD 53.8 million in housing support for 96 grants and 178 loans to Sharjah citizen residents in July 2019 (Emirates News Agency, 2019b).

Debt relief is also a tool of social development policy, as well as a major fiscal outlay. In 2016, the UAE encouraged private banks to restructure nearly USD 2 billion of outstanding small and medium-sized business loans before a new bankruptcy law went into effect (Kassem, 2016). And a 2019 commitment to pay the personal debt of 3,000 citizens reflects a major instance of state patronage not in line with a diversification agenda or liberalised economy (RT, 2019). The pressures of an economy that is not growing quickly or showing signs of acceleration require government intervention, but in very specific and targeted ways. Debt relief, housing benefits, and business loans are tools to manage possible discontent.

Economic diversification policy

In policies aimed specifically at shifting the investment climate and encouraging development of the non-oil economy, the UAE has been forward-looking and ahead of its peers in attracting FDI and institutionalising a liberal economic environment. But that is not necessarily changing the trajectory of FDI, as globally there is some disappointment in FDI inflows to emerging markets over the last ten years, especially since the global financial crisis. FDI inflow to the GCC has been largely flat since 2011 (JP Morgan Chase, 2019). According to research by JP Morgan Chase (2019), net FDI inflow to the UAE has been about 0.6 per cent of GDP since 2011, with no change in the post-oil boom of 2015–18, despite policy initiatives to attract it. In October 2018, the UAE issued a new law to set up an FDI unit in the Ministry of Economy to be responsible for FDI policies and tracking its flow, as well as streamlining registration of investors (Bridge, 2018).

The emirate of Dubai has repeatedly made concessions to attract foreign investors, including the expansion of its free zones and investment in new fintech and innovation centres within its Dubai International Financial Centre (Martin, 2019). The Dubai World Trade Centre made similar efforts to reduce licensing fees as part of the emirate's 'ease of doing business' initiative in 2018 (Emirates News Agency, 2018). Abu Dhabi has increasingly tried to compete with the private-sector business climate of Dubai, with a range of programmes and incentives to build its own financial centre and set of free zones in the Dubai model. In May 2018, Abu Dhabi's Department of Economic Development announced that companies based in the emirate's free zones could operate outside of the zones with a dual licence, a departure from Gulf practices of protecting local businesses from foreign-owned competition (Abbas, 2018). Prior to the essential erasure of the line between free zones and normal emirate borders, the UAE cabinet had approved a decision to allow 100 per cent ownership of certain kinds of firms outside of free zone borders (Ernst & Young, 2018). The sectoral limits continue to exclude the oil industry, any military or defence technology, and some social services, though the education and health sectors have largely been liberalised. The competitive nature of Dubai–Abu Dhabi liberalisation efforts have helped the UAE to be more open than some of its GCC peers in terms of foreign ownership and investment rules. However, in the summer of 2019 Bahrain broke the taboo and allowed foreign investment into its oil discovery and production, something no other GCC state has tried (Sertin, 2019).

Conclusion

Economic diversification has taken on new urgency after the decline in oil prices of 2014. Oil prices will likely be lower for longer, and new consumer bases in Asia will determine the demand curve for Gulf energy products in the years to come. However, the demands for growth in the non-oil economy also coincide with demands for regional security and political stability. The UAE and its GCC peers have carefully navigated a period in which political and economic demands increasingly conflict. The policy landscape is wider, more competitive, and more creative as a result of these challenges. It is also true that citizen and foreign residents have become dual constituencies, especially in the case of the UAE, in which so much of the economy depends on expatriate labour and consumption. Moreover, the domestic politics of the UAE have undergone profound changes, as the federal system is stressed with increasing inequality between emirates, competition for resources, and new institutional challenges of resource sharing and distribution. Taxes will necessitate better reporting and transparency in federal budget authorities and also give a better understanding of citizen and foreign resident economic activity. The problem of sustainability and climate change will loom large in the coming decades, affecting oil demand and revenue, but also the liveable standards of rising temperatures and energy consumption at home.

The most compelling challenges have been to the state itself – to justify its place in a changing global economy and to present to citizens a plan for reform and change without upsetting a political system with few valves for dissent or negotiation. Therefore, policy flexibility and shifts, with the capacity for fiscal stimulus and reversals, have been essential. The current period is one in which expectations will be high, but growth could continue to be low. For the emirate of Dubai, its own forward thinking in diversification may be punished, as diminished oil revenues and foreign policy continue to dominate policymaking choices at the federal level. Nevertheless, compared to its peers, the UAE is well positioned to weather the storm.

Notes

1 In his book *The economy of the Gulf states*, Gray (2019) elaborates the concept of late rentierism and state capitalism as pillars of the organisation of Gulf economies. I rely on his concept of late rentierism to describe how some policy shifts and even 'reforms' may entrench economic dependence on energy resources.

References

Abbas, W. (2018), 'Abu Dhabi launches first phase of dual licence initiative', Zawya, 15 September. https://bit.ly/2MNtFij (accessed 18 November 2019).

Abu Basha, M. (2019), 'UAE economics: Abu Dhabi – Oil sector pushes GDP growth to three-year high, but masks weak non-oil backdrop', EFG Hermes Holding S.A.E., 31 July. https://bit.ly/38cC4U8 (accessed 19 November 2019).

Abu Omar, A. (2020), 'UAE gets first rating from Fitch ahead of federal bond sale', Bloomberg, 11 November. https://bloom.bg/3tllDNw (accessed 11 March 2021).

Barbuscia, D. (2018a), 'Abu Dhabi's Mubadala sets final price guidance for 10-yr dollar bond-document', Reuters, 31 October. https://reut.rs/3ql0y3w (accessed 18 November 2019).

Barbuscia, D. (2018b), 'Emirate of Sharjah sells $1 bln sukuk – official', Reuters, 8 March. https://reut.rs/30lFAqF (accessed 18 November 2019).

Barbuscia, D. (2019), 'Mubadala signs $2 billion revolving loan', Reuters, 2 May. https://reut.rs/3qlA9CR (accessed 11 February 2020).

Bridge, S. (2018), 'UAE unveils new plan to boost foreign direct investment', Arabian Business, 30 October. https://bit.ly/2PsJEmS (accessed 18 November 2019).

Cape, J. (2018), 'Three down, three to go in the GCC as Bahrain set to roll out VAT', *The National*, 19 December. https://bit.ly/3bnsUGt (accessed 18 November 2019).

Carvalho, S., and D. Barbuscia (2018), 'Abu Dhabi meets bond investors in non-deal roadshow, sources say', Reuters, 6 December. https://reut.rs/3ea1ZzB (accessed 18 November 2019).

Duncan, G. (2018), 'Municipality fee for villas and apartments in Abu Dhabi to be set at 5 per cent', *The National*, 27 June. https://bit.ly/3kQDybQ (accessed 18 November 2019).

Elsheshtawy, Y., and O. Al Bastaki (2011), 'The Dubai experience: Mass transit in the Arabian Peninsula', The International Forum on Urbanism. https://bit.ly/3th3PSV (accessed 18 November 2019).

Emirates News Agency (2018),'Dubai World Trade Centre announces up to 70% reduction on free zone licencing and incorporation fees', Zawya, 25 June. https://bit.ly/3892R3p (accessed 18 November 2019).

Emirates News Agency (2019a), 'UAE approves VAT revenues distribution between federal and local governments', Zawya, 30 May. https://bit.ly/2PqEBmX (accessed 18 November 2019).

Emirates News Agency (2019b), 'Sharjah approves $53.79mln for housing support beneficiaries', Zawya, 2 July. https://bit.ly/2OpwqXA (accessed 18 November 2019).

Ernst & Young (2018), 'UAE announces significant changes to foreign ownership and visa rules', 24 May. https://go.ey.com/3uWLSeF (accessed 18 November 2019).

Fitch Solutions (2019), 'MENA: Mixed progress on GCC diversification', MENA Macro Intelligence Monthly Outlook Presentation, May. https://bit.ly/3bk4w8t (accessed 11 February 2020).

Goldman Sachs Global Investment Research (2018), 'EM sovereign credit – main risk to supply technical stems from oil exporters (in particular the GCC region)', Goldman Sachs Economics Research EM in Focus 2, 7 December.

Gray, M. (2019), *The economy of the Gulf states*. Newcastle upon Tyne: Agenda Publishing.

Hickin, P. (2018), 'Energy reform is a price worth paying', *The National*, 29 June. https://bit.ly/2MNuKGT (accessed 18 November 2019).

JP Morgan Chase (2019), 'EMEA emerging market focus: FDI trends', JP Morgan Chase Economic Research Global Data Watch, 14 June. London: JP Morgan.

Kassem, M. (2016), 'UAE banks bail out 1,700 SMEs and corporations with Dh7 billion in loans restructured', *The National*, 19 September. https://bit.ly/3ebfGy9 (accessed 18 November 2019).

Kerr, S. (2019), 'Dubai fears the end of its "build it and they will come" model', *Financial Times*, 1 April. https://on.ft.com/3ePiek5 (accessed 18 November 2019).

Krane, J., and F. J. Monaldi (2017), 'Oil prices, political instability, and energy subsidy reform in MENA oil exporters', James A. Baker III Institute for Public Policy of Rice University. https://bit.ly/3c7UXIQ (accessed 18 November 2019).

Lohade, N. (2017), 'Abu Dhabi launches $10 billion bond sale', *The Wall Street Journal*, 3 October. https://on.wsj.com/2Oqvqmd (accessed 18 November 2019).

Martin, M. (2019), 'What slump? Dubai finance hub plans Canary Wharf-sized expansion', Bloomberg, 8 January. https://bloom.bg/3xPm5Xp (accessed 18 November 2019).

MEED (2019), 'Beyond 2020: The outlook for Middle East construction', *MEED Business Review*, 4:6.

Meredith, S. (2019), 'Oil prices will stay "relatively benign" despite escalating Iran tensions, Morgan Stanley says', CNBC, 22 July. https://cnb.cx/30gDCrG (accessed 19 November 2019).

Mubadala (2018), 'RDIF and Mubadala celebrate six-years of partnership', 24 May. https://bit.ly/3rkK3pE (accessed 19 November 2019).

Nasser, R. (2019), 'Underwhelming gains: Economic activity in the Middle East', HSBC Bank Middle East Ltd, 31 July.

Nehme, D. (2019a), 'ADNOC, China's Wanhua Chemical sign deal potentially worth $12 billion', Reuters, 23 July. https://reut.rs/2O30Uil (accessed 28 January 2020).

Nehme, D. (2019b), 'UAE announces $8.7bln housing plan for citizens-Dubai Ruler's tweet', Zawya, 25 February. https://bit.ly/3qh65In (accessed 18 November 2019).

Odeh, L., and A. Abu Omar (2019), 'Another freebie gone as Abu Dhabi set to start road-toll system', Bloomberg, 25 July. https://bloom.bg/2QF1Wm5 (accessed 18 November 2019).

RT (2019), 'UAE pledges to wipe out $100mn in debt for over 3,000 of its citizens', 25 February. www.rt.com/business/452344-uae-clear-million-loans/ (accessed 18 November 2019).

Sanderson, D., and S. Khan (2019), 'Ghadan 21: Abu Dhabi to boost business and ecotourism with major new reforms', *The National*, 25 June. https://bit.ly/2O3uaFI (accessed 18 November 2019).

Sdralevich, C. A., R. Sab, Y. Zouhar, and G. Albertin (2014), 'Subsidy reform in the Middle East and North Africa: Recent progress and challenges ahead', International Monetary Fund, 9 July. https://bit.ly/38d5zFh (accessed 18 November 2019).

Sertin, C. (2019), 'Bahrain will allow foreign companies 100% ownership of oil and gas projects', Oil and Gas Middle East, 9 June. https://bit.ly/3efgrWW (accessed 19 November 2019).

Sharif, A., and Z. Fattah (2019), 'Abu Dhabi commercial bank to cut about 2,000 jobs after merger', Bloomberg, 3 July. https://bloom.bg/3nFMAcY (accessed 18 November 2019).

Smith, D. G., T. Cerdan, L. Drissi, and S. M. Al-Amin (2018), 'United Arab Emirates issues national public debt law', Squire Patton Boggs, October. https://bit.ly/3eoA3rZ (accessed 18 November 2019).

Standard Chartered Bank (2019), 'Economies: Middle East, North Africa and Pakistan', Standard Chartered Bank Global Focus: Economic Outlook Q3 2019, July.

The Economist (2017), 'The ambitious United Arab Emirates: The Gulf's "little Sparta"', 6 April. https://econ.st/3riSRfC (accessed 19 November 2019).

The National (2016), 'New departure fee for Abu Dhabi airport passengers', 15 June. https://bit.ly/3kLEsGm (accessed 18 November 2019).

The National (2019), 'Abu Dhabi is in the vanguard of economic diversification', 25 June. https://bit.ly/3c4r2kP (accessed 18 November 2019).

UAE Government (2019), 'Government accelerators', The United Arab Emirates Government Portal, 24 September. https://bit.ly/2NUMmkV (accessed 18 November 2019).

Vahdat, A., and A. Batrawy (2019), 'UAE and Iran hold rare talks in Tehran on maritime security', The Associated Press, 31 July. https://apnews.com/49c6da1c33fd45bbaf1b14836ee5e2ec (accessed 19 November 2019).

Vertin, Z. (2019), 'Red Sea rivalries: The Gulf states are playing a dangerous game in the Horn of Africa', *Foreign Affairs*, 15 January. https://fam.ag/3gUsFFI (accessed 19 November 2019).

Warrier, R. (2019), 'UAE cabinet decides to amend, waive fees for 1,500 federal services', Construction Week Online, 29 May. https://bit.ly/3c0Odg4 (accessed 18 November 2019).

Young, K. E. (2013), 'The emerging interventionists of the GCC', London School of Economics and Political Science, LSE Middle East Centre Paper Series, 2. https://eprints.lse.ac.uk/55079/ (accessed 18 November 2019).

Young, K. E. (2017), 'UAE fiscal policy: Shining light on federal resource sharing', The Arab Gulf States Institute in Washington, 20 October. https://bit.ly/3qh8liP (accessed 18 November 2019).

Young, K. E. (2018), 'A home of one's own: Subsidized housing as a key lever of Gulf domestic policy', The Arab Gulf States Institute in Washington, 15 June. https://bit.ly/3roeuLM (accessed 18 November 2019).

Young, K. E. (2019a), 'Is debt answer to Middle East economic troubles?', Al-Monitor, 9 April. https://bit.ly/3rkWNN3 (accessed 19 November 2019).

Young, K. E. (2019b), 'The Gulf's eastward turn: The logic of Gulf-China economic ties', American Enterprise Institute, 14 February. https://bit.ly/2OkvJPj (accessed 19 November 2019).

Young, K. E. (2019c), 'Can Saudi Arabia afford a stimulus?', Al-Monitor, 6 March. https://bit.ly/38dR97J (accessed 18 November 2019).

Zriqat, T. (2016), 'Sharjah residents rush to attest tenancy contracts ahead of price hike', *The National*, 31 July. https://bit.ly/30j31Rq (accessed 18 November 2019).

8

Egypt's twisted hydrocarbon dependency: A case of persistent semi-rentierism

Amr Adly

Introduction

Whereas most of this volume focuses on the impact of lower oil prices on the Middle East and North Africa's (MENA) oil-rich and predominantly rentier economies, this chapter examines Egypt as a peculiar case of semi-rentierism with a resilient dependency on hydrocarbon rents in the external and state sectors. Semi-rentierism is an asymmetrical dependency of generally resource-poor economies on hydrocarbon-based rents in the mode of insertion in the global division of labour through trade and capital flows, as well as in state revenues. Despite becoming a net oil importer in 2006, and a net natural gas importer in 2012, Egypt remains heavily dependent on crude oil and gas for its foreign currency earnings. Directly, crude oil and natural gas still represent 35–40 per cent of total merchandise exports and receive around 60 per cent of net foreign direct investment (FDI) inflows. Indirectly, the Egyptian economy depends on workers' remittances, namely, from migrant workers in oil-rich Gulf Cooperation Council (GCC) countries in addition to large capital transfers in the form of in-kind and cash aid and concessionary loans and public and private foreign investment from oil-rich Arab nations.

Unlike extractive industries in oil-rich MENA countries, these industries have never constituted a sizeable percentage of Egypt's gross domestic product (GDP). In fact, the structure of the Egyptian economy is among the most diversified in the region. The structure of GDP shows that most output and employment is created outside of the energy sector (see Table 8.1).

Despite being itself poor in hydrocarbons, Egypt, like MENA more generally, has been and continues to be heavily dependent on crude oil and natural gas in its insertion into the global division of labour. The mode of insertion refers simply to what Egypt competitively offers to the world for exchange, which is referred to here as the external sector. So far, the Egyptian economy has been by and large a factor-based economy, which depends both directly and indirectly on cheap energy as its comparative advantage.

Table 8.1 GDP structure – Egypt, 2006, 2010, and 2017 (per cent).

	2006	2010	2017 (2016 prices)
Agriculture	14.07	13.26	11.45
Manufacturing	16.11	15.63	16.66
Hydrocarbons	15.02	13.81	12.41
Suez Canal	3.40	3.31	2.34
Other	51.40	53.99	57.14

Source: Central Bank of Egypt (2018a). Author's calculation.

By the same token, in recent years, the Egyptian state has remained reliant on non-tax revenues that accrue largely from hydrocarbon sales in addition to the Suez Canal fees and external aid from oil-rich allies. This has been the case despite the dwindling ratio of these rent-based revenues to total revenue as well as to GDP, indicating an inability to institutionally adapt to declining rents by raising tax revenues, another feature of persistent semi-rentierism.

This chapter explores the political economy of Egypt's twisted hydrocarbon dependency, which has paradoxically restrained the country's ability to benefit from cheaper international energy prices since 2014, despite being a net energy importer. Being a semi-rentier state and economy, Egypt has been facing problems adapting its external and state sectors to declining rents, much like rentier economies, despite not possessing such natural riches in the first place. Herein lies the potential contribution of this chapter, as it sheds light on the hardships facing semi-rentier economies in diversifying even after they have become net energy importers and amid lower international energy prices.

The origins of Egypt's semi-rentierism

After the discovery of oil on its lands in the early twentieth century, Egypt was traditionally a net oil-exporting nation. This lasted until 2006, when the country for the first time in its contemporary history became a net oil importer. The impact of this change was mitigated temporarily with an increase in natural gas production almost during the same time (Springborg, 2012: 297). By 2012, Egypt had become a net energy importer, including both oil and gas. Consumption overtook production. Since then, energy imports (of both oil and natural gas products) have made up almost one-fifth of the country's import bill (Trading Economics, 2018).

Now a net importer, soaring international energy prices between 2008 and 2014 widened the Egyptian trade and budget deficits, as the government maintained a universal fuel subsidy programme that covered households and producers alike, including big domestic and multinational companies in energy-intensive industries like iron and steel, aluminium, cement, glass, and fertilisers (Moerenhout, 2018). The situation was exacerbated further by the security issues and political turmoil that followed the 2011 popular uprising, which soon translated into energy shortages and electricity outages from 2012 to 2014.

The sheer decline in international energy prices since 2014, combined with the affirmation of Egypt's position as a net energy importer, should have theoretically enabled or at least eased its diversification away from dependency on crude energy in its external sector and as a source of state revenues. Nonetheless, this did not happen on any significant scale.

I argue that Egypt's twisted and persistent hydrocarbon dependency can be explained by two predominant and largely interrelated factors: the first is institutional, while the other is geo-economic. The institutional explanation builds on an already well-developed body of literature traced back to Karl's (1997) classic *The paradox of plenty*, which brought institutionalism into the loop of studying the developmental and political-economic repercussions of rentierism, used here interchangeably with rent dependency. Institutions are sticky. They do not readily change in response to economic shocks or downturns, nor do they adjust automatically and functionally to changing economic conditions like lower energy prices in the short and medium terms. This is especially the case when earlier institutional arrangements serve certain distributional coalitions that are likely to mobilise in order to sustain them even when rents are dwindling or at least not mobilise to bring about the alternative institutional arrangements necessary for diversification (Knight, 1992; Goldberg, 2004). Moreover, past experiences and functions determine what these state institutions can do by shaping their bureaucratic capacities, networks of coordination with private businesses and societal groups, as well as defining the horizons for state intervention. Practical regulatory legacies are likely to persist for some time even if their earlier functions have lost relevance and new needs arise on the macro level.

In Egypt, institutions governing the policy areas linking state and economic actors proved disabling for coordinating the diversification efforts of the external sector (i.e. international trade and capital flows) away from dependency on hydrocarbon-based rents. The same handicap could be noticed in state finances, where declining rent-based revenues led to a decline in total revenues amid an inability to build the capacities necessary for extracting tax revenue from the economy. These missing institutional capacities come under two broad categories: absolute capacities in reference to the more or less Weberian traits of autonomy and internal coherence of bureaucratic bodies in charge of formulating and pursuing economic policies (Amsden, 1992; Evans, 2012); and relative capacities that allude to the formal, informal, and semi-formal channels of communication, cooperation, and coordination between the state and private economic actors in a manner that enables the coordination of long-term plans for upgrading (Shafer, 1994).

State institutions do not emerge in a vacuum. They are rather the result of broader power relations that link the state to various societal groups at different junctures. What I assert in this context is that semi-rentierism, defined in terms of the over-represented weight of hydrocarbon-based rents in the external and state sectors, makes it hard for absolute and relative capacities in areas relevant for economic and revenue diversification to emerge or to evolve even when the overall weight of such resources in the economy is declining. Sustained rent dependency weakens the state institutions necessary for regulating and extracting resources from the economy as well as those required for state–business coordination of diversification or upgrading of the external sector. They instead reproduce other capacities that are allocative in essence and that involve rent-seeking and distribution from the state to society rather than vice versa.

The second factor behind Egypt's persistent semi-rentierism is geo-economic regional dynamics. Geo-economics is an international political economy concept that revolves around power as the source and consequence of economic actions (Baru, 2012). It found its material translation in multiple channels of oil-rent recycling from the GCC to Egypt in a way that mitigated the overall impact of the latter becoming a net energy importer through much of the 2010s. These dynamics kept the Egyptian economy and state dependent on higher oil prices through aid packages, cheap credit and investment, and flows of workers' remittances, even when the country itself was becoming a net importer. Needless to say, this muted the impact of Egypt's changing situation and generated mixed signals for policymakers and major stakeholders on the need for diversification away from crude energy.

The chapter proceeds hence by providing a condensed review of the extant literature on the ordeal of rentier economies and the difficulties of diversifying in an age of low energy prices, while trying to keep the focus on the category of semi-rentier economies and states such as Egypt. The fourth section explores both arguments by applying them to the specificities of Egypt's political economy through taking a historical outlook on the country's energy production and consumption with a special focus on the post-2014 period. The section tackles in some detail the evolution of Egypt's external sector and the over-representation of hydrocarbon-based rents in it by looking into trade and FDI and capital inflows, including workers' remittances and intergovernmental loans and credit. The subsequent section focuses on the state sector by exploring the increasingly complicated presence of hydrocarbons on both the revenue and expenditure sides. The final section concludes.

State of the art: Literature review

Rentierism as an institutional disability

Following the two oil shocks of 1973 and 1979, political economists started addressing the impact of rent on the political and economic evolution of countries rich in natural resources. Generally speaking, the literature underlined a strong association between under-development and rent dependency. Economically, scholars talked about a resource curse where countries with the biggest access to rent are the ones most likely to project poor growth records in the long term (Chaudhry, 1989; Ross, 2003; Deacon, 2011). The Dutch disease thesis was among the first to suggest an economic mechanism explaining such unexpected outcome (Bruno and Sachs, 1982). According to the theory, rent as windfall income leads to currency appreciation, encourages public and private consumption, and undermines the competition of productive sectors. Early studies caught a link between the rise in rent-based income and the reallocation of resources away from productive sectors like industry and agriculture into other markets characterised by high speculation, like construction and financial intermediation, which served as venues for rent recycling. Politically, most rentier states were authoritarian (Wantchekon, 2002; Schlumberger, 2008; Ross, 2011). A straightforward explanation was that rentier states were too autonomous from their societies to be democratic or accountable to their peoples. In a rentier state, there was no room for the classical formula of 'no taxation without representation', for the state was in no need of taxation to start with (Schwarz, 2008). Instead, it was societal groups that were in dire need of seeking rent from the state rather than vice versa. This could be delivered in return for loyalty through patronage, a bloated public sector, and generous subsidies (Hertog, 2016).

By the 1990s, institutionalism made its way into rentier-state literature. Karl (1997) was among the first to invoke an institutional mechanism in explaining the economic and political repercussions of rent dependency. According to her, rentier states could not easily adjust to the decline in rents. Instead, most found themselves at the mercy of cycles of boom and bust that they could not control, as shown by the glut of 1986 and its devastating impact on the finances of a great many oil-rich countries in the Global South at the time. Karl argued that the answer was institutional, as rentier states lacked the institutional capacity to regulate their economies or to extract resources from them through taxation. Conversely, their institutional capacities developed for the allocation of resources that accrued directly to them in boon years. When prices were low and resources scarce, these states found themselves stuck with these lopsided institutional capacities, on the one hand, and captured by social coalitions of beneficiaries that had grown on the state body during the heydays, on the other. In confirmation of Karl's thesis, none of the major oil or gas producers worldwide could hitherto diversify away from dependency on their extractive industries since the 1970s, despite being subjected to the vagaries of high and low international energy prices.

This chapter finds this argument relevant not just for the flagrant cases of rent-dependent countries like the oil-rich MENA states but also in a case of semi-rentierism like that of Egypt. Semi-rentierism was brought up as early as Luciani and Beblawi's (1987) classic work. The term, however, has not been defined in a clear-cut manner. Whereas the criterion they used was the ratio of 'rents' to GDP, this ignored the complexities of the external and state sectors through which relations with the international economic order are defined. Egypt is an interesting case of semi-rentierism from this angle. Extractive industries are by no means huge in relation to the economy as a whole. Semi-rentierism reveals itself the most with the primacy of crude energy in accessing foreign currency earnings, either directly for the state treasury or for the economy as a whole, as explained earlier (Adly, 2016).

The other dimension of semi-rentierism is the significant share of rent-based resources in state revenues. External rents (including oil and gas revenues but also locational rents – like the Suez Canal – foreign aid, and cheap credit) have constituted around 45 per cent of total revenues through the past fifteen years (Schwarz, 2008; Soliman, 2011). Even though they were never as predominant as in rentier states, they still retained some considerable weight, especially when it came to foreign currency earnings that accrued directly to the state coffers. This is precisely the case given the limited ability of the state to tax other private foreign currency earning sectors like tourism, export-oriented manufacturing, and agricultural activities.

Moreover, these rent-based revenues seem indispensable due to a general inability to increase tax revenues amid a decline in rent-based ones. Despite the fact that rents have been declining as a percentage of GDP since the late 1990s, the Egyptian state has not been capable of compensating for this by building effective taxation capacities. Hence, the decline in rent-based revenues has often translated into an overall decrease or stagnation in total state revenues as a percentage of GDP (Soliman, 2011).

Geo-economics of rent recycling

Most literature on rentierism has subscribed to methodical nationalism, where the emphasis was put almost exclusively on how national governments obtained and managed the rent that accrued to them. There has been little focus on how the generation of rent-based revenues has regional, rather than merely national, political-economic dynamics as well, with some notable exceptions (Richter, 2007: 178; Hanieh, 2015). Dependency on rents is a regional feature of MENA that has puzzlingly characterised almost all countries regardless of their natural endowments. Rents, especially those generated from oil and gas, get recycled, as mentioned earlier, creating primary, secondary, and even tertiary beneficiaries among MENA countries (and beyond, as in the case of South and Southeast Asian labour in the GCC). I argue that these regional factors have been critical in sustaining Egypt's twisted dependency on hydrocarbons by numbing the incentives for developing the institutions necessary for diversification among major policymakers and stakeholders. Regional rent recycling hence impeded the institutional changes required for pushing for diversification.

MENA's geo-economics has persistently corresponded to the position occupied in the global division of labour since the end of the Second World War. The average share of fuel exports in total merchandise exports for MENA as a whole has been estimated at 65.7 per cent in 2014. This is hardly different from the situation in 1981, where it stood at two-thirds of the total (Farsoun, 1988: 156; World Bank, 2018a). These figures indicate that very little progress in economic diversification has been achieved over nearly three decades.

Egypt's geo-economic mode of insertion into the regional political economy militated significantly against the incentives and capacity to erect the institutions necessary for diversifying its external sector. MENA's geo-economic dynamics refer to the exact position that Egypt has occupied since the first oil shock in 1973, which led to the rise of primarily non-trade flows with the energy-rich Gulf countries (and to a lesser extent Iraq in the 1980s as well as pre-2011 Libya). These were flows of capital and labour that took the form of intergovernmental aid and credit, private and public investment from capital surplus countries in Egypt and other capital-poor economies, and flows of remittances corresponding to the export of semi- and low-skilled labour from the latter to the former (Henwood, 1993; Galal and Hoekman, 2003). Three mechanisms have been involved:

(1) Selective access to labour markets in oil-rich and often population-scarce countries, which was instrumentally used to create regional alignments by rewarding and punishing oil-poor/labour-abundant countries (Fargues *et al.*, 2015). Egypt's receipt of remittances from the GCC is a case in point that is discussed in more detail in the following subsections.
(2) Intergovernmental aid and credit: a typical example would be the post-1990 reconstruction of Lebanon after fifteen years of civil war (Corm, 1998; Daher, 2016). Another example is the readmittance of Egypt into the Arab League within the same context, followed by a massive debt relief underwritten by the Gulf monarchies and the USA (Ikram, 2007: 150). Contrastingly, Yemen was punished twice for siding with Iraq. Economically, the GCC countries halted aid and massively expelled Yemeni workers. Politically, Saudi Arabia supported (the ultimately failed) secessionist attempt by the South in 1994 in retaliation against Ali Abdullah Saleh's regime. Aid flows from the oil-rich GCC monarchies proved decisive in shaping national political outcomes in the aftermath of the 2011 revolutions. Egypt once again is a case in point. The country received a massive USD 23 billion of in-kind and cash assistance from Saudi Arabia, the United Arab Emirates (UAE), and Kuwait as budget support between 2013 and 2015 (Reuters, 2015). The generous financial support came in the wake of the 2013 military takeover and served as part and parcel of a broader political and diplomatic effort to stabilise and legitimise the new regime in Egypt both domestically and internationally. The trend was resumed after the July 2013 takeover.
(3) Intra-regional investment flows: the third mechanism of intra-regional rent recycling has been private, public, and semi-public investment from oil-rich into oil-poor Arab countries. Attracting part of the capital surpluses in oil-rich countries has served as a major objective in the development strategies of countries like Egypt, Turkey, and Tunisia since the 1970s and 1980s. It has also been instrumental for the fixing of their balance of payments deficits. Whereas these investment flows have been generally private and hence driven by profit considerations, they have rarely been completely devoid of political factors.

The brief review of literature has shown that the long-term developmental impacts of rentierism are often explained in institutional terms, referring to how the state is related to societal actors, including the most relevant in non-hydrocarbon economic sectors. The following sections will explore the Egyptian case in more depth with a focus on the post-2014 period by applying the argument developed above on Egypt's external sector and its state sector, that is, the weight of hydrocarbon-based rents in public revenues.

The external sector: Egypt as a semi-rentier economy

Exports

Figure 8.1 shows the changing composition of Egyptian merchandise exports from 1980 to 2017. Fuels and minerals have consistently retained a sizeable share of total exports. Their share had declined, however, from nearly two-thirds in the first half of the 1980s to around a half of total merchandise exports throughout the 1990s and until 2005–6, when it increased thanks to the large natural gas deposits discovered at that time, which rendered Egypt a net gas exporter for several years. As Egypt's gas boom proved to be short-lived due to exaggerated reserves and rising consumption, the share of fuels and minerals declined sharply in 2009 to around one-third of total exports and continued to decline until it reached 25 per cent in 2017, probably for the first time in Egypt's contemporary economic history.

The declining share of fuels and minerals in Egyptian merchandise exports gives the impression of a successful diversification, especially since manufactured goods raised their share in total exports consistently through the 1990s and 2000s. However, this can be quite deceiving, as the increase in the relative weight of manufactured goods did not translate into a higher growth rate of total exports. This means that this diversification was a mathematical representation of decreasing fuels and minerals exports rather than a real increase in the value of manufactured goods. Figure 8.2 shows the long-term growth of total exports and the annual growth rates of the share of fuels and manufactured goods from 1980 to 2017.

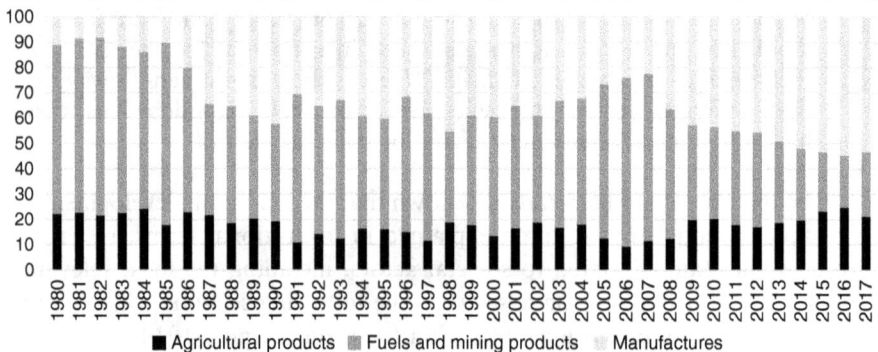

Figure 8.1 Composition of exports – Egypt, 1980–2017 (percentage of total exports). Source: World Trade Organization (2020). Author's calculation.

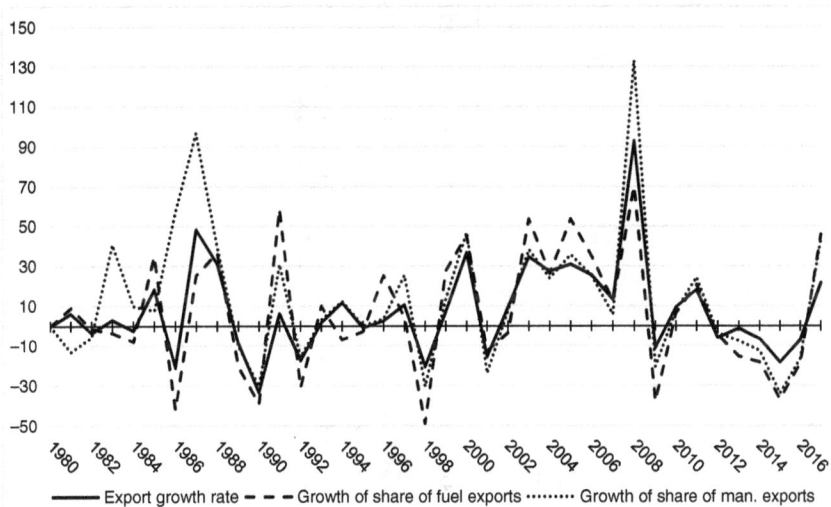

Figure 8.2 Growth rate of merchandise exports and share of manufacturing and fuel exports – Egypt, 1980–2017 (per cent). Source: World Trade Organization (2020). Author's calculation.

Figure 8.2 suggests two remarks. First, the rise in the share of manufactured goods has consistently been higher than the growth rate of total exports. This means that the expansion in the share was a reflection of the decline in the share of fuels and minerals rather than a real increase in the former. The second is that the almost perfect mimicking of growth rates of fuels exports by total exports throughout the years of analysis indicate that increasing and decreasing values of fuels exports – depending on international prices and local production and consumption – remained the main determinant of total merchandise exports.

Another restraint on the diversification impact of the increasing share of manufactured exports is the rising weight of energy-intensive manufactured goods. Industries like iron and steel, aluminium, glass, chemicals including petrochemicals, fertilisers, and cement are all energy intensive and have hence been heavily dependent on cheap supplies of fuels in order to gain a comparative advantage in international markets. Table 8.2 shows the progression of different shares of manufactured goods in total manufactured goods exports from 1990 to 2017. It is noticeable that the share of textiles and clothing has witnessed a steep decline. These are typically labour-intensive sectors that Egypt has historically had comparative advantage in. Conversely, iron and steel and the broad category of chemicals, which includes a number of energy-intensive industries like cement and fertilisers, have increased their

Table 8.2 Breakdown of manufactured exports by sector – Egypt, 1990–2017 (per cent).

	1990	1991	1992	1993	1994	1995	1996	1997	1998	1999	2000	2001	2002	2003	2004	2005	2006	2007	2008	2009	2010	2011	2012	2013	2014	2015	2016	2017
Iron and steel	5.50	5.31	12.84	12.84	9.84	11.53	6.18	7.23	8.03	5.54	7.32	10.06	14.24	19.99	22.13	22.29	28.72	24.62	13.12	6.22	8.05	9.61	6.05	7.18	4.66	2.73	4.38	6.78
Chemicals	8.34	14.85	15.07	10.88	11.75	14.48	15.85	13.26	18.91	20.31	17.19	23.82	18.44	24.47	17.32	21.37	25.17	24.65	31.10	32.00	32.18	33.89	34.93	34.74	32.85	28.33	31.81	34.51
Textiles	37.58	44.39	36.74	38.73	45.45	41.07	38.05	35.70	31.34	27.31	22.80	21.30	15.22	14.82	11.62	10.85	8.69	9.45	8.02	10.31	11.78	11.32	10.68	10.96	11.52	12.86	11.22	9.95
Clothing	9.77	15.47	15.16	17.75	16.96	18.23	21.40	17.35	23.67	21.38	17.36	17.60	12.60	12.42	10.03	7.34	5.07	5.95	8.17	13.36	11.64	11.55	9.84	10.05	9.61	11.95	10.65	10.83

Source: World Trade Organization (2020). Author's calculation.

share. It is also noteworthy that the share of the sectors picked up around 2005–6, which is the period that coincided with Egypt's large natural gas discoveries and the generous subsidies that were extended to the industrial sector at the time (Adly, 2012: 208–9). This can be taken roughly as a form of disguised hydrocarbon dependency lurking behind what looked like a successful diversification into higher value-added manufactured products.

These long-term trends of failed diversification away from hydrocarbon exports have impacted Egypt's road to economic recovery following the adoption of the International Monetary Fund (IMF) programme in late 2016. According to the World Bank's 'Egypt's economic monitor', Egypt's non-hydrocarbon exports have not shown a strong positive reaction to the sharp depreciation of the Egyptian pound since November 2016 compared to other countries of the same income level as Egypt. '[A] significantly larger depreciation in the Egyptian pound was only followed by an export increase of 16 per cent in about a year-time' (World Bank, 2019: 29). Much of the increase in overall exports had to do with the recent big natural gas discoveries in the Mediterranean. Gas and oil exports increased by 29.9 per cent in 2017–18 (constituting 31.6 per cent of total exports), whereas non-oil exports increased just at a rate of 9.7 per cent (making up 68.4 per cent of total merchandise exports) (Central Bank of Egypt, 2018b: 1–2). Once again, this leads us back to the status of the productive sectors in Egypt and their ability to contribute to redefining the country's position in the global division of labour.

It is noteworthy that the recent natural gas discoveries will not convert Egypt back into a net energy exporter, even if they reflect positively on the Egyptian trade deficit by reducing the country's dependence on imported natural gas. Official estimates claim the offshore field is expected to save Egypt some USD 1 billion annually in gas imports, but against a backdrop of a total import bill of around USD 62 billion. Liquefied Natural Gas (LNG) imports accounted for 3.6 per cent of the country's total import bill by the end of 2017. The savings are significant but by no means a game changer, given that other fossil fuels comprise nearly 15 per cent of imports (Trading Economics, 2018).

Foreign direct investment

Egypt's dependency on hydrocarbons in the external sector is not confined to merchandise exports. It extends to the composition of FDI, which has been historically tilted towards the oil and gas sectors (Hanafy, 2015: 21). FDI composition is quite significant in at least two main regards: the first is its contribution to the flow of foreign currency and hence its impact on the current account and the balance of payments. The other aspect is its potential productive role through technology and skill transfer and enhancing exports through backward and forward linkages. FDI has been a cornerstone in Egypt's development strategy since the adoption of *infitāḥ* in 1974. It remains the case. The government revamped the investment law in 2017, furnishing very generous subsidies and incentives to foreign investors. It was hoped that the sharp depreciation in the value of the Egyptian pound in November 2016 would lure foreign companies into investing in labour-intensive activities. Little of this has actually materialised given the long-standing hydrocarbon dependency in attracting FDI.

In 2016–17, about 50 per cent of FDI inflows went to the oil and gas sector (*Egypt Today*, 2017). The share of the oil and gas sector in total FDI inflows was a staggering 84 per cent during the first quarter of 2017 (Egypt Oil and Gas, 2017).

Intuitively, this sector has no backward or forward linkages with the rest of the economy given its capital- and technology-intensive character, which also makes it barely contribute to job creation. This means that little technology or skill transfer has ever taken place, not to mention the opportunity missed of having FDI positively impact manufactured exports by integrating local producers and suppliers into global value chains. Conversely, the share of the productive sectors in FDI inflows in Egypt has been historically marginal. According to data offered by the Central Bank of Egypt (2018a, 2018c, 2018d) for 2017–18, the share of manufacturing and agriculture was 10.4 and 0.1 per cent, respectively. Construction received 5.7 per cent, whereas tourism, communications, and real estate obtained 0.3, 2.2, and 2.5 per cent, respectively. The rather meagre share of total FDI in construction and real estate also highlights the Egyptian government's problematic strategy in basing a recovery on massive public and private investment in real estate and housing. It shows just how false the government's claim is that large construction projects would attract correspondingly large FDI. Despite an expected boost to vertical supply chain industries – e.g. cement, iron, steel, and real estate services – it will unlikely contribute to bettering Egypt's external position. Construction, financial, and non-financial services are largely non-tradeables, resulting in minimal impact at best on improving Egypt's balance of payments position. These sector boosts will neither compensate for the large import bill nor increase exports.

Intergovernmental credit and aid

Egypt's dependency on recycled oil rent from the GCC countries was exacerbated after 2011 through multiple channels, some of which were public while others were private. Public transfers included intergovernmental loans and aid whereas private flows took usually the form of investments and, more prominently, workers' remittances underlying the importance of accessing GCC labour markets.

Under the rule of the Muslim Brothers (2012–13), Egypt started its path of external borrowing from political allies, namely, Qatar, and to a lesser extent Turkey, and the government in Libya (Tripoli), which at the time was dominated by the Muslim Brothers. The Muslim Brothers' government received around USD 10 billion during its one-year rule (Badawi, 2013). The trend was resumed after the July 2013 takeover by Saudi Arabia, the UAE, and Kuwait, who offered Egypt a massive USD 23 billion in budget support (Reuters, 2015). This sum did not include the financing of arms purchases from France and Russia, put by some at an additional USD 10 billion (France 24, 2015; Colonna, 2018).

Table 8.3 demonstrates the dramatic rise in the share of external grants in total state revenues between 2012 and 2014, which corresponds to the interval when GCC governments came to the aid of those ruling Egypt in attempts to influence the final outcome after the 2011 revolution. The increase in intergovernmental aid was more pronounced in the external sector, as it offered the Egyptian government much needed US dollars to pay for imports and meet other foreign obligations.

However, Egypt's intensified dependence on the GCC came at the wrong time. In 2014, international oil prices collapsed. So far, they have not returned to pre-2014 levels and are not expected to do so in the foreseeable future. Collapsing oil prices have put considerable strain on the GCC budgets, namely, that of Saudi Arabia, with a relatively large population to sustain, an open-ended military venture in Yemen since 2015, and uncertainties about the crown succession. Yet, massive Gulf aid was sustained through these challenging times, showing the primacy of political and geopolitical factors. The flow of aid declined following the decline in oil prices in 2014 and could be taken as an explanation for Egypt resorting to the IMF by the end of 2016. However, the dense links between Egypt and Saudi Arabia and the UAE proved beneficial even with the IMF deal, as they remained de facto underwriters of the Egyptian government's international obligations.

GCC-emanating private investments and workers' remittances

Hydrocarbon rent recycling has not been confined to intergovernmental channels. Private flows have also been another traditional venue of their redistribution on a regional scale with direct repercussions for Egypt's access to foreign currency and hence to the current account and balance of payments position. These private flows assumed two principal forms: private foreign investment originating in oil-rich GCC countries and workers' remittances flowing from them to the Egyptian economy.

Table 8.3 Share of external grants in state revenue – Egypt, 2006–17 (per cent).

	2006	2007	2008	2009	2010	2011	2012	2013	2014	2015	2016	2017
External grants as share of total revenues in %	1.89	0.59	2.77	1.43	0.76	2.90	1.29	18.45	4.72	0.72	2.34	0.22

Source: Central Bank of Egypt (2018c). Author's calculation.

Table 8.4 demonstrates the evolution of the share of GCC-emanating foreign investment in net FDI inflows into Egypt. The trends are somehow clear. The overall share of GCC countries witnessed a considerable increase after the 2008 positive oil shock that created massive reserves in the oil-rich countries, with international oil prices skyrocketing. The GCC countries would sustain their share through 2014, when the decline in oil prices would translate into a downward trend for their share. This is not, however, the only determinant. Politics has been a force to reckon with, especially following the 2011 revolution and the turmoil that ensued in its aftermath both domestically and regionally. This is a reminder of the semi-public charac-teristics of private investment in GCC countries, where oil rents are either directly controlled by some state agencies, as sovereign funds that make private investments, or are channelled through politically connected business-men. In either case, profit-making motivations are often deeply entangled with geopolitical considerations. This has been primarily the case with Qatari and UAE investments in Egypt after 2011.

Table 8.4 demonstrates that Qatari and UAE-originating FDI have gone in opposite directions. The share of Qatari FDI picked up considerably after 2011 and peaked in 2012 under the brief rule of the Muslim Brothers. It is noteworthy that this was happening amid a decline in total FDI net inflows due to intensifying political turmoil, indicating a political rationale behind the boost. Qatari FDI inflows then plummeted suddenly after the toppling of the Muslim Brothers and have stagnated ever since. To the exact contrary, the UAE share kept declining from 2009 through 2013 and then witnessed a sharp increase after 2013. The Saudi share in net FDI inflows shows a similar trend, albeit milder. This can only be interpreted in the light of political alignments with certain actors in Egypt that created opportunities and mitigated risks for different GCC investors.

It is also noteworthy that the increase in the share of GCC-emanating FDI in Egypt became most visible after 2011, which, as indicated earlier, was a period of FDI contraction due to heightened political and security risks. The political and economic turmoil that followed the revolution translated into declining net FDI inflows, precisely from the USA and the EU, where most FDI inflows traditionally originated. This was the context in which the relative weight of GCC-originating FDI increased considerably, manifesting yet another form of Egypt's twisted hydrocarbon dependency.

Another paramount form of private-capital inflows is workers' remittances, which have masked Egypt's overwhelming dependency on oil rents despite it being a net energy importer. By the same token, GCC countries and Jordan were the origin of 70.8 per cent of total remittances that officially flowed into the Egyptian economy in 2017 (World Bank, 2019). These remittances have been quite significant for Egypt's access to foreign currency since the

Table 8.4 Share of GCC countries in net FDI inflows – Egypt, 2004–17 (per cent).

	2004	2005	2006	2007	2008	2009	2010	2011	2012	2013	2014	2015	2016	2017
Saudi Arabia	0.83	1.62	1.85	2.76	6.34	4.79	9.43	6.04	5.11	6.81	10.17	4.51	4.33	3.84
Qatar	0.05	0.10	0.02	1.40	0.65	1.04	8.75	0.88	10.01	2.61	1.48	2.81	2.14	2.14
UAE	1.04	1.03	27.59	5.49	12.79	4.49	18.77	14.06	12.80	9.60	21.67	19.17	10.55	13.93
GCC	4.92	5.02	29.86	22.05	21.62	14.20	43.19	26.73	36.45	28.86	40.74	31.84	21.50	23.82

Source: Central Bank of Egypt (2018d). Author's calculation.

1970s. Access to oil-rich (and population-scarce) countries' labour markets has also been crucial for relieving the Egyptian labour market of mounting population pressure.

Historically, Egypt has been a principal recipient of remittances. It is estimated to have received a massive USD 166.56 billion in remittances during the period between 1989 and 2012, or an average of USD 5 billion annually, a sum far greater than that received by developing countries of similar or greater population sizes such as Turkey or Brazil (World Bank, 2017). Figure 8.3 shows the relative weight of remittances from 1977 through to 2017. After a relative decline in the 1990s, the share of remittances picked up from 2008, with the jump in international oil prices. Remittances grew in absolute and relative terms after 2011 until they hit a record 10 per cent of GDP in 2017, which is the highest since 1991. In 2017, Egypt received a massive USD 21 billion in remittances (Refaat, 2018), which is four times as big as the Suez Canal revenues and around 70 per cent of total exports. The great rise in the share of remittances in GDP is most likely reflective of the depreciation in the value of the Egyptian pound following the devaluation of November 2016.

Workers' remittances continued to flow to Egypt despite international oil prices plunging in the summer of 2014. Egypt received USD 19.2 billion in remittances in 2015, predominantly from the GCC countries. Figure 8.3 shows that remittances stood at USD 16.7 billion in 2016 and USD 18.3 in 2017 and were projected at USD 23.4 billion for 2018 (with no final official figures out yet). So far, the Saudis on their part have hitherto largely exempted Egyptian workers from the Saudisation plans adopted by Crown Prince Muhammad bin Salman. While worker-exporting countries, mainly

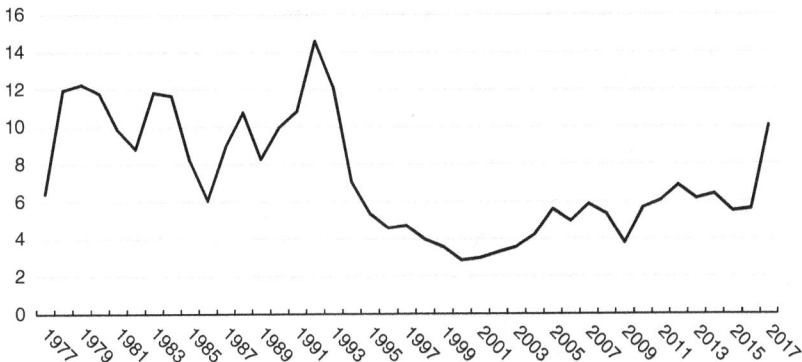

Figure 8.3 Personal remittances received – Egypt (percentage of GDP).
Source: World Bank (2018b).

in South Asia, suffered the most due to mass layoffs and a corresponding decline in remittances, Egyptians have escaped unscathed. It seems that the Saudi government is well aware of the potential negative impact on Egypt's economy in the event of reduced amounts of remittance transfers, bearing in mind their large share in foreign currency earnings. This has to be seen within the broader geopolitical context, which requires a stable Egypt with an allied regime in power (Abu Basha, 2018: 17).

The above subsections have shown in some detail the regional geo-economic mechanisms of oil-rent recycling that sustained and reproduced Egypt's twisted dependency on hydrocarbon-based rent, despite becoming itself a net energy importer. These mechanisms are political-economic in character, as they were closely related to Egypt's geopolitical situation in MENA before and after the 2011 Arab popular uprisings. They also demonstrated that Egypt's mode of insertion into the global division of labour and the basic features of its external sector are shaped by its insertion into the region's geo-economics. Egypt's heavy dependence on remittances, loans and aid, and investments flowing from the GCC, combined with its general inability to diversify its exports and FDI composition away from oil and natural gas, contributed to the current institutional impasse. The country is becoming more dependent on a dwindling resource in its external sector. It has not been capable of taking the greatest advantage of lower international oil prices since 2014. The signals were distorted by the rent that was regionally recycled, silencing much of the impact on the economy. These factors have numbed the incentives among policymakers and stakeholders to erect the institutional arrangements necessary for coordinating the diversification efforts with non-energy sectors, as explained earlier.

The state sector: Egypt as a semi-rentier state

The over-representation of hydrocarbon-based rents in Egypt's external sector confirms its status as a semi-rentier economy. By the same token, hydrocarbon-based rents (together with locational and geostrategic rents like the Suez Canal fees and external grants) have retained a significant share in state revenues, giving the Egyptian state persistent semi-rentier traits (Richter, 2007: 189–90). Overall, rent-based revenues averaged between 50 and 60 per cent of total revenues in the period 2006–12. Despite a relative decline in their weight, they still retained an average 45 per cent of state revenue in the period between 2014 and 2017 (see Table 8.5).

Hydrocarbons appeared prominently in Egypt's public finances in the twenty-first century. Being a net energy importer as of 2012 (and a net oil importer since 2006) made Egypt vulnerable to international oil prices.

Table 8.5 Revenue indicators – Egypt, 2006–18 (per cent).

	2006	2007	2008	2009	2010	2011	2012	2013	2014	2015*	2016	2017
Egyptian General Petroleum Company (EGPC): Total revenue	17.70	21.92	19.33	19.03	18.32	14.07	16.00	13.07	11.41	9.18	6.41	6.54
Suez Canal Authority (SCA): Total revenue	10.25	10.19	8.31	7.31	8.66	8.00	7.07	6.24	6.06	6.03	6.84	5.76
Non-tax and disguised rents: Total revenue	59.31	60.17	56.10	56.09	50.61	50.80	50.86	43.06	47.62	38.22	45.05	40.06

Source: Central Bank of Egypt (2018c). Author's calculation.
* Data for 2015 includes only the budget sector. The figures for all other years refer to total government budget including the budget sector, the National Investment Bank, and the Social Insurance Fund.

Indeed, a growing energy import bill increased the pressure on Egypt's dwindling foreign reserves, which fell from USD 37 billion in 2010 to USD 15.9 billion in December 2012 (BBC Arabic, 2012). A significant percentage of diesel, petrol, and butane was imported due to Egypt's limited refining capacities. The depletion of foreign reserves put downward pressure on the national currency, leading to higher inflation rates and a parallel foreign exchange market combined with a general economic slowdown. This combination also contributed significantly to Egypt's budget deficit because of universal fuel subsidy programmes inherited from Husni Mubarak's time. The budget deficit jumped from an already high rate of 8.1 per cent of GDP in 2010 to 11.6 per cent in 2015 (Trading Economics, 2017). By 2014, immediately before the plunge in international oil prices, energy subsidies made 21 per cent of total state expenditure (Egypt Oil and Gas, 2018).

Egypt was temporarily relieved due to the sharp collapse in international oil prices in mid-2014 combined with large in-kind and cash aid from oil-rich GCC allies, which helped meet Egypt's needs for imported energy at a relative low cost. The government simultaneously started cutting energy subsidies in 2014. The adoption of the IMF deal in November 2016 was followed by four rounds of energy subsidy slashing (Egypt Oil and Gas, 2018). Indeed,

> the fuels subsidy bill has decreased from 3.3 per cent of GDP in 2016/17 to a projected 2.7 per cent of GDP in 2017/18. Even though the overall share of energy subsidies in government expenditure declined, the net impact was relatively offset by the sharp depreciation of the Egyptian pound versus the dollar in a way that translated into a bigger cost for importing needed energy. (*Egypt Today*, 2018)

As with the external sector, Egypt's relations with hydrocarbon rents remained complicated in the state sector. Hydrocarbon-based revenues (and other locational and geopolitical rents) have also remained significant on the revenue side, albeit with a downward trend. It is noteworthy that as of 2006, hydrocarbon exports and Suez Canal revenues have been paid as a form of tax. The Egyptian General Petroleum Corporation (EGPC) and the Suez Canal Authority (SCA) started transferring money to the treasury in two ways: the first was the more familiar non-tax transfer, which handled these revenues as property income to the state, whereas the other counted them as tax revenues paid by companies that both public authorities owned. It is not clear why this reclassification happened under the Ahmed Nazif cabinet (2004–11). This led to a numerical boost in tax revenues as a result of disguising what were essentially non-tax rents by relabelling them as corporate taxes. What is even more ironic is that the EGPC and the SCA so-called taxes together constituted around two-thirds of corporate tax revenues collected by the Egyptian government.

These disguised rents showed the limited capacity of the Egyptian state to collect genuine income taxes from capital holders in the private sector that came to supply more than 70 per cent of Egypt's non-hydrocarbon GDP as early as 1990 (Central Bank of Egypt, 2017, quoted in Adly, 2017: 6). This is yet another example of an institutional area that was not adequately developed in time to substitute for declining rent revenues by raising taxes. This is typical of rentier states whose capacities evolve in the direction of allocating rather than collecting tax revenue from their economies (Soliman, 2011). However, it has also been the case with Egypt, a prototypical semi-rentier state.

Tables 8.5 and 8.6 demonstrate the evolving share of total state revenues from the EGPC and SCA (2006–17). Table 8.5 presents their share in total state revenues; it also estimates the share of non-tax revenues plus the disguised rents collected as taxes from the two public authorities.

There is a clear downward trend with EGPC revenues. They decreased dramatically from a peak of 22 per cent in 2007 to around 6 per cent in 2017. This trend is expected to be partly reversed with the recent surge in the production and export of natural gas, but not in any drastic or long-lasting manner, as explained earlier. The case of the SCA is as telling, despite not being hydrocarbon based. It is, however, noteworthy that the revenue that accrues from these two authorities is like no other, given that it is in USD, which makes it qualitatively quite important by constituting a direct and rather stable source of hard currency that flows directly to the treasury.

Table 8.6 demonstrates the evolution of the tax and non-tax revenues from EGPC and SCA as a percentage of tax revenue during the same time interval, 2006–17. The decline is obvious from 2007 onwards. The steep fall in the share of EGPC after 2013 must have been the result of lower international oil prices.

The share of non-tax revenues (plus disguised rents) has also witnessed a decline. This, however, did not mean successful revenue diversification away from rent dependency into more taxation. In fact, the overall ratio of tax revenue to the GDP declined between 2008 and 2012 and stagnated afterwards (see Table 8.7). This means that the decrease in rent-based revenue and other non-tax resources was not compensated for by large increases in tax revenues, once again confirming the sluggish institutional adjustment that faces rentier and semi-rentier states. In 2015, the ratio of tax revenues to GDP – a strong indicator of the capacity of the state to extract resources from society – was lower than that of 1996 and almost equal to that of 2002, showing basically long-term stagnation.

Table 8.6 Rent-based revenues in total state revenues – Egypt, 2006–17 (per cent).

	2006	2007	2008	2009	2010	2011	2012	2013	2014	2015*	2016	2017
Egyptian General Petroleum Company (EGPC): Tax revenue	31.83	39.76	34.17	33.86	28.80	23.67	25.73	26.07	20.07	12.81	10.48	9.53
Suez Canal Authority (SCA): Tax revenue	18.43	18.49	14.68	13.00	13.62	13.46	11.36	12.45	10.66	8.42	11.19	8.40

Source: Central Bank of Egypt (2018c). Author's calculation.
* Data for 2015 includes only the budget sector. The figures for all other years refer to total government budget including the budget sector, the National Investment Bank, and the Social Insurance Fund.

Table 8.7 The ratio of tax revenues to GDP – Egypt, 1996–2015 (per cent).

	1996	1997	1998	1999	2000	2001	2002	2003	2004	2005	2006	2007	2008	2009	2010	2011	2012	2013	2014	2015
Tax revenue as share of GDP in %	17.46	15.96	n/a	n/a	n/a	n/a	13.41	13.35	13.84	14.07	15.83	15.35	15.32	15.66	14.13	14.01	12.38	13.50	12.22	12.52

Source: World Bank (2020).

Declining total revenue and constant or increasing expenditure led to a heavier dependence on borrowing from the early 1990s until the present. Egypt's public debt, still predominantly domestic despite the recently soaring foreign debt, rose sharply as a percentage of GDP from 73.7 per cent in 2010 to 85 per cent in 2015, then further to 92 per cent of GDP in 2017 (Trading Economics, 2019).

Conclusion

This chapter focused on Egypt as a case of a semi-rentier economy that shows a diversified GDP structure while sustaining heavy dependency on hydrocarbon-based rents in its external and state sectors. It has argued that in a seeming age of low energy prices, Egypt should be a winner, as it became a net energy importer at least in 2012. However, specific institutional and regional geo-economic factors have shown the situation to be much more complex than mere economic analysis would suggest. The Egyptian external sector is facing big challenges to diversify away from this twisted dependency on hydrocarbon-based rents. In the same vein, moving away from rent-based into tax-based revenues is proving to be as challenging. The Egyptian state lacks the needed institutional arrangements for diversification in both sectors. Part of this is traced to its continued direct and indirect access to secondary rent through multiple channels of rent recycling that created mixed signals and weakened the incentives to bear the political and economic costs of building the required institutions in non-energy sectors.

Egypt is in dire need of building competitive advantage in non-extractive industries like manufacturing, agriculture, and high-skilled services that might attract FDI capitalising on the country's cheap factors of production, especially following the flotation of the pound in 2016. Egypt can also capitalise on its trade arrangements with the EU market, with whom it has an association agreement that came into effect in 2004. Egypt can become a hub for Asian producers that do not enjoy the same access to the EU markets. This has already started but has not reached a large scale yet. It requires a coordinated industrial policy that would link trade and investment policies with infrastructure plans and vocational education.

However, none of the above prescriptions can find their way into reality without a conscious and rigorous process of institutional capacity-building in policy areas most relevant for diversification of Egypt's external sector and public revenues away from depending on rents. These in turn are not likely to materialise without the dismantling or the disintegration of the sociopolitical coalitions producing and reproducing the allocative institutional settings and the emergence of broader coalitions bent on establishing the prerequisites for a more productive and less rent-dependent economy. Simply put, if Egypt is a net energy importer, it should act like one.

References

Abu Basha, M. (2018), '4Q17 BOP chartbook: CAD stable at three-year low as remittances keep breaking record high', EFG Hermes, 23 April. https://bit.ly/3e9Zg8M (accessed 15 March 2019).

Adly, A. (2012), *State reform and development in the Middle East: Turkey and Egypt in the post-liberalization era*. London: Routledge.

Adly, A. (2016), 'Egypt's oil dependency and political discontent', Carnegie Middle East Center, 2 August. https://bit.ly/3ubIlIw (accessed 15 March 2019).

Adly, A. (2017), 'Too big to fail: Egypt's large enterprises after the 2011 uprising', Carnegie Middle East Center, March. https://bit.ly/2Pr5o2v (accessed 15 March 2019).

Amsden, A. H. (1992), *Asia's next giant: South Korea and late industrialization*. Oxford: Oxford University Press.

Badawi, N. (2013), 'Following Morsi's ouster, Qatar's support to Egypt in question', *Daily News Egypt*, 14 July. https://bit.ly/2PxR9ZP (accessed 15 March 2019).

Baru, S. (2012), 'Geo-economics and strategy', *Survival*, 54:3, 47–58.

BBC Arabic (2012), 'Al-Markazi al-Masri: Ihtiyati al-naqd fi mustawa harij wa-Mursi yashidu bi-l-numu' [Central Bank of Egypt: Foreign reserves reach a critical level and Mursi praises economic growth], 29 December. https://bbc.in/3bltkNv (accessed 8 September 2019).

Bruno, M., and J. Sachs (1982), 'Energy and resource allocation: A dynamic model of the "Dutch disease"', *The Review of Economic Studies*, 49:5, 845–59.

Central Bank of Egypt (2018a), 'Structure of GDP (factor cost-constant prices)'. https://bit.ly/3xJ6AQH (accessed 9 September 2019).

Central Bank of Egypt (2018b), 'Egypt external position quarterly report', 52, 2015–16.

Central Bank of Egypt (2018c), 'Consolidated fiscal operations of the general government'. https://bit.ly/38aCLNO (accessed 9 September 2019).

Central Bank of Egypt (2018d), 'Net foreign investment by country'. https://bit.ly/3ebBjOD (accessed 9 September 2019).

Chaudhry, K. A. (1989), 'The price of wealth: Business and state in labor remittance and oil economies', *International Organization*, 43:1, 101–45.

Colonna, J. (2018), 'SIPRI: Egypt's arms imports skyrocket amidst greater security threats', *Egypt Today*, 2 March. https://bit.ly/30fSaYG (accessed 15 February 2017).

Corm, G. (1998), 'Reconstructing Lebanon's economy', in N. Shafik (ed.), *Economic challenges facing Middle Eastern and North African countries*. London: Palgrave Macmillan, pp. 116–35.

Daher, J. (2016), *Hezbollah: The political economy of Lebanon's party of God*. London: Pluto Press.

Deacon, R. T. (2011), 'The political economy of the natural resource curse: A survey of theory and evidence', *Foundations and Trends in Microeconomics*, 7:2, 111–208.

Egypt Oil and Gas (2017), '84% of Q1 FDI flows to energy sector', 11 December. https://egyptoil-gas.com/news/84-of-1st-qtr-fdi-flows-to-energy-sector (accessed 8 September 2019).

Egypt Oil and Gas (2018), 'Energy subsidy cuts the hardest part of Egypt's economic reform', 18 June. https://bit.ly/2O8iq4C (accessed 8 September 2019).

Egypt Today (2017), 'Egypt's oil and gas sector requires majority of FDI: Report', 17 August. https://bit.ly/3v02HFk (accessed 9 September 2019).

Egypt Today (2018), 'Egyptian energy subsidy costs increase 34% in H1 of 2017/18: Min', 7 February. https://bit.ly/2Orbsre (accessed 9 September 2019).

Evans, P. B. (2012), *Embedded autonomy: States and industrial transformation.* Princeton: Princeton University Press.

Fargues, P., F. De Bel-Air, and N. M. Shah (2015), 'Addressing irregular migration in the Gulf States', GLMM Policy Brief 1, 1–6.

Farsoun, S. K. (1988), 'Oil, state and social structure in the Middle East', *Arab Studies Quarterly*, 10:2, 155–75.

France 24 (2015), 'Egypt, France to conclude 5.2 billion Euro deal for rafale jets', 16 February. https://bit.ly/3bjgrDz (accessed 2 February 2018).

Galal, A., and B. M. Hoekman (eds) (2003), *Arab economic integration: Between hope and reality.* Washington, DC: Brookings Institution Press.

Goldberg, E. (2004), *Trade, reputation, and child labor in twentieth-century Egypt.* Berlin: Springer.

Hanafy, S. (2015), 'Sectoral FDI and economic growth: Evidence from Egyptian governorates', Joint Discussion Paper Series in Economics, 37.

Hanieh, A. (2015), 'Shifting priorities or business as usual? Continuity and change in the post-2011 IMF and World Bank engagement with Tunisia, Morocco and Egypt', *British Journal of Middle Eastern Studies*, 42:1, 119–34.

Henwood, D. (1993), 'Global economic integration: The missing Middle East', *Middle East Report*, 184, 7–8, 31.

Hertog, S. (2016), 'Rent distribution, labour markets and development in high rent countries', London School of Economics and Political Science, Kuwait Programme on Development, Governance and Globalisation in the Gulf States, Research Papers, 40.

Ikram, K. (2007), *The Egyptian economy, 1952–2000: Performance policies and political issues.* London: Routledge.

Karl, T. L. (1997), *The paradox of plenty: Oil booms and petro-states.* California: University of California Press.

Knight, J. (1992), *Institutions and social conflict.* Cambridge: Cambridge University Press.

Luciani, G., and H. Beblawi (eds) (1987), *The rentier state.* London: Croom Helm.

Moerenhout, T. (2018), 'Reforming Egypt's fossil fuel subsidies in the context of a changing social contract', in J. Skovgaard and H. van Asselt (eds), *The politics of fossil fuel subsidies and their reform.* Cambridge: Cambridge University Press, pp. 262–85.

Refaat, T. (2018), 'Remittances of Egyptians abroad hit $2.14 billion in 10 months', Ahram Online, 20 December. https://bit.ly/3bl8mOs (accessed 9 September 2019).

Reuters (2015), 'Egypt got $23 billion in aid from Gulf in 18 months – minister', 2 March. https://reut.rs/3c2m8Vu (accessed 9 February 2018).

Richter, T. (2007), 'The political economy of regime maintenance in Egypt: Linking external resources and domestic legitimation', in O. Schlumberger (ed.), *Debating Arab authoritarianism: Dynamics and durability in nondemocratic regimes.* Redwood City: Stanford University Press, pp. 177–93.

Ross, M. L. (2003), 'The natural resource curse: How wealth can make you poor', in I. Bannon and P. Collier (eds), *Natural resources and violent conflict: Options and actions.* Washington, DC: World Bank, pp. 17–42.

Ross, M. L. (2011), 'Will oil drown the Arab Spring? Democracy and the resource curse', *Foreign Affairs*, 90:5, 2–7.

Schlumberger, O. (2008), 'Structural reform, economic order, and development: Patrimonial capitalism', *Review of International Political Economy*, 15:4, 622–49.

Schwarz, R. (2008), 'The political economy of state-formation in the Arab Middle East: Rentier states, economic reforms, and democratization', *Review of International Political Economy*, 15:4, 599–621.

Shafer, D. M. (1994), *Winners and losers: How sectors shape the developmental prospects of states*. Ithaca: Cornell University Press.

Soliman, S. (2011), *The autumn of dictatorship: Fiscal crisis and political change in Egypt under Mubarak*. Stanford: Stanford University Press.

Springborg, R. (2012), 'Gas and oil in Egypt's development', in R. E. Looney (ed.), *Handbook of oil politics*. London: Routledge, pp. 295–311.

Trading Economics (2017), 'Egypt government budget 2002–2018'. https://tradingeconomics.com/egypt/government-budget (accessed 9 September 2019).

Trading Economics (2018), 'Egypt imports by category'. https://tradingeconomics.com/egypt/imports-by-category (accessed 8 February 2018).

Trading Economics (2019), 'Egypt government debt to GDP'. https://tradingeconomics.com/egypt/government-debt-to-gdp (accessed 11 February 2019).

Wantchekon, L. (2002), 'Why do resource dependent countries have authoritarian governments?', *Journal of African Finance and Economic Development*, 5:2, 57–77.

World Bank (2017), 'Personal remittances received: Brazil, Egypt, and Turkey (current US dollars)', World Bank Data. https://data.worldbank.org/indicator/BX.TRF.PWKR.CD.DT?view=chart (accessed 12 June 2017).

World Bank (2018a), 'Fuel exports (% of merchandise exports), the Middle East and North Africa', World Bank Data. https://bit.ly/30f6yjN (accessed 10 September 2019).

World Bank (2018b), 'Egypt: Remittance received as a % of GDP', World Bank Data. https://bit.ly/3kPFJfr (accessed 11 September 2019).

World Bank (2019), 'Migration and remittances data'. https://bit.ly/3rmk01d (accessed 8 October 2019).

World Bank (2020), 'Tax revenue (% of GDP) – Egypt, Arab Republic', World Bank Data. https://data.worldbank.org/indicator/GC.TAX.TOTL.GD.ZS?locations=EG (accessed 23 April 2021).

World Trade Organization (2020), 'Egypt total merchandise exports by commodity', World Trade Organization Time Series. https://bit.ly/3ecGQEY (accessed 9 September 2019).

9

Oil and turmoil: Jordan's adjustment challenges amid local and regional change

Riad al Khouri and Emily Silcock

Introduction[1]

Cheap oil would seem to be a blessing for Jordan, as the country imports almost all its energy needs. Petroleum represented the Kingdom's largest merchandise import in 2017 (Observatory of Economic Complexity, 2019). Both the International Monetary Fund (IMF) and the Central Bank of Jordan (CBJ) predicted that low oil prices would help spur fiscal reforms and reduce Jordan's public debt (CBJ, 2015; IMF, 2015).

In turn, high oil prices can be blamed for aggravating Jordan's fiscal deficits (El-Said and Becker, 2001; Knowles, 2004). Two mechanisms act as important links between high oil prices and Jordan's fiscal imbalances. First, aid from oil-exporting Gulf Cooperation Council (GCC) states in the 1970s helped fund Jordan's large public sector, which accounts for 26 per cent of employment and more than 55 per cent of the gross domestic product (GDP), one of the highest ratios in the world (Fanek, 2017). Second, high oil prices amplify remittances and employment opportunities for Jordanians in GCC states. Much like direct oil income, aid and remittances can undermine states' capacity to raise taxes and cut spending (Chaudhry, 1997; Malik, 2017; Sowell, 2017). These 'unearned' revenues abet profligate spending and disincentivise tax collection when times are good, leaving in place a massive public sector without the means to fund itself, when times are bad. As such, rents from aid and remittances deepen recipient states' economic and political dependence on donor countries. This dependence can constrain a recipient state's foreign policy and place external interests over domestic ones (Ryan, 2002).

This chapter investigates the impact of low oil prices from 2014 to 2018 on Jordan. We assess whether cheap oil facilitated fiscal reform, defined in terms of reducing public-sector debts. We also examine whether cheap oil, by reducing aid from oil-exporting GCC states, weakened GCC pressure on Jordanian foreign policy. One might expect cheap oil to help the Jordanian state slash fiscal deficits. Savings on the fuel bill can offset the economic burdens of higher taxes and spending cuts, helping to make fiscal reform more politically bearable. The reduction in aid from oil-exporting states in times of cheap oil could also pressure the Jordanian state to cut spending and raise taxes. The reduction of GCC aid following low oil prices should also reduce Jordan's economic and political dependency on GCC states. Low oil prices could therefore coincide with greater fiscal reform and foreign policy autonomy.

Yet there is mixed evidence of these expectations. Cheap oil dramatically lowered Jordan's energy import bill, and lower oil prices did correlate with a reduction in foreign aid from GCC states. Remittance flows, however, remained relatively constant from 2014 to 2018. In terms of political outcomes, cheap oil was not accompanied by fiscal reform in Jordan. Public-sector expenditures surpassed growth in domestic revenues, further expanding Jordan's current account deficits during this period. In foreign policy, however, Jordan demonstrated greater resistance to mainstream GCC pressure in policies towards Qatar, Syria, and Yemen. Yet, such foreign policy develop-ments remained short-lived. The scale of Jordan's debt crisis, coupled with the depth of public discontent over fiscal reform programmes, ultimately compromised Jordan's fiscal reform and pushed Amman to accept more aid from GCC states after unprecedented tax protests in 2018.

The next section surveys the history of Jordanian fiscal and foreign policy. It highlights the centrality of aid and remittances in linking the two realms of policy. The third section illustrates how energy costs, levels of foreign aid, and remittances to Jordan have evolved since the global decline in oil prices in the summer of 2014. The fourth section then assesses Jordan's fiscal and foreign policy during the cheap oil era. We find that low oil prices did not correlate with lower fiscal deficits. In foreign policy, however, Jordan has displayed greater autonomy from the GCC. The section then outlines why this change should be temporary. We describe Jordan's 2018 tax protests, which resulted in more foreign aid from GCC states. We expect this aid to sustain Jordan's fiscal deficits and compromise its foreign policy. The chapter concludes by explaining why widespread predictions of the effect of the reduction in oil prices were wrong. It proposes that regimes dependent on aid-based rents may be especially resistant to economic reform.

A brief history of oil, aid, debt, and reform in Jordan

Aid and oil in Jordan

Foreign aid has subsidised the Jordanian state since its founding in the 1920s. Jordan's strategic location – between Israel, Iraq, the Kingdom of

Saudi Arabia, Palestine, and Syria – has made the country's survival indispensable to regional and global powers. Over the past three decades, Jordan's location and coordination have been ideal for dealing with the Israeli–Palestinian conflict, waging the 'war on terror', invading Iraq, and discreetly providing support to Western intervention in Syria (Yom and Al-Momani, 2008). Jordan's strategic importance to the global and regional powers undergirds its privileged access to foreign aid. This strategic position has led Jordan to become one of the largest middle-income country recipients of foreign grants in the world, even prior to the beginning of the Syrian crisis and humanitarian financial appeals made since. Maintaining access to aid is a pillar of Jordanian foreign policy (Ryan, 2002: 49).

Aid and oil are linked in Jordan. Though the country benefits from many donors, oil-exporting Arab states have played a vital role in providing aid since the 1970s, when oil prices rose. During this era, aid constituted 16 per cent of Jordan's GDP (Jordan Strategy Forum, 2018: 12). Saudi Arabia has been the leading aid-giver since 2003, although assistance from other GCC states is also significant. Aid from the USA is also important, though not more so than that of Saudi Arabia. An important milestone in this process is that, from 2011, aid from GCC states, particularly from Saudi Arabia, helped the Jordanian economy and the Hashimite regime weather Arab Spring protests (Beck and Hüser, 2015). In late 2011, as the Jordanian government at the time was contending with a large budget deficit, the Kingdom of Saudi Arabia and oil-rich neighbours decided to extend USD 5 billion in financial aid for development schemes in Jordan over five years (Ayesh, 2018). By undermining oil-exporting donors' financial surpluses, cheap oil jeopardises Jordan's access to foreign aid.

At the same time, cheap oil has also constituted a form of aid to Jordan. Prior to 2003, the Kingdom had been supplied with cheap oil from Iraq. Baghdad had covered all of Jordan's requirements for oil during the previous thirteen years under a special arrangement, with one half supplied at preferential prices and the other as an outright gift. However, following the 2003 Western invasion of Iraq, Jordan lost this important regional ally. Needing to recover its access to oil, Jordan then strengthened relations with the Kingdom of Saudi Arabia to offset losses in energy resources. Yet, although the Saudis did come to Amman's rescue, there was no agreement to sell oil to Jordan at the same low prices that it enjoyed from Iraq. Saudi Arabia said that it and the United Arab Emirates (UAE) would replace the oil that Jordan received from Iraq, but at market prices. Oil could have come from elsewhere, but geographical proximity resulting in reduced transport burdens, and greater reliability of supply, made the Saudi and Emirati sources more attractive (Schenker, 2016; Robins, 2019).

Aid rentierism, debt, and the crisis in Jordan since 2018

In many cases, aid can help countries accumulate capital, escape poverty traps, and kickstart development (Arndt *et al.*, 2015). However, in some cases, sustained aid can have detrimental effects on the political economy of a country (Moss *et al.*, 2006; Rajan and Subramanian, 2007). Jordan appears to some extent to be an example of the latter. The country has displayed 'surprisingly poor growth coincident with having relatively good economic policies and inflows of aid' (Burnside and Dollar, 2004: 783). Aid income, like other external rents, allows the possibility of countries becoming 'rentier states' (Malik, 2017). The literature on aid rentierism argues that aid can undermine the country's institutions in two ways. First, aid encourages the state to buy support from the electorate, by giving them protected markets, subsidies, public employment, and tax breaks. This can impede sustainable growth, particularly by distorting relations between the state and the private sector. Second, it reduces the extent to which the government relies on taxes, thus lowering incentives for the state to develop the domestic economy by cultivating taxable domestic surplus and for it to invest in or establish effective private-sector regulation (Peters, 2009).

The development of Jordan as an aid rentier is connected to oil prices and has led to high levels of Jordanian debt. Higher oil prices in the 1970s flushed the Kingdom with aid to fund a sprawling and highly interventionist public sector (Knowles, 2004: 55). When oil prices plunged in the late 1980s, the Jordanian state lost the means to fund its massive public sector. In 1989, Jordan's debts were two and a half times greater than its GDP (Sowell, 2017: 4). Jordan's debt woes prompted its first structural adjustment programme with the IMF. Jordan's debt then stabilised in the 1990s and 2000s, dropping to 58 per cent of GDP in 2008.

However, Jordan continued to receive high levels of aid, and the same pattern is re-emerging. Jordan's aid receipts have been used for budget support, supporting the Jordanian regime by helping to fund the military and the civil bureaucracy. Based on the latest published official statistics, the country's civil bureaucracy employed around 265,000 people in 2017, up from 195,000 in 2008, an increase of about 36 per cent. This remarkable number must be seen against the light of conventional wisdom among Western donors and international financial institutions about the need to shrink the public sector. On the other hand, the whole Jordanian economy included 1,022,000 employed persons in 2017, up from 770,000 in 2008, a rise of only 33 per cent (Jordanian Department of Statistics, 2019). The share of the civil bureaucracy in the employed workforce has therefore gone up, even as the government pays lip service to shedding state jobs. The public sector now accounts for 26 per cent of employment, and more than 55 per cent of GDP, one of the highest ratios in the world (Fanek, 2017). In other words, as the private sector does not deliver on job creation, the state is using aid to act as employer, bloating the civil bureaucracy (Knowles, 2004).

This mounting level of government spending, coupled with slow growth following the 2008 financial crisis and the Arab Spring protests, pushed Jordan's debt-to-GDP ratio to 93.4 per cent in 2015, as oil prices plummeted (Sowell, 2017: 4). Yet the government was unable to spur GDP growth via a business sector and workforce alike that sought state favours and handouts, those in turn leading to the taking on of more public debt. Such patterns of weak production and heightened consumption continue today, making worse the country's finances in general.

Against this background of mounting economic pressure, in 2016, the Jordanian government and the IMF once again agreed to an Extended Fund Facility (EFF) programme to lower debt levels and fiscal imbalances to maintain the country's credit standing. Under the EFF, in January 2018, Jordan announced tax hikes crucial to gradually reducing record public debt. The package included removing exemptions on general sales tax and raising the previous 4–8 per cent rates on some goods to 10 per cent. This affected many items, ranging from Internet and electricity to soft drinks and stationery, with special taxes imposed on tobacco and premium gasoline. Farming inputs like animal feed, insecticides, and fertilisers, which were previously exempt from taxation, became tax-liable, raising production costs by up to 10 per cent, some of which were passed on to consumers. At the same time, Jordan announced the end of subsidies, such as those to bread. The intention was to generate over USD 760 million in extra revenue (IMF, 2016).

Most of the economic measures in the 2018 budget bill came into force promptly, but the impact of the measures was softer, as King Abdullah intervened to grant an exemption from the sales taxes on medicines. Yet, the measures worsened the plight of poorer Jordanians, and triggered some civil unrest (The New Arab, 2018). Much of the reform heightened public criticism of the regime's economic policy including through attacks against the government and on the prime minister personally, which is quite rare in Jordan (ANSA Med, 2018).

Important industrial action and other protests continued in Jordan throughout 2020, mainly in relation to complaints about the rising cost of living, coupled with the government's parlous fiscal condition. The latest of these actions took place in September and October 2019 in the form of a major successful strike by teachers in the public sector, resulting in the government raising teacher wages. A conclusion to this process of rising economic complaints, resulting in further political and fiscal tension, does not appear to be near.

Cheap oil and energy costs, foreign aid, and remittances

Cheap oil, caused by the oil price decline of 2014, should have five immediate implications for Jordan's political economy. It should lower Jordan's energy bill, reduce aid, lower remittance levels, reduce exports to oil-exporting states, and lower receipts from GCC tourists. The following subsections assess these expectations.

Effect of oil price changes on Jordan's energy bill

Jordan imports most of its fuel needs, with local gas extraction and renewable energy production making very modest contributions. Unsurprisingly, low oil prices drastically reduced spending on oil in Jordan. The cost of Jordan's oil imports decreased by 30 per cent from 2014 to 2018, netting a USD 1.8 trillion saving in energy costs (CBJ, 2019). Figure 9.1 illustrates the strong correlation between global oil prices and the aggregate value of crude petroleum imports to Jordan.

Both private consumers and government entities benefit from cheaper oil. The Jordanian Central Bank estimated in 2015 that, all else being equal, a

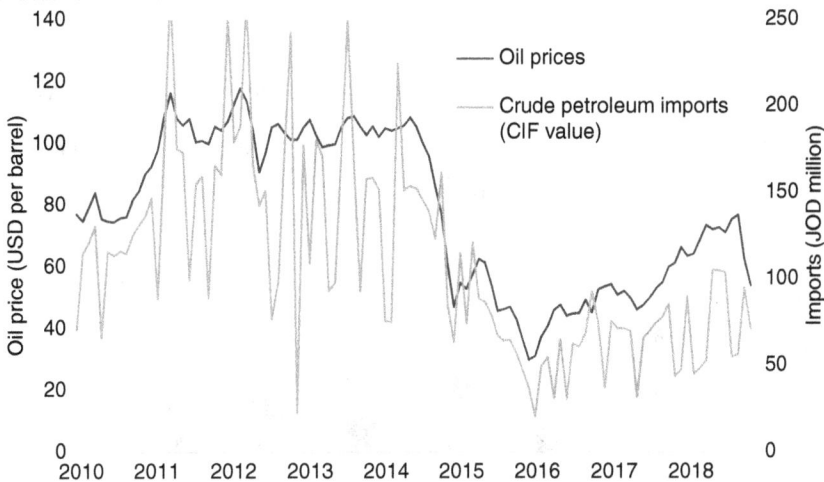

Figure 9.1 Value of Jordan's crude petroleum imports, 2014–18 (in JOD million). 1 Jordanian dinar (JOD) is approximately equal to USD 1.4. Source: CBJ (2019).

global oil price of USD 50 per barrel would trim Jordan's current account deficit from 7.5 to 4.7 per cent of GDP (CBJ, 2015: 10). The IMF cast cheap oil as a boon for fiscal reform in Jordan (IMF, 2015). A cheaper energy bill could alleviate painful but necessary cuts in public-sector spending.

Theoretically, lower energy prices should also promote growth in energy-importing countries by lowering production costs (Ali Al-Zeaud, 2014; Moawad Ahmed, 2016). However, economists have found mixed evidence of the negative association between oil prices and growth in Jordan. Berument *et al.* (2010) and Abu Asab (2017) find that negative oil price shocks have little effect on GDP, whereas Mohaddes and Raissi (2013) and Ali Al-Zeaud (2014) find they are associated with an increase in GDP. Relatedly, Maghyereh *et al.* (2019) find that an increase in the uncertainty surrounding oil prices is associated with a decrease in Jordanian GDP. Other economic linkages between Jordan and oil-exporting states, like foreign aid, remittances, and export markets, may account for these mixed results.

Jordanian foreign financial support and oil prices

Low oil prices also correlate with less aid from GCC states. Figure 9.2 graphs the price of oil and the volume of external grants to Jordan across donors. Grants from GCC states almost halved from 2014 to 2018 (Ayesh, 2018). Aid from the USA and the EU, however, has remained robust. This is unsurprising; confronted by the dire economic ramifications of cheap oil,

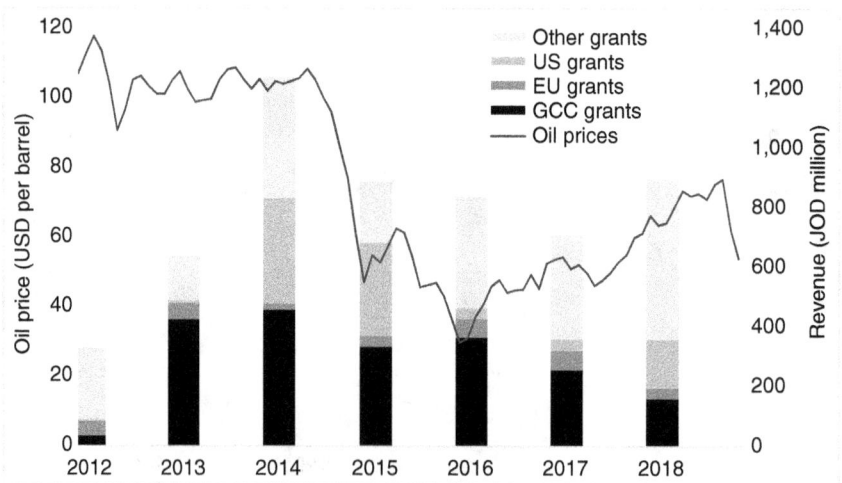

Figure 9.2 Jordanian central government external grants (in JOD million).
Source: CBJ (2019).

oil-exporting countries will have less disposable income to donate to Jordan, whereas oil prices make little difference to aid allocations of the USA and the EU.

However, foreign aid is not purely driven by changes in oil prices; it clearly has a political dimension as well as an economic one. Saudi–Jordan relations are complicated, including when it comes to aid and oil supplies. High Jordanian aid receipts in 2011 and 2014 appear to have been politically motivated, as they were aimed at preserving regional stability in reaction to Arab Spring protests in Jordan.

However, by 2015, Riyadh's financial aid to Amman slowed to a trickle after the oil price drop. This seems to suggest that in the immediate aftermath of the price crash and later decrease in oil revenues, harsh economic realities may have forced Saudi Arabia to scale back ambitious foreign policy goals. Yet a political element entered this equation, with the undeclared, but nevertheless prominent, idea that aid slowed or even stopped because, despite Saudi pressures, Jordan was reluctant to supply military and other support, including providing training, to Syrian insurgent groups. This was in pursuit of Jordan's long-standing policy of neutrality in regional conflicts, after a pro-Iraq stance in the 1991 Gulf conflict resulted in hardship for many Jordanians, including their mass expulsion from Kuwait. However, it is difficult to disentangle these two elements of Saudi decision-making, which is quite opaque and tough to analyse, as it runs on informal, secretive, and personal, non-transparent lines (Schenker, 2016).

By 2016, however, relations between Amman and Riyadh were gradually improving, partly because it became clearer that the Saudi policy in Syria of backing anti-government forces was failing. Relations were reinforced by a visit by Jordan's King Abdullah to Riyadh in April of that year, which brought the announcement of a new joint Saudi–Jordan Coordination Council designed to deepen strategic relations and enhance cooperation in all fields. At the Jordanian monarch's earlier meeting with the then deputy Crown prince of Saudi Arabia, Muhammad bin Salman, the two sides had already named investment sectors, with a focus on Jordan's Aqaba Special Economic Zone (something of a Jordanian investment flagship) and the opening of new opportunities for Jordan's exports to Saudi Arabia. Jordan then issued a new investment law to govern large-scale infrastructure and development projects that are open to Arab and foreign investment institutions, which would pave the way for major Saudi investment (Yom, 2017). Saudi Arabia's support to Jordan, despite modest apparent political changes, may be evidence that aid receipts from Riyadh might also have depended on economic factors, such as a change in oil prices, which somewhat recovered in 2016, rather than on purely political considerations. However, Schenker (2016) proposes a political explanation, arguing that Amman is likely to have substantively committed itself to Riyadh in ways not made public, in return for financial relief from Saudi Arabia.

In 2018, despite the decline in oil prices, and frightened by the increased instability inside Jordan, the Saudi Fund for Development and the Jordanian government signed a deal to reschedule the USD 114 million that Amman owed it. The debt included nineteen due loans for repayment over twenty years with a grace period of five years (Tayseer, 2018). In June of the same year, the Kingdom of Saudi Arabia, the UAE, and Kuwait announced a USD 2.5 billion aid package for Jordan (Furlow and Borgognone, 2018). This aid package consisted primarily of loans in the form of deposits with the CBJ and funding for infrastructure investment, not outright grants; as a result, these GCC partners could withhold deposits and delay or even cancel investments to censure and pressure Jordan if necessary (Furlow and Borgognone, 2018). In March 2019, the Kingdom of Saudi Arabia transferred over USD 300 million to Jordan in the latest aid package designed to ease the financial strain in Amman (Ma'ayeh, 2019).

Saudi Arabia, despite differences with Jordan, thus continued to supply Amman with aid, as the preservation of the Jordanian monarchy remains of major interest to Riyadh (Riedel, 2019). Despite the decline in oil price, Saudi Arabia restarted and continued to finance parts of the gap Jordan encountered. Yet, overall there is no unambiguous answer to the question of the extent to which changes in oil prices have led to changes in foreign financial support to Jordan. The 2018–19 increase in financial support from the Saudis seems clearly politically motivated, whereas the fall in aid in 2014–15 is more likely to be attributable to oil price shocks. Even if we cannot directly attribute all changes in aid to changes in oil price, it is likely that declining oil prices lead to greater uncertainty in aid receipts, which has a similar outcome to aid simply being lower; in both cases the Jordanian government is less able to count on aid in the future, and therefore cannot make plans that include this revenue stream (Connable, 2015).

Effect of oil price changes on Jordanian remittances

In addition to curbing aid from GCC states, cheap oil might also weaken remittance levels and employment opportunities for Jordanians in the GCC. As at 2015, there were about 1 million Jordanians working outside Jordan, of which 80 per cent were in GCC countries. In 2013–15, remittances from Jordanians in Saudi Arabia were on average over 38 per cent of remittance flows into Jordan, making it the single biggest source of remittances (De Bel-Air, 2016; Jordan Strategy Forum, 2018; Momani, 2018).

Given Jordan's high levels of unemployment, these expatriates and their remittances supply an important socio-economic safety valve that helps to

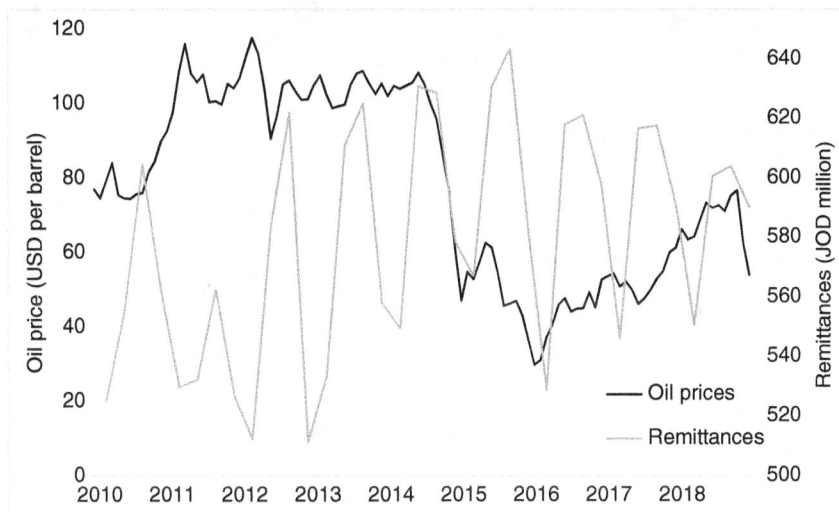

Figure 9.3 Value of workers' remittances – Jordan, 2014–18 (in JOD million).
Source: CBJ (2019).

support local incomes. Low oil prices undermine oil-exporting states' economies, which could jeopardise Jordanian employment opportunities in the GCC and remittance flows. However, Figure 9.3 demonstrates that remittances have changed little since 2014, perhaps because employment, especially in Saudi Arabia, changed very little in response to the change in oil prices. Cheap oil may thus have had little direct effect on remittances to Jordan. That is perhaps because Jordanian expatriate human resources in Saudi Arabia and other GCC states tend to be more skilled and so less dispensable in a downturn. Thus, Jordanian doctors, engineers, lawyers, and similar higher-grade labour would continue working and making transfers home, while unskilled workers from other countries would be laid off in a downturn caused by cheaper oil.

Effect of oil price changes on Jordan's exports

Cheap oil may undermine Jordanian exports to oil-exporting markets. Figure 9.4 shows Jordanian exports to the GCC states and Iraq during 2015–16. Exports to the GCC countries appear unaffected by changes in oil prices, whereas exports to Iraq appear to be highly correlated with oil prices. However, the changes in exports to Iraq may also be driven by difficulties at the Iraqi–Jordanian border from mid-2014, as well as unrest inside Iraq, leading to a reduction in Jordan's business with its north-east neighbour.

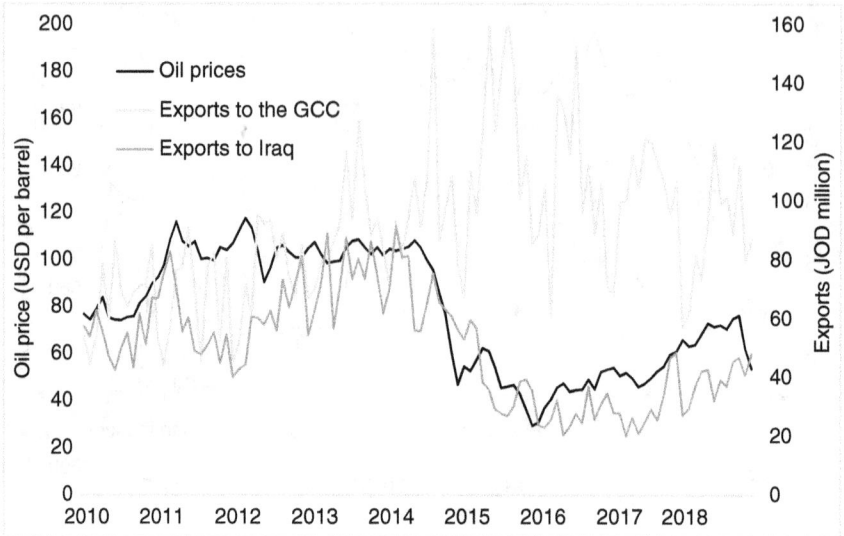

Figure 9.4 Value of Jordanian merchandise exports to selected oil-producing Arab countries, 2014–18 (in JOD million). Source: CBJ (2019).

The eruption of the Islamic State in the northern and western parts of Iraq in June 2014 had a negative impact on the Iraqi economy, leading to fewer imports, as well as making transport in the country and across western borders parlous, sometimes even bringing trade with Jordan to a standstill.

Effect of oil price changes on Jordanian tourism

Finally, tourism is a significant source of external revenue for Jordan. The income from GCC-country tourists visiting Jordan fell somewhat in 2015–16 following on the oil price decline, as seen in Figure 9.5. There is no evidence that this decline in tourism was politically or otherwise non-economically motivated. For example, the number of tourists coming to Jordan was unaffected by GCC-member governments telling their nationals not to visit Amman or Petra, or by the Jordanian government frowning on or impeding them from coming. Therefore, it is possible to explain declining tourism receipts in 2015–16 by the oil price drop. However, Jordan was able to gain a much higher income from tourists from the GCC countries in 2017–18; oil prices recovered somewhat then, but it is notable that tourist revenues in 2018 are much higher than in 2014, despite oil prices being lower. This is partly due to a general upgrading of Jordan's tourism sector, making it more attractive to visitors from GCC countries.

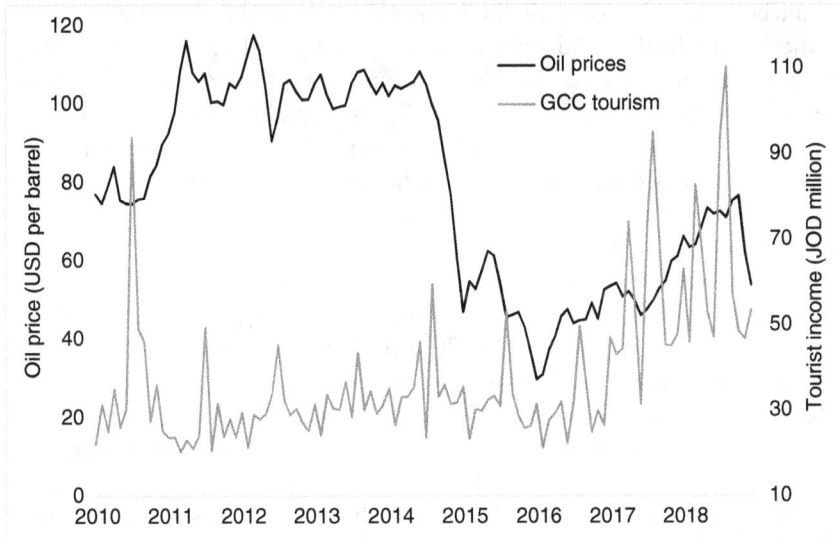

Figure 9.5 Receipts from GCC-country tourists – Jordan, 2014–18 (in JOD million). Source: Ministry of Tourism and Antiquities (2019).

Overall, the effect of oil prices on aid receipts and energy costs stand out as the predominant effect of the fall in oil prices on the Jordanian economy. There is little correlation between oil prices and remittances, exports, or tourism receipts. Unsurprisingly, oil prices seem strongly correlated with the Jordanian energy bill. Oil prices are not the sole driver of changes in aid receipts throughout the period but seem to be responsible for at least the fall in aid from 2014 to 2015, as well as an increase in uncertainty about future aid receipts. Other channels/factors of influence also operate to link oil price changes to the domestic Jordanian economy, but the main one is clearly the bill for energy imports.

Cheap oil and fiscal reform and foreign policy autonomy

The previous section illustrated that low oil prices correlated with cheaper energy imports and a reduction in aid to Jordan from GCC countries. This section assesses the political implications of these developments on fiscal reform and foreign policy autonomy. We begin by discussing the effects on the Jordanian government budget, which was affected by the change of oil prices predominantly via the decrease in aid receipts. This led to a decline in capital expenditure and an increased need to raise domestic revenue. It might be hoped that over the long term this would reduce the level of aid rentierism in Jordan and help to develop the social contract by making the government more accountable to taxpayers. However, since 2014, we see no evidence of this (Moore, 2017). If anything, the social contract eroded as the Jordanian government attempted to continue spending as it was before the decline in aid receipts; its inability to do so has meant that the state delivered less to Jordanians, while at the same time trying to tax them more.

Second, we discuss the effect of the change in oil prices on Jordan's relationship with regional powers, particularly the Kingdom of Saudi Arabia. To some extent, the decline in GCC aid allowed Jordan to be less at the whim of its benefactors. Jordan resisted toeing the Saudi line in the Syrian and Israeli–Palestinian conflicts. However, Riyadh still holds significant sway over Jordan, not least because of the importance of remittances from Jordanian workers in Saudi Arabia to the Jordanian economy.

The recent crisis in Jordan shows how these two effects are intertwined. Jordanian protests, starting in 2018, were sparked by the government's plans to reform income tax, attempting to be more self-sufficient. In order to stabilise the political situation in the wake of protests, Jordan accepted USD 2.5 billion from the Kingdom of Saudi Arabia and the UAE, having the inverse effect of making them more dependent on their regional allies (Furlow and Borgognone, 2018). This exemplifies how Jordan has been facing the dilemma between raising taxes and accepting help from regional powers, along with the required political concessions. It may seem that the decrease in oil prices and the subsequent drop in aid are therefore responsible for Jordan's current crisis. However, this dilemma existed long before 2014; the decline in oil prices simply reduced the state's room for manoeuvre, exacerbating pre-existing difficulties. Jordan's problems would not be solved by aid being restored to pre-2014 levels, as that does not allow escape from this dilemma. Instead it would be necessary to take measures to rebuild the social contract, particularly by tackling mismanagement and corruption.

Oil prices and fiscal reform

One might expect cheap oil to help the Jordanian state reduce fiscal deficits. Savings at the pump offset the economic burdens of higher taxes and spending cuts, making fiscal reform more politically bearable. The IMF's 2015 final review for its stand-by arrangement with Jordan predicted that low oil prices would aid fiscal reforms and help stabilise Jordan's public debt (IMF, 2015: 1). It was hoped that lower aid would encourage Jordan to generate more of its own revenue and cut back on high levels of public expenditure. These expectations were mostly wrong.

Table 9.1 Jordanian central government budget indicators, 2014–18 (in JOD million).

Year	Expenditure			Revenues and grants			Overall balance	
	Capital expenditure	Current expenditure	Total expenditure	External grants	Domestic revenues	Total revenue	Excluding grants	Including grants
2014	1,137.5	6,713.6	7,851.1	1,236.5	6,031.1	7,267.6	−1,820.0	−583.5
2015	1,098.2	6,624.5	7,722.7	886.2	5,910.6	6,796.8	−1,812.1	−925.9
2016	1,029.1	6,919.1	7,948.2	836.0	6,233.6	7,069.6	−1,714.6	−878.6
2017	1,060.2	7,113.0	8,173.2	707.9	6,717.4	7,425.3	−1,455.6	−747.9
2018	947.7	7,619.6	8,567.3	894.7	6,944.9	7,839.6	−1,622.3	−747.6

Source: CBJ (2019).

Low oil prices did correlate with higher levels of domestic revenue. Domestic revenue, the amount of revenue the Jordanian government collected, excluding external grants, increased 15 per cent from JOD 6 billion to 6.9 billion from 2014 to 2018 (CBJ, 2019), as shown in Table 9.1. However, contrary to prevailing expectations, public-sector spending increased as well. Public-sector spending rose 9 per cent from 2014 to 2018. Crucially, current expenditures, which represent public-sector spending on subsidies, pensions, social assistance, and public-sector wages, constituted the bulk of growth in expenditure spending. Levels of capital expenditures, public-sector spending in growth enhancing assets like infrastructure, public services, and utilities, decreased by 16 per cent from 2014 to 2018. Public-sector expenditures outgrew increases in domestic revenue, meaning Jordan's current account deficit as a percentage of GDP increased from 7.2 per cent in 2014 to 10.6 per cent in 2017 (World Bank, 2019).

Why was the Jordanian government unable to capitalise on low oil prices to curb public-sector spending? A singularly important statistic about Jordan is that about 29 per cent of males in the country's labour force work in the public sector, the bulk of them in the armed and security forces (Jordanian Department of Statistics, 2019). State pensions are also absorbing a growing share of state expenditures. Kirk Sowell, a leading expert on Jordan's political economy, estimates that security and pension costs constitute 40 per cent of government spending (Sowell, 2017: 2). Military and public-sector employees are a vital political constituency for the Hashimite regime (Baylouny, 2008). Furthermore, many of Jordan's pension and social security obligations predate the cheap oil era. Oil price fluctuations are irrelevant to these 'locked-in' current expenditures. As a result, all the government's spending cuts targeted capital expenditures. Thus, while benefitting from cheap oil prices, Jordanians were paying higher taxes while enjoying fewer government services (Moore, 2017). As will be discussed in the subsection after the next, on the crisis in Jordan since 2018, this proved politically untenable.

Oil prices and foreign policy autonomy

Cheap oil may have had a bigger impact on Jordanian foreign policy. Decreased aid following the oil price drop from 2014 allowed Jordan to be somewhat less beholden to its benefactors. Jordan's resistance to Saudi pressures to intervene in Syria (Stratfor, 2012) best illustrates this autonomy. Jordan has maintained rhetorical support for GCC and Arab League resolutions on the Syrian uprising but did not back direct foreign intervention. Jordan's resistance to intervention reflects the Kingdom's policy of regional neutrality since it refrained from participating in the Gulf War (1990–91).

Jordan also resisted GCC pressures to boycott Qatar. In June 2017, Jordan recalled its ambassador to Qatar, siding with the regional axis of the Kingdom of Saudi Arabia, which imposed a land, sea, and air blockade of the gas-rich state. However, in July 2019, Jordan re-established diplomatic ties by naming an ambassador to Qatar, despite pressure from the Kingdom of Saudi Arabia and the UAE (Younes, 2019). Amman was also able to resist pressure from the Kingdom of Saudi Arabia in Yemen, withdrawing support for the Saudi-led coalition forces (Kilani, 2019) and otherwise playing Riyadh off against Doha, also a source of external funding for Jordan (Younes, 2019).

Donors such as the Kingdom of Saudi Arabia and the USA also tried to pressure Jordan to join their Israel–Palestine peace plan, the 'Deal of the Century', which Amman resisted. Jordan made no concessions demanded by the USA and its allies that would be strongly unacceptable to most Jordanians, and many others in the Arab world. Amman instead aligned its Middle East policy with underlying changes in the region. Safeguarding Palestinian rights, Jordan's King Abdullah has said there can be no peace without Jerusalem as Palestine's capital (*Haaretz*, 2018). Such stances affirm Jordan's position on Palestine. For Amman to waver on this would be politically impossible, but Jordan's position has displeased some, regionally and internationally, with blandishments offered for Amman to budge.

However, lower oil prices by no means meant that Jordan escaped from GCC influence. The Kingdom of Saudi Arabia particularly is the largest source of incoming Jordanian personal remittances, and its importance is increasing. In 2013–15, remittances from Jordanians in the Kingdom of Saudi Arabia reached an average of over 38 per cent of remittance flows into Jordan, from only about 25 per cent in 2011–12. Given that a pro-Iraq stance in the 1990–91 Gulf conflict caused mass expulsion of Jordanians from Kuwait, Jordan worries that failure to kowtow to the Kingdom of Saudi Arabia could lead to a similar outcome, which would mean Jordan losing its main remittance source (De Bel-Air, 2016; Jordan Strategy Forum, 2018). Further, following recent instability in Jordan, Amman turned to the Kingdom of Saudi Arabia for funding, meaning reduction in Saudi influence over Jordan may have been only temporary, as further discussed in the next subsection.

The crisis in Jordan since 2018

The crisis in Jordan since 2018 exemplifies how the problems of Jordanian budgets and reliance on regional powers, discussed in the previous two subsections, are intertwined. The state's attempts to break away from the influence of regional powers led to it adopting the IMF's EFF programme for financial autonomy, which ended up causing turmoil. Among significant protests, demonstrators in early 2018 unusually called on Prime Minister Hani Mulqi to step down, which he eventually did, with reformist Omar Razzaz succeeding. Under him, in early 2019, parliament passed the EFF income tax provision. In October 2018, not long after the dismissal of Mulqi's government, rallies demanded comprehensive reform. Among demonstrators' slogans was 'no to the hegemony of the Royal Court', criticism previously taboo in the country (Harel, 2018). Further, recent protests included Jordanian tribal groups traditionally loyal to the ruling Hashimite royal family. Now, joining opposition ranks, tribal figures demanded restitution of funds from privatised public institutions and called for action against corrupt figures, some Hashimite linked. There has also been lower respect for state symbols, accompanied by increases in the intensity and scale of confrontations with security forces, news about which is heavily censored. The king's response has been appeasement, which further strains state finances.

In order to stabilise the political situation in the wake of this unrest, in 2018 Jordan accepted financial support from the Kingdom of Saudi Arabia, the UAE, and Kuwait, as described above. This exemplifies how Jordan is on the horns of a dilemma. Amman cannot do without Gulf money, which has historically contributed to Jordan's stability, and attempts to raise domestic income to reduce reliance on regional powers failed, as Jordan was unable to cut huge current expenditure and justify the necessity to raise taxes. At the same time, Jordan cannot blindly accept outside financial support, as political strings attached could undermine the state.

Aid uncertainty after the oil price fall contributed to protests (Ayesh, 2018). However, this dilemma existed pre-2014; as explained above, Jordan's aid rentier position caused governance erosion, with widespread mismanagement and corruption. The decline in oil prices reduced state room for manoeuvre between the two horns of the dilemma, exacerbating pre-existing problems, but oil price decline is not responsible for Jordan's current predicament. The current Jordanian crisis is more related to decades of aid rentierism than to the aid decrease driven by oil price changes.

Conclusion

Despite widespread predictions that oil price decreases would benefit the Jordanian economy, including through the state budget, we find little evidence that this has been the case. The savings from cheap oil pale in significance compared to the cuts in aid from the GCC, spurred by the GCC's own decline in oil revenues. Part of the reason for this is that GCC aid, unlike energy savings, is much more targeted to politically important communities; the benefits of cheap oil are dispersed, whereas the costs are predominantly borne by the groups on which the Jordanian regime relies to ensure its stability, many of which are in the employ of the state. Predictions of Jordan's response to the change in oil prices have simply considered Jordan as an oil-importing economy and have not considered the further complexities of Jordan's relationship with oil, and more importantly, oil producers.

As we have explored, the change in oil prices had a direct effect on Jordan's energy bill and aid receipts. We find that there was little effect on Jordan's income from remittances, exports, or tourism. However, it is necessary to go beyond these initial changes in order to see the full impact of the change in oil prices on Jordan. It was anticipated that a fall in aid could help Jordan to raise its own revenues and to become more accountable to its population (CBJ, 2015; IMF, 2015). In doing so, it was hoped that Jordan would reduce its level of public expenditure, particularly by tackling its bloated bureaucracy. However, Jordan was not able to do this; it was not politically feasible for Jordan to backtrack on decades of public expenditure which had been funded by aid receipts.

However, the reduction in aid did help Jordan to somewhat be less at the whim of regional benefactors. Even so, this development was short-lived, as Jordan turned to the Saudis for support in the wake of the current crisis in Jordan. Lower aid from the GCC did not mean that Amman was any less reliant on Riyadh when Jordanians were faced with difficulties.

Despite this connection between aid, oil prices, and the current crisis in Jordan, the crisis cannot be solely attributed to the oil price fall. If oil prices had stayed stable, it is possible that Jordanian protests would not have broken out in the summer of 2018. However, it is likely that the feeling of malaise among the Jordanian public would have erupted at some point. The oil price decline was more a spark than an underlying cause of this unrest. Cheap oil, and the subsequent fall in aid, served to throw the pre-existing problems in the Jordanian political economy into sharp relief.

The stability of Jordan's economy thus continues to rely on realisation of the multilateral and bilateral commitments made to support it, with all the issues this entails in raising the country's already high debt or in the political price to be paid for foreign grants, as well as the possible impact of fluctuating oil prices on outside financial support. It goes without saying that this is a fragile, if not to say unsustainable, situation. What is sometimes ignored, however, is that the heightened expectations of Jordanian consumers add difficulty to the problem, as people become accustomed to demanding more without the ability to pay for it.

As seen above, before the latest round of reforms, people were frustrated with government efforts to help create private-sector jobs and dampen the cost of living. The poor economic conditions and a stymied political process combine to produce pessimism. The demonstrations protesting the country's situation reflect growing antagonism. Jordan's dire economic situation means that popular concerns among Jordanians focus on the economy, and it remains to be seen how these latest reforms will make a big difference. In such a situation, questions of greater or lesser aid and other financial inflows based indirectly or otherwise on levels of oil prices become a less important focus than the rentier mentality that is at the heart of the system. Then again, it can be that it is the very aid – much of it from oil producers – that somehow or other promotes rentierism.

Notes

1   The authors are greatly indebted to Dr Steve Monroe, Postdoctoral Research Fellow at the Harvard Kennedy School's Middle East Initiative, for substantive assistance with this chapter. Thanks also to Will Crass and to Abdullah Shah for their research support, as well as to Dr Oroub Elabed for valuable suggestions concerning source material. However, neither they nor anyone else bears responsibility for any of the chapter's errors or omissions, which are entirely the authors' fault.

References

Abu Asab, N. (2017), 'Asymmetric oil price shocks and economic activity in developing oil-importing economies: The case of Jordan', *International Journal of Economics and Financial Issues*, 7:2, 118–24.

Ali Al-Zeaud, H. (2014), 'An investigation of Granger causality between oil-price, inflation and economic growth in Jordan', *Global Journal of Management and Business Research*, 14:6, 33–41.

ANSA Med (2018), 'Jordan to cut bread subsidies, risk of protests', 3 January. https://bit.ly/3vCqJ8X (accessed 6 November 2019).

Arndt, C., S. Jones, and F. Tarp (2015), 'Assessing foreign aid's long-run contribution to growth and development', *World Development*, 69, 6–18.

Ayesh, M. (2018), 'Why Jordan needs Saudi Arabia', Middle East Eye, 12 June. www.middleeasteye.net/opinion/why-jordan-needs-saudi-arabia (accessed 15 October 2019).

Baylouny, A. M. (2008), 'Militarizing welfare: Neo-liberalism and Jordanian policy', *Middle East Journal*, 62:2, 277–303.

Beck, M., and S. Hüser (2015), 'Jordan and the "Arab Spring": No challenge, no change?', *Middle East Critique*, 24:1, 83–97.

Berument, M. H., N. B. Ceylan, and N. Dogan (2010), 'The impact of oil price shocks on the economic growth of selected MENA countries', *The Energy Journal*, 30, 149–76.

Burnside, C., and D. Dollar (2004), 'Aid, policies, and growth: Reply', *American Economic Review*, 94:3, 781–4.

CBJ (2015), 'The effect of the decline of oil prices on the Jordanian economy', February. https://bit.ly/3vedKv4 (accessed 10 November 2019).

CBJ (2019), 'Statistical database'. www.cbj.gov.jo (accessed 5 November 2019).

Chaudhry, K. A. (1997), *The price of wealth: Economies and institutions in the Middle East*. New York: Cornell University Press.

Connable, B. (2015), *From negativity to positive stability: How the Syrian refugee crisis can improve Jordan's outlook*. Santa Monica: RAND Corporation.

De Bel-Air, F. (2016), 'Migration profile: Jordan', European University Institute: Migration Policy Centre, 6.

El-Said, H., and K. Becker (eds) (2001), *Management and international business issues in Jordan: The potential of an Arab Singapore?* Binghamton: Haworth Press.

Fanek, F. (2017), 'Public sector still the leader', *The Jordan Times*, 13 August. http://jordantimes.com/opinion/fahed-fanek/public-sector-still-leader (accessed 10 October 2019).

Furlow, R., and S. Borgognone (2018), 'Gulf designs on Jordan's foreign policy', Carnegie Endowment: Sada, 17 July. https://carnegieendowment.org/sada/76854 (accessed 14 July 2019).

Haaretz (2018), 'Jordanian king says no peace without Jerusalem as capital of Palestine, days after meeting Netanyahu', 21 June. https://bit.ly/3bucr33 (accessed 15 November 2019).

Harel, Z. (2018), 'Growing calls in Jordan to enact political reforms limit king's powers', The Middle East Media Research Institute, Inquiry and Analysis Series, 1427. https://bit.ly/3cmrbQI (accessed 10 October 2019).

IMF (2015), 'Jordan: Seventh and final review under the stand-by arrangement and proposal for post-program monitoring', Country Report, 225. https://bit.ly/3v6gc6N (accessed 10 November 2019).

IMF (2016), 'IMF executive board approves US$723 million extended arrangement under the Extended Fund Facility for Jordan', 25 August. https://bit.ly/38mpQZ9 (accessed 16 July 2019).

Jordanian Department of Statistics (2019), 'Statistical yearbook 2019'. http://dosweb.dos.gov.jo (accessed 5 November 2019).

Jordan Strategy Forum (2018), 'Jordanian expatriates in the Gulf'. https://bit.ly/3t0B1yk (accessed 3 October 2019).

Kilani, A. (2019), 'Has Jordan distanced itself from Saudi Arabia?', Lobe Log, 6 September. https://lobelog.com/has-jordan-distanced-itself-from-saudi-arabia/ (accessed 10 November 2019).

Knowles, W. M. (2004), *Jordan since 1989: A study in political economy*. London: I.B. Tauris.

Ma'ayeh, S. (2019), 'Saudi transfers another 334 million to Jordan in latest assistance package', *The National*, 20 March. https://bit.ly/3t9Emeu (accessed 1 September 2019).

Maghyereh, A. I., B. Awartani, and O. D. Sweidan (2019), 'Oil price uncertainty and real output growth: New evidence from selected oil-importing countries in the Middle East', *Empirical Economics*, 56:5, 1601–21.

Malik, A. (2017), 'Rethinking the rentier curse', in G. Luciani (ed.), *Combining economic and political development: The experience of MENA*. Boston: Brill, pp. 41–57.

Ministry of Tourism and Antiquities (2019), 'Statistics'. www.tourism.jo/Contents/Statistics.aspx (accessed 5 November 2019).

Moawad Ahmed, S. (2016), 'The impact of oil prices on the economic growth and development in the MENA countries', Middle East Studies Association, 89073.

Mohaddes, K., and M. Raissi (2013), 'Oil prices, external income, and growth: Lessons from Jordan', *Review of Middle East Economics and Finance*, 9:2, 99–131.

Momani, A. (2018), 'Measurements of Jordanian abroad and non-Jordanians in Jordan', OECD International Forum on Migration Statistics, 15–16 January. www.oecd.org/migration/forum-migration-statistics/2.B-3-Ahmed-A-Momani.pdf (accessed 6 November 2019).

Moore, P. W. (2017), 'The fiscal politics of rebellious grievance in the Arab world: Egypt and Jordan in comparative perspective', *The Journal of Development Studies*, 53:10, 1634–49.

Moss, T., G. Petersson, and N. van de Walle (2006), 'An aid-institution paradox? A review essay on aid dependency and state building in sub-Saharan Africa', Centre for Global Development, Working Paper, 74.

Observatory of Economic Complexity (2019), 'Jordan', Observatory of Economic Complexity. https://oec.world/en/profile/country/jor/#Imports (accessed 10 November 2019).

Peters, A. M. (2009), 'Special relationships, dollars, and development: U.S. aid and institutions in Egypt, Jordan, South Korea, and Taiwan'. PhD dissertation, University of Virginia.

Rajan, R. G., and A. Subramanian (2007), 'Does aid affect governance?', *American Economic Review*, 97, 322–7.

Riedel, B. (2019), 'Jordan's King Abdullah is facing new risks from his own friends', Brookings, 14 June. https://brook.gs/2O9ja9X (accessed 15 October 2019).

Robins, P. (2019), *A history of Jordan*. Cambridge: Cambridge University Press.

Ryan, C. R. (2002), *Jordan in transition: From Hussein to Abdullah*. Boulder: Lynne Rienner.

Schenker, D. (2016), 'Promised Saudi support to Jordan: At what price?', The Washington Institute for Near East Policy, 9 May. https://bit.ly/3tb3kuh (accessed 4 September 2019).

Sowell, K. H. (2017), 'Jordan's reform imperative: Fiscal challenges, public policy and the constraints of political economy', Utica Risk Services.

Stratfor (2012), 'The costs of Jordan's energy insecurity', Stratfor Worldview, 22 May. https://worldview.stratfor.com/article/costs-jordans-energy-insecurity (accessed 14 July 2019).

Tayseer, M. (2018), 'Saudi fund for development reschedules Jordan's debt', Bloomberg, 16 December. https://bloom.bg/3nGYljg (accessed 6 November 2019).

The New Arab (2018), 'Riots break out in Jordan over bread price hikes', 5 February. https://bit.ly/3rB1ylB (accessed 13 October 2019).

World Bank (2019), 'Current account balance (% of GDP) – Jordan', World Bank Data. https://data.worldbank.org/indicator/BN.CAB.XOKA.GD.ZS?locations=JO (accessed 25 November 2019).

Yom, S. (2017), 'The art of the hedge: Jordan between Washington, Riyadh, and Doha', Middle East Eye, 19 June. https://bit.ly/3eqJAP5 (accessed 15 October 2019).

Yom, S., and M. H. Al-Momani (2008), 'The international dimensions of authoritarian regime stability: Jordan in the post-cold war era', *Arab Studies Quarterly*, 30:1, 39–60.

Younes, A. (2019), 'Inching away from Saudi-UAE axis, Jordan restores ties with Qatar', Al Jazeera, 9 July. https://bit.ly/2OFAfrE (accessed 10 November 2019).

10

Lower oil prices since 2014: Good news or bad news for the Lebanese economy?

Mohamad B. Karaki

Introduction

Lebanon is well known to be sensitive to imported commodity prices. While the manufacturing sector is relatively small (around 8 per cent of gross domestic product (GDP)), oil and energy-related imports constitute a sizeable share of total imports. In 2018, the value of fuel minerals and oil was equivalent to 20 per cent of total imports. During that year, 86 per cent of imported mineral fuel and oil were allocated to Electricité du Liban (EDL), a government-owned enterprise that continues to be the main electricity producer in the country. These imports represent more than 16 per cent of the general government expenditures.

Previous studies have found that a reduction in oil prices tends to stimulate oil-importing economies and negatively affects oil-producing countries. For instance, in Saudi Arabia, the largest oil producer in the Middle East, the government registered huge budget deficits in 2015 due to the losses in oil revenues. In addition, the recent oil price decline has reduced Saudi Arabia's foreign exchange reserves substantially. Kilian (2017) found that if the drops in net foreign assets continued at the same rate as the one that occurred during the 2015–16 period, Saudi Arabia would have entirely lost its foreign exchange reserves by 2020. The USA, however, has benefitted from this recent oil price decline episode. Baumeister and Kilian (2016) found that the decrease in oil prices led to an increase in US real GDP by 0.7 per cent. They attribute this increase in GDP to an increase in real consumption and a reduction in the petroleum trade deficit.

In this chapter, I study the effect of lower oil prices on the Lebanese economy. Although a reduction in oil prices tends to bring good news for oil-importing economies, understanding the effect of lower oil prices on the Lebanese economies is not straightforward. Given that Lebanon heavily relies on remittances coming from Gulf Cooperation Council (GCC) countries, little is known about how a reduction in oil prices will affect the Lebanese economy. Furthermore, there is a lot of ambiguity on how oil price declines will affect the Lebanese economy in the future, when Lebanon becomes an oil-producing country.

The aim of this study is to shed light on the impact of the 2014–16 oil price decline on the Lebanese economy. To proceed with this investigation, I examine data from Banque du Liban (BDL), government agencies, international organisations, and other reputable institutions. On the one hand, I find evidence that the reduction in oil prices during that period had a positive effect on the Lebanese economy in different aspects. Despite the various negative shocks that hit the Lebanese economy during that period, manufacturing output increased, consumption spending increased, remittances inflows increased, and the trade deficit and the budget deficit dropped. On the other hand, the decrease in oil prices during that period had a negative effect on the oil and gas sector. In its current embryonic stage, this sector suffered from a decrease in the number of bidding companies. Moreover, I shed light on the Dutch disease and the different possible measures that could be implemented to avoid it in the future when Lebanon becomes an oil-exporting country.

The remainder of this chapter is organised as follows. First, I review and evaluate the relevance of the different transmission channels through which oil prices affect the Lebanese economy. Second, I discuss how lower oil prices contributed to the 2015 deflation. Third, I present an overview of the Lebanese oil and gas sector and discuss the measures that need to be taken to avoid the Dutch disease. Finally, I close this chapter with concluding remarks.

How do oil prices affect the Lebanese economy?

Since the mid-1970s stagflation, oil prices have been viewed as a major source of economic fluctuations. Early papers have found that an unexpected increase in the real price of oil can cause large economic downturns for an oil-importing economy (see e.g. Hamilton, 1983; Mork, 1989). Furthermore, by the early 2000s, there was a consensus among economists that higher oil prices have recessionary effects on oil-importing economies whereas unexpected declines in the real price of oil have little impact on economic activity (see e.g. Mork, 1989; Hamilton, 1996, 2003; Davis and Haltiwanger, 2001). This view has been challenged by Kilian and Vigfusson (2011), who found the effect of oil price increases and oil price decreases to be symmetric. This symmetry implies that when oil prices increase, for instance, by 10 per cent, the economy will contract by an amount that is equivalent to the stimulating effect triggered by an oil price decrease of 10 per cent. Along with the change in the understanding that oil price declines have on the economy, there has been an important development in the understanding of how oil prices affect the economy. Specifically, while before the 2000s most economists believed that oil prices primarily affect the economy through a change in the cost of production, the modern view nowadays is that oil prices largely operate by affecting the demand side of the economy. This chapter begins by discussing the different channels through which lower oil prices affect the Lebanese economy.

The cost channel

Standard macroeconomic textbooks suggest that for an oil-importing country, a sudden decline in the real price of oil reduces the cost of production. Specifically, industries that rely on oil-related products to produce goods and services such as transportation and rubber and plastics will now produce goods and services at a lower cost.

As a result, this overall reduction in the economy's cost of production leads to an increase in the economy's level of output and a reduction in inflation. Table 10.1 reveals that, compared to 2012 and 2013, the manufacturing share of GDP increased in 2014 and 2015. For instance, the share of manufacturing of food products in GDP increased by 0.19 per cent in 2014 and 0.11 per cent in 2015. Moreover, Table 10.1 shows that, relative to the 2012–13 period, the share of chemicals and rubber and plastics in GDP in Lebanon also increased in the 2014–15 period. Furthermore, the share of transport in GDP increased by 0.18 per cent in 2014 and 0.17 per cent in 2015. These findings reveal that industries that rely on oil as an input of production expanded following the onset of the oil price decline period in 2014.

Demand side channels

Most economists now agree that oil prices primarily affect the economy through a change in the demand side of the economy (see Edelstein and Kilian, 2009; Kilian, 2014; Herrera, 2018; Herrera *et al.*, 2019). Moreover, nowadays, policymakers also acknowledge that oil prices mainly affect the economy through demand side channels. For instance, Ben Bernanke (2006) underscored that higher oil prices have a negative effect on consumption expenditure. Also, Yellen (2016) stressed that the 2014 oil price decline allowed households in the USA to increase their consumption spending primarily through an increase in their purchasing power.

The most notable demand channel that is known to be the main transmission channel through which oil prices affect the economy is the discretionary

Table 10.1 Contribution of manufacturing and transport industries as a percentage to GDP in Lebanon.

Industry	2012	2013	2014	2015	2016	2017
Manufacturing of food products	1.43	1.46	1.65	1.76	2.10	2.05
Beverages and tobacco manufacturing	0.81	0.82	0.87	0.88	0.96	0.96
Textile and leather manufacturing	0.37	0.35	0.39	0.47	0.46	0.48
Wood and paper manufacturing; printing	0.61	0.62	0.61	0.61	0.60	0.57
Chemicals, rubber, and plastics manufacturing	0.63	0.68	0.72	0.82	0.81	0.84
Non-metallic mineral manufacturing	0.87	0.96	0.84	0.90	0.97	0.89
Metal products, machinery, and equipment	2.30	2.35	2.21	2.29	2.13	2.10
Other manufacturing	0.35	0.43	0.46	0.51	0.59	0.67
Transport	2.95	3.07	3.25	3.42	3.30	3.16

Source: Central Administration of Statistics (2018).

income (leftover income for consumption) channel. This channel implies that as oil prices drop, petrol prices tend to decrease. As a result, the discretionary income will increase, and that leads to an increase in consumption spending (see Edelstein and Kilian, 2009; Baumeister *et al.*, 2018).

To evaluate whether this channel was an important transmission channel for the 2014–16 oil price decline, I first track the domestic petrol price in the Lebanese market during that period. Table 10.2 shows that petrol prices fell sharply after 2014. In fact, between 2014 and 2015, petrol prices fell by 40.97 per cent. This decline in petrol prices coincided with an increase in consumption spending during that period. Table 10.3 reveals that the consumption levels in the 2014–16 period were larger than the consumption level during the 2011–13 period. For instance, compared to 2013, consumer spending increased by more than 7 per cent in 2014. Overall, the data implies that the decrease in oil prices lowered petrol prices in the Lebanese market and possibly increased consumption spending, which helped to boost economic activity.

Table 10.2 Petrol prices in Lebanon (98 octane) (in LBP).

Year	June 2014	December 2015
Price per litre	1,745	1,030

Source: IPT (2019).

Table 10.3 Household consumption expenditure (in LBP billion) in Lebanon.

Year	2011	2012	2013	2014	2015	2016	2017
Consumption expenditure	53,384	58,723	60,403	65,214	64,675	68,232	73,963

Source: Central Administration of Statistics (2018).

Table 10.4 Private savings as a percentage of GDP in Lebanon.

Year	2013	2014	2015	2016
Savings	1,906	−1,617	2,810	994

Source: Central Administration of Statistics (2018).

Another demand channel is the uncertainty channel. This channel suggests that when oil prices change unexpectedly, whether upwards or downwards, households tend to postpone their spending on durable goods items and increase their precautionary saving.

The data on saving reported in Table 10.4 reveals that in 2014, the private saving level decreased very sharply. Then in 2015, private saving increased importantly and then it fell back by more than 64 per cent in 2016. These very sharp ups and downs in private saving cannot be solely attributed to the 2014–16 oil price decline because the Lebanese economy was hit by a variety of negative demand shocks during that period. These different demand shocks, which directly affect economic activity and household saving, can be summarised below:

Terrorism

Compared to the 2012–14 period, Lebanon experienced a surge in terrorist attacks from the start of 2014. The average terrorism index jumped from 5.09 in the 2011–13 period to 6.03 in the 2014–16 period (see Table 10.5).

Table 10.5 Terrorism index.

Year	2010	2011	2012	2013	2014	2015	2016	2017
Terrorism index	4.88	4.58	4.48	6.21	6.38	6.07	5.64	5.15

Source: Institute for Economics and Peace (2019).

In general, the increase in terrorism negatively affected business and consumer confidence and reduced economic activity (Sandler and Enders, 2008).

Garbage crisis

In July 2015, a major garbage crisis erupted in Lebanon. The managing company for collecting waste back then, Sukleen, suspended its collection services following the closure of the Naʿma landfill. As a result, a series of protests occurred in 2015 and 2016, which negatively affected economic activity (Iskandarani, 2019).

Presidential vacancy

Lebanon had no president from May 2014 until October 2016. The failure to elect a head of state for more than two years increased economic uncertainty, given that presidential vacuums in the past had often been associated with security threats that claimed the lives of many civilians. It also caused more political divisions among the population, which often translated into internal conflicts between the supporters of opposing political groups (Mourad, 2014).

The increase in the number of Syrian refugees

In April 2013, the number of registered Syrian refugees was only 298,165; however, this number increased beyond 1,000,000 starting in May 2014 (UNHCR, 2019). Many news analysts and policymakers have claimed that the influx of Syrian refugees has not only increased Lebanese youth unemployment importantly but also reduced the wage rate in the agriculture and construction sectors (see e.g. LCPS, 2016).[1]

Overall, it would be hard to pin down the effect of the 2014–16 oil price decline on saving by examining the data. Different demand shocks occurred during that period, and these shocks triggered random changes in individuals' willingness to readjust their saving level. Thus, evaluating the contribution that oil prices had on saving during the 2014–16 period would require estimating a structural vector autoregressive model and conducting a historical decomposition exercise. We leave that for future research.

Table 10.6 Volume of car sales to Lebanon.

Year	2013	2014	2015	2016	2017	2018
Volume	37,500	40,300	40,700	37,800	39,881	35,301

Source: CEIC (2019).

The operating cost channel

One other demand side channel is the operating cost channel. This channel states that when oil prices decline, petrol prices fall; therefore, there is a drop in the cost of operating goods that rely on energy, leading to an increase in consumption of these goods. Specifically, this channel implies that lower oil prices tend to increase consumption spending on cars, especially large sport utility vehicles (SUVs). Table 10.6 reveals that, compared to other years, the total market volume for car sales in Lebanon reached its highest and second highest level in 2014 and 2015, respectively. This data reveals that Lebanese citizens were motivated to buy more vehicles during these years, possibly due to low petrol prices.

The budget channel

In this section, I study the impact of the 2014–15 oil price decline on the Lebanese budget deficit. This channel is crucial given that Lebanon is now facing very serious fiscal challenges. In fact, in 2018, Lebanon ranked third worldwide in its debt-to-GDP ratio at 151 per cent (see Figure 10.1).

Before evaluating how lower oil prices can affect the Lebanese budget, some historical contextualisation is needed to understand why Lebanon has a very high level of debt to GDP. Since the start of the civil war in 1975, Lebanon has been dealing with important fiscal challenges. Dibeh (2002) provides a comprehensive discussion on the political economy behind the large budget deficits and the currency crisis that erupted during that period. In fact, since 1978, the Lebanese government has been incurring a large amount of deficit to maintain the operations of government institutions, leading to a surge in debt accumulation. In a state of war, Dibeh (2002) claims that the Lebanese government started to service its debt by increasingly relying on a Banque du Liban balance sheet. These measures fuelled inflation and led to a severe collapse of the Lebanese pound (LBP) towards the end of the civil war. From the early 1990s until the 2000s, and while the government was implementing the reconstruction plan, the budget deficit widened progressively, as the government was relying on domestic credit at a high

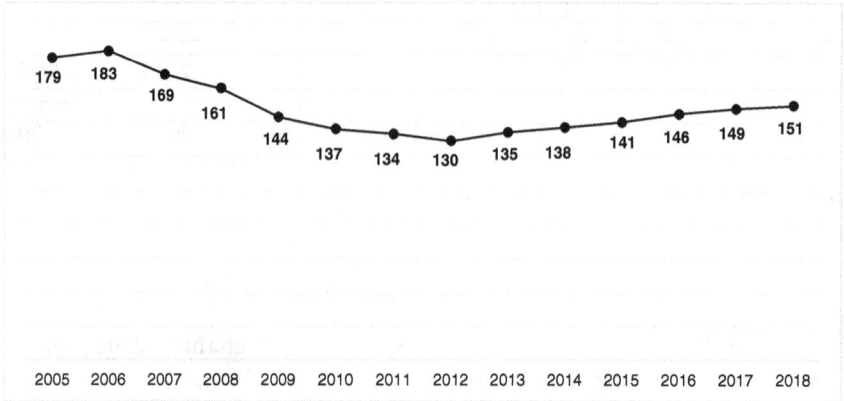

Figure 10.1 Debt-to-GDP ratio (per cent) in Lebanon. Source: Republic of Lebanon Ministry of Finance Public Debt Directorate (2019).

interest rate (see Neaime, 2004). For instance, the three-month treasury bill rate was 34.18 per cent in September 1992 and remained above 10 per cent until November 2002. The accumulation of deficits led to a debt-to-GDP ratio of 183 per cent in December 2006. The reconstruction after the war along with the subsidised interest rate on housing loans by Banque du Liban helped to stimulate GDP. As a result, the debt-to-GDP ratio dropped to 130 per cent in 2012. Yet in recent years, several factors have worsened the fiscal health of the Lebanese economy. First, the influx of Syrian refugees slowed down GDP growth in Lebanon (LCPS, 2016). Second, the emergence of the garbage crisis in 2015 placed additional strain on economic activity. Third, the major salary adjustment of government employees in July 2017 contributed to the widening of the budget deficit.

There is a clear consensus among researchers and news analysts that the inefficiency of EDL, the government-owned enterprise that generates and distributes electricity in Lebanon, is one main structural reason behind Lebanon's budget deficit. On average, EDL incurs annual losses that range between USD 1.5 and 2 billion. The main reason behind these losses is the fact that the cost of generating electricity is higher than the tariffs collected. Obeid (2018) states that the average tariff per kilowatt-hour (kWh) consumed is 9.5 cents, which has remained fixed since 1996 (McDowall, 2019), whereas the cost to generate each kWh ranges between USD 0.17 and 0.23 depending on oil prices. In 2018, government expenditure on EDL constituted around 16 per cent. Many news analysts have repeatedly claimed that the large subsidies that the government gives to EDL are still in place due to corruption and political interest (Abdelnour, 2003; Al-Jack, 2018).

Table 10.7 Electricité du Liban (EDL) expenditure and production
(in LBP million).

Year	EDL budget expenditure (in LBP millions)	growth rate (%)	Electricity production (millions of kWh)	growth rate (%)
2014	3,157,034	n/a	12,522	n/a
2015	1,710,875	−46	12,520	0
2016	1,397,274	−18	13,130	5
2017	2,001,625	43	15,030	14
2018	2,647,265	32	15,245	1

Source: Banque du Liban (2019) and Republic of Lebanon Ministry of Finance (2019).

Did the decrease in the price of oil during the 2014–16 period help in lowering the Lebanese government's spending on EDL? Table 10.7 reveals that, following the 2014 oil price decline, the government expenditure on EDL fell by 61 per cent in 2015 and 20 per cent in 2016. Given that electricity production increased by 4.86 per cent between 2014 and 2016 (see Table 10.7), these figures imply that oil prices helped to reduce the Lebanese government budget deficit. In fact, as indicated on EDL's website, more than 98 per cent of the government transfers to EDL are direct expenditures on petroleum products.

The trade channel

Fluctuations in oil prices have a direct effect on Lebanon's trade balance because Lebanon mainly relies on energy imports to satisfy local demand. In fact, data from the US Energy Information Administration (2020) reveals that in 2017, Lebanon ranked 59 out of 214 countries in total consumption of petroleum products with 153,000 barrels per day. When oil prices increase, the value of imported fuel oil also increases, which results in an increase in the trade deficit. Persistent trade deficits can pause an important threat on Lebanon's fixed exchange rate with the US dollar that has been in place since 1997. Did the oil price decline that started in 2014 reduce Lebanon's trade deficit?

Before evaluating the impact of the 2014 oil price decline on the trade balance, let us first elaborate on the link between trade deficits and the LBP value using the recent trade figures for the Lebanese economy. In 2018, Lebanon's trade deficit was around USD 17 billion, which is equivalent to 30 per cent of GDP. This deficit in the trade balance led to a current account deficit of around USD 12 billion. Moreover, in 2018, the deficit in the current account balance was larger than the financial account and capital account, resulting in a balance of payment deficit. In 2018, the balance of payment recorded an unusual deficit of USD 4.8 billion. Compared to 2017 (when the value of the balance of payment deficit was USD 156 million), the balance of payment deficit increased by 2,976 per cent. Given that Lebanon has a fixed exchange rate regime, a balance of payment deficit is only possible if commercial banks and the central bank start drawing out of their foreign reserves. In 2018, the USD 4.8 billion balance of payment deficit was due to a USD 2.3 billion decline in BDL's net foreign assets and a USD 2.5 billion drop in commercial banks' net foreign assets (Barakat *et al.*, 2018). These large reductions in net foreign assets are alarming and can pose an important threat to the stability of the Lebanese pound. In fact, large drops in an economy's net foreign assets are known to be a major reason behind speculative attacks.

The decrease in oil prices that started in June 2014 had a positive effect on Lebanon's trade balance. According to the Republic of Lebanon Ministry of Finance's (2014) International Trade Monthly Bulletin, the value of imported mineral fuel and oil declined by 5 per cent despite an increase in volume by 2.1 per cent, and the trade deficit in 2014 was 0.6 per cent lower than the trade deficit recorded in 2013. In 2015, the trade deficit shrank by 12 per cent compared to the deficit registered in the previous year (Republic of Lebanon Ministry of Finance, 2015). This trade deficit reduction was mainly due to the 30 per cent reduction in mineral fuel and oil imports. In fact, lower oil prices were the primary reason behind the reduction in the value of imported mineral fuel and oil, especially given the fact that the volume of such imports increased by 6 per cent compared to 2014. Overall, the data clearly reveal that lower oil prices helped to reduce the trade deficit and show that lower oil prices positively affect economic activity through a trade channel.

The remittances channel

Lebanon is well known to be highly dependent on remittance inflows. Figure 10.2 reveals that following 2002, the share of remittance inflows in GDP was larger than 10 per cent. Given that remittance inflows are crucial to maintaining the stability of the banking sector and the LBP value (Union of Arab Banks, 2016), and because some of these remittances come from the Lebanese diaspora that work in GCC countries, it is therefore important to evaluate how the 2014 oil price decline affected remittance inflows to Lebanon.

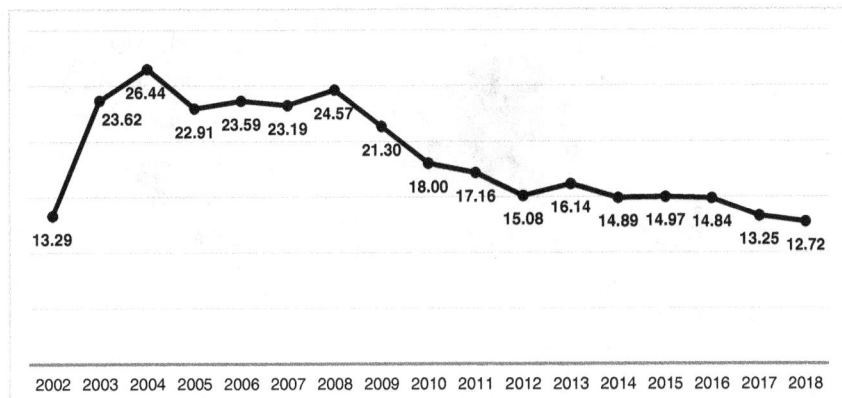

Figure 10.2 Contribution of remittance inflows to GDP (per cent) in Lebanon.
Source: World Bank (2019).

Table 10.8 Remittance inflows (in USD billion) to Lebanon.

Year	2013	2014	2015	2016	2017	2018
Inflows	6.600	6.253	6.619	6.930	6.484	6.397

Source: Banque du Liban (2019).

Before we proceed, it is essential first to discuss the relevance of remittances to the economic well-being of households, the banking sector, and the stability of the Lebanese pound. First, as remittance inflows from non-resident households to resident households increase, resident households will have an increase in their purchasing power. Consequently, resident households will be able to increase their consumption expenditure on goods and services. Second, remittance inflows are the backbone of the Lebanese banking sector because 27 per cent of remittances flowing to Lebanon go into bank deposits (Credit Libanais, 2017). These US dollar deposits have been the engine of the BDL financial engineering framework, which allowed commercial banks to gather a considerable amount of interest from BDL and provided BDL with an increase in foreign currency reserves, which is very much needed to preserve the LBP value (Chaya, 2019).

How did the 2014–16 oil price decrease affect remittance inflows to the Lebanese economy? Table 10.8 reveals that remittance inflows dropped by 5.26 per cent in 2014 and then increased by 5.85 per cent and 4.70 per cent, respectively. To better understand the pattern of remittance inflows

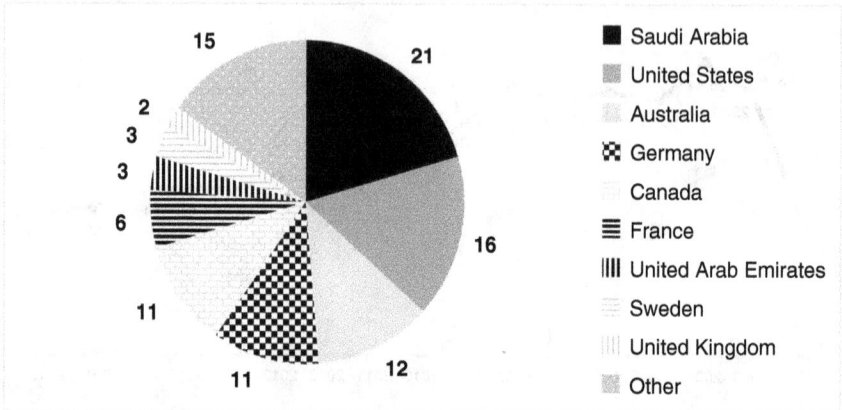

Figure 10.3 Percentage of remittance inflows to Lebanon in 2017. Note: 'Other' refers to a group of 204 countries each of which has a share of remittance inflows to Lebanon of less than 2 per cent. Source: World Bank (2017).

during that period, we need to first detect the top remitting countries to the Lebanese economy and see how these countries were affected by the oil price decline. Figure 10.3 reveals that 21 per cent of the remittance inflows to Lebanon come from Saudi Arabia, which makes Saudi Arabia the top remitting country to Lebanon. Given that Saudi Arabia is a major oil-exporting country that suffered from a sharp reduction in its oil revenues due to the oil price decline in 2014 (Kilian, 2017), remittances sent from Saudi Arabia to Lebanon decreased (Union of Arab Banks, 2016). While Saudi Arabia tends to be negatively affected by lower oil prices, several top remitting countries to Lebanon benefitted from the 2014–16 oil price decline period. For instance, the USA added jobs (Herrera *et al.*, 2017) and experienced an increase in real GDP (Baumeister and Kilian, 2016). Moreover, a statement on monetary policy published by the Reserve Bank of Australia states that lower oil prices have benefitted the Australian economy (Reserve Bank of Australia, 2015). In addition, Canada's central bank governor Stephen Poloz stated that the Canadian economy is now more resilient to lower oil prices given that the oil and gas sector only constitutes 3.5 per cent of Canada's total output (CBC, 2015). Furthermore, Bergmann (2019) showed that Germany's GDP growth increases following a reduction in oil prices.

Overall, the data that I collected on remittances from BDL reveals that, while inflow of remittances decreased in 2014, the level of remittances increased in 2015 and 2016, respectively. This increase can possibly stem from the benefit that lower oil prices bring to the USA, Australia, and Germany.

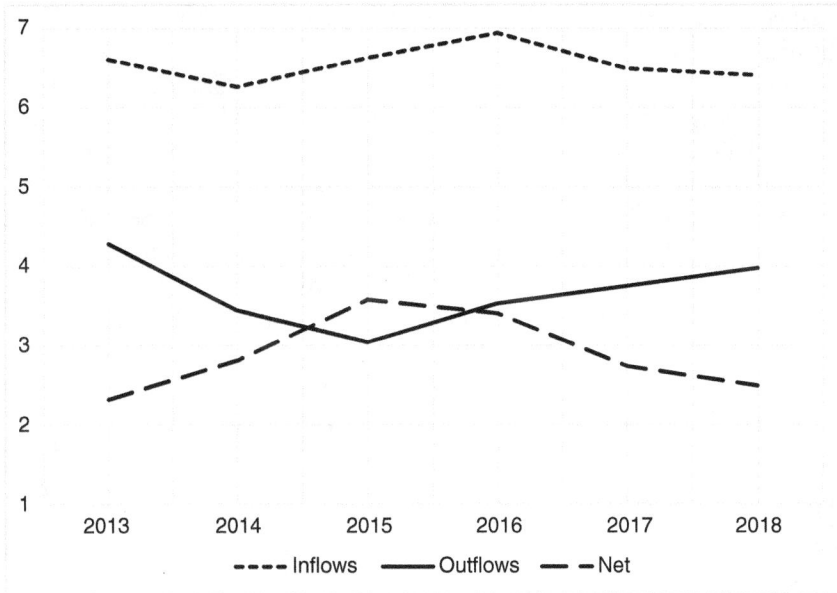

Figure 10.4 Inflowing, outflowing, and net remittances (in USD billion) for Lebanon. Source: Banque du Liban (2019).

While remittance inflows are important for the Lebanese economy, what is more important is the net inflow of remittances.[2] For instance, what is really relevant for the BDL is to have an increase in the net new dollar inflows to the economy. Figure 10.4 reveals that the net remittance inflows increased from 2013 to 2015 mainly due to a reduction in remittance outflows. This reduction in remittance outflows could be due to the reduction in GDP growth from 2.6 per cent in 2013 to 2 per cent and 0.8 per cent in 2014 and 2015, respectively. This slowdown in Lebanon's economic activity can possibly explain why foreigners living in Lebanon sent less money to their relatives outside Lebanon during that period.

The deflation of 2015–16

During the 2014–16 oil price decline period, the Lebanese economy experienced a notable deflation. According to the Central Administration of Statistics (2019), the inflation rate was −3.75 per cent and −0.82 per cent in 2015 and 2016, respectively. Was this deflation primarily caused by the decrease in oil prices or by the series of negative demand shocks that hit the Lebanese economy during that period? What are the repercussions of this deflation on the economy?

The term deflation is often linked to the 1929–33 Great Depression in the USA or to Japan's deflationary period between 1991 and 2010. These two examples of deflation were primarily caused by a reduction in aggregate demand leading to a sharp slowdown in economic activity. Deflation can put the economy in a vicious circle when the primary cause behind it is a reduction in aggregate demand. As spending decreases, deflation increases the real debt burden for firms and households. As a result, households reduce their consumption spending and firms decrease their investment projects, which further reduces economic activity.

Bordo *et al.* (2004) claim that deflation can be good or bad depending on the cause behind it. Specifically, they state that 'good' deflation occurs when the increase in the economy's ability to supply goods and services exceeds the reduction in the economy's aggregate demand. They refer to 'bad' deflation as the reduction in the overall price level in the economy arising from a reduction in aggregate demand that exceeds any increase in the economy's aggregate supply.

During the 2014–16 period, the Lebanese economy's aggregate supply increased following the reduction in oil prices. Moreover, as previously discussed, Lebanon faced a series of negative demand shocks during that period, such as the presidential vacuum, the increase in the number of Syrian refugees, the terrorist attacks, and the garbage crisis. The data on nominal GDP from the Central Administration of Statistics (2019) reveals that nominal GDP increased by 3.41 per cent between 2014 and 2015.[3] Thus, it is evident that the increase in aggregate supply caused by the decline in the price of oil overtook the series of negative aggregate demand shocks that occurred during that period. As a result, the 2015 deflation seems to have been a 'good' deflation for the Lebanese economy.

The oil and gas industry following the 2014 oil price decline and future adjustment policies

In this section, I will analyse the effect of oil prices on Lebanon's oil and gas sector. First, I will provide a brief historical overview of this embryonic sector. Second, I will discuss how the 2014–16 oil price decline and the four-year licensing delay from 2013 to 2017 have affected this sector. Finally, I will shed light on the future economic problems that oil price fluctuations would bring as Lebanon becomes an oil-exporting country.

In December 2009, a new hope emerged for Lebanon when Noble Energy discovered enormous oil and gas reserves in the Levant Basin region. Located in the Eastern Mediterranean basin, this region consists of the seabed off the coast of Lebanon, Syria, Cyprus, and Israel. In 2010, the reserves were estimated to be around 1.7 billion barrels of crude oil and at least 122 trillion cubic feet of natural gas (Energy Information Administration, 2013; Credit Libanais, 2015). Several important steps were taken by the Lebanese government to proceed with the oil and gas exploration. Among these were the passing of the Offshore Petroleum Resources Law in August 2010, the appointment of the Petroleum Administration in December 2012, and the launch of the bidding process by the Ministry of Water and Energy.

The fluctuations in oil prices had a direct effect on the development of the oil and gas explorations. After the start of the first qualification stage in 2013, more than fifty companies were eligible to place their bids (Azar, 2017). Among these companies, forty-six were highly interested in exploring Lebanon's seabed for potential oil and gas fields. Yet, political instability led to a substantial delay in initiating the bidding process. Specifically, the licensing process was delayed for four years largely due to two reasons. First, the government was unable to pass the decree related to dividing the oil and gas zone into ten blocks. Second, the government failed to reach a consensus on how the extracted resources would be shared between the companies and the government (*The Economist*, 2017). In fact, the inability of the government to make effective decisions was mainly due to the strong divide between politicians who supported/opposed the Syrian regime. This strong divide also led to sharp disagreements on how to address the influx of Syrian refugees to Lebanon, which eventually led to a political deadlock and a presidential vacancy from 2014 to 2016. In fact, it was only after General Michel Aoun became president in October 2016 that the serious discussions regarding oil and gas were back on the table. However, after this four-year delay, Lebanon found itself with very few bidders due to the sharp decrease in oil prices during the 2014–16 period (see Azar, 2017).

Recently, the oil and gas industry has seen some important development. In 2018, Lebanon managed to sign its first contract for offshore oil and gas exploration of two blocks (blocks 4 and 9) with three companies: France's Total, Italy's Eni, and Russia's Novatek. Moreover, Lebanon has taken active steps to pursue oil and gas exploration in the other blocks. For instance, the Lebanese government has been working on a cooperative agreement with Cyprus to pursue oil and gas extractions in common fields. Moreover, the minister of water and energy, who took office in January 2019, Nada Boustani, has also declared that the Lebanese government is working on a regional cooperative agreement with Egypt to reactivate the Arab Gas Pipeline (Lebanon Gas and Oil, 2019).

Despite the different steps that the government has taken to pursue oil and gas exploration, pundits are concerned that the high level of corruption, political instability, and the high debt-to-GDP ratio in the country could drag the Lebanese economy towards the pre-resource curse (LOGI, 2018). The pre-resource curse would occur if the highly positive expectations among policymakers regarding the future oil revenues that could be brought in end up pushing the government to borrow excessively and spend more in advance. Several countries have fallen into the pre-resource curse. Among these countries are Ghana and Mozambique. For instance, in Ghana, following the oil discoveries in 2007 and 2010, the government borrowed heavily. Excessive borrowing leads to an increase in domestic market interest rates due to the reduction in the country's national saving level, which deteriorates investment and leads to a slowdown in economic growth. Compared to an average GDP growth of 7 per cent between 2003 and 2013, the average growth rate in Ghana fell to 4 per cent during the 2014–16 period (Cust and Mihalyi, 2017).

Although the Lebanese government is highly indebted, and despite the high level of corruption in the country, the chances that the Lebanese government would fall into the pre-resource curse trap remains low. Following the April 2018 CEDRE conference (Conférence Économique pour le Développement par les Réformes et avec les Entreprises), Lebanon secured around USD 10.2 billion in loans at an interest rate of around 1.5 per cent and USD 860 million in grants. The main requirement to unlock these loans and grants is to reduce the budget deficit by 5 per cent over the next five years. Lebanon has started implementing important austerity measures to reduce the budget deficit. These measures include a reduction in spending on the electricity sector and an increase in indirect taxes. A policy paper by the Republic of Lebanon Ministry of Energy and Water (2019) discusses the steps that the Lebanese government intends to take to reduce the financial deficit caused by EDL. Among these steps are improving the collection of electricity bills, raising the tariff, decreasing operating costs, and cooperating with the private sector to increase electricity generation while relying more on natural gas rather than fossil fuels. Thus, the fact that the government must commit to low budget deficits to access the CEDRE funds implies that the odds of Lebanon falling into the pre-resource curse trap may possibly be low. Moreover, in January 2017, Lebanon submitted a request to join the Extractive Industries Transparency Initiative (EITI). This membership brings several benefits that can greatly reduce the chances of the Lebanese government falling into the pre-resource curse trap. First, upon signing up for EITI membership, the government would be obliged to disclose the revenues it receives from the oil and gas sector. In addition, companies that are handling the extraction process would also be required to disclose information about their payment activities. This increase in transparency could boost the citizens' trust in the government and reduce corruption.

Second, with Lebanon joining the EITI, citizens would have a tool in their hand to evaluate the performance of the government; thus, they would be able to hold the government accountable should the oil and gas sector be mismanaged. Third, the EITI requires civil society to have an active role in making decisions on rules and regulations in the sector. Note here that there is an important disagreement in the literature related to the relationship between civil society and corruption. While one group of researchers, often called the optimists, believe that civil societies can reduce corruption by disseminating public awareness, constantly evaluating the role of public institutions, and being involved in policy advocacy related to anti-corruption, another group of researchers, often called the sceptics, believe that civil societies are corrupt because they lack transparency, suffer from poor financial supervision, and tend to have little autonomy from their donors and the government (Villanueva, 2019).

Another challenge that news analysts are highly concerned about following the oil and gas extraction process is the Dutch disease. This economic anomaly is mainly observed in resource-rich economies. To better understand how Lebanon could be subject to the Dutch disease, suppose that Lebanon decides to continue its fixed exchange rate policy in the future. As Lebanon starts to export oil and gas to other countries, an inflow of US dollar payments comes to the Lebanese economy. The US dollars brought into Lebanon are eventually converted to Lebanese pounds, leading to an increase in the money supply. As the money supply rises, the demand for goods and services in the economy increases, leading to an increase in overall prices in Lebanon. This increase in the inflation rate appreciates the Lebanese real exchange rate, which makes the domestic production less competitive on the global level. Consequently, as other industries are unable to export goods and services to other countries, businesses start contracting and laying off workers leading to a slowdown in economic activity and an increase in unemployment.

While there is a growing fear of the Dutch disease in countries that are pursuing oil and gas exploration, Sachs (2007) claims that such fears can be overblown if an adequate national development strategy for the oil revenues is put in place. He argues that the Dutch disease mainly occurs in oil-rich countries that allocate their oil revenues for public consumption rather than public investment. In fact, Sachs (2007) suggests that if oil revenues are allocated towards public investment such as developing the country's infrastructure, then workers' productivity in both traded and non-traded sectors will increase. As a result, even if the real exchange rate appreciates, the non-oil sector will continue to expand due to the increase in labour productivity triggered by the increase in public investment spending. Thus, Sachs (2007) claims that an oil-exporting country can avoid the Dutch disease if a serious development strategy is implemented. Such a strategy should be focused on allocating the oil revenues towards long-term development strategies. According to Sachs (2007), the measures needed to achieve long-term development are as follows. First, the oil wealth must be allocated towards public investment in order to increase the provision of public and merit goods. As the poor gain access to better public goods and to more merit goods, the society as a whole benefits from both social and political stability. Second, policymakers should have a clear budget plan that maintains the stability of the economy in the short run as well as in the long run. In other words, the budget framework should account for the fact that oil prices are constantly in a state of flux, which implies that oil revenues can be unpredictable. In addition, public investment should mainly complement rather than substitute for private investment. In other words, public investment should focus on public and merit goods, whereas private investment should focus on the provision of private (excludable) goods and services. Moreover, Sachs (2007) claims that the spending on investment should be conducted within a fixed time frame so that the positive spillover effects of public investment are assimilated smoothly in the economy. For instance, as the government increases its spending on public hospitals, an increase in human capital should be moving in tandem. If not, then a very fast increase in spending on public hospitals can lead to economic inefficiencies.

A recent report by the Lebanese Oil and Gas Initiative (LOGI) states that Lebanon should follow the Norwegian example by putting all the oil revenues into a sovereign wealth fund and set a fiscal rule where government spending would be equal to the tax revenue and a fraction of the oil revenues (see LOGI, 2018). While this policy recommendation by LOGI (2018) is partly appropriate given the high level of corruption in Lebanon, a more effective policy for the allocation of Lebanon's oil revenues should incorporate elements of Sachs' (2007) measures, the promotion of economic development and long-term economic growth. In 2019, Lebanon continued to suffer from poor infrastructure (Saghir, 2019) and low quality of public education (Loo and Magaziner, 2017). Work by Calderón and Servén (2004) found that spending on infrastructure increases economic growth and reduces income inequality in developing countries. Moreover, Hanushek (2013) has shown that improving the quality of education is crucial for long-term growth in developing countries. Thus, while intergenerational equity is important, Lebanon cannot develop and achieve improvements in standards of living without improving the infrastructure and the quality of education.

To summarise, at its early stage, the oil and gas sector is negatively affected by lower oil prices due to the reduction in the number of bidding companies. As the bidding process begins for other blocks, low oil prices can continue to play against the prosperity of this sector. In the future, as

Lebanon becomes an oil-producing country, lower oil prices will also have a negative effect on future oil revenues. Moreover, international empirical evidence reveals that oil-producing countries tend to be subjected to the Dutch disease, and the only way to avoid this economic anomaly is by adopting an effective national development strategy. This strategy, however, should be carefully designed to avoid corruption and selective distribution by politicians.

Conclusion

The Lebanese economy is known to be sensitive to oil price fluctuations. The imports of mineral oil and fuel constitute a sizeable portion of total imports. This chapter provided a discussion on the effect of lower oil prices on the Lebanese economy following the large oil price declines that occurred in the 2014–16 period.

First, I reviewed the different transmission channels through which oil prices affect the macroeconomy. The data reveals that lower oil prices have increased the share of manufacturing in total output. In fact, the decrease in the cost of production in the economy translated into an increase in output and a negative inflation rate in 2015. These changes in the macroeconomic indicators reveal that the cost channel played an important role in the transmission of oil prices to the Lebanese economy. Moreover, during the 2014–16 oil price decline period, petrol prices fell, and despite the various negative demand shocks that hit the Lebanese economy during that period, consumption expenditure increased. Furthermore, car sales to Lebanon were higher in 2014 and 2015 compared to other years, which can possibly indicate that the operating cost channel also played a role in the transmission of oil prices. In addition, the decrease in oil prices helped to reduce the budget deficit, due to a reduction in government allocation of funds to EDL despite an increase in electricity production. The decline in oil prices has also reduced the trade deficit due to a reduction in the value of imported mineral oil and fuels. Last but not least, remittance inflows did not fall during the oil price decline period despite the negative repercussions that the low oil prices had on Saudi Arabia. Instead, remittance inflows increased because at least 50 per cent of these remittances come from countries that tend to benefit from lower oil prices.

Second, I discussed the 2015 deflation and the different causes behind it. Specifically, I reviewed whether the deflation was driven by the reduction in oil prices or the series of negative aggregate demand shocks that hit the economy during that period. In 2015, the Lebanese economy was hit by a series of negative aggregate demand shocks such as the high number of terrorist attacks, the garbage crisis, the government gridlock, and the surge in the number of Syrian refugees. All these demand shocks negatively affected consumer and business confidence, leading to a slowdown in economic activity and a reduction in inflation. In addition, the decrease in oil prices increased the economy's aggregate supply, due to the reduction in the cost of production, and increased the economy's aggregate demand due to the increase in consumption associated with higher discretionary income. In 2015, the resulting effects of these shocks on the economy were an increase in GDP and a decrease in the overall price level. This outcome most likely implies that the increase in aggregate supply associated with the oil price decline was the primary cause behind the 2015 deflation, and that this deflation is classified as good deflation based on work by Bordo *et al.* (2004).

Third, I looked into the effect of oil prices on the oil and gas sector. At its current early stage, lower oil prices have damaging effects on the oil and gas sector as the number of bidding companies interested to pursue the oil and gas explorations would fall. Currently, there is a low chance that Lebanon would fall into the pre-resource curse trap due to the unanimous interest among Lebanese politicians to unlock the CEDRE funds. However, pundits are worried of the Dutch disease when Lebanon finally becomes an oil-exporting country.

Overall, lower oil prices tend to benefit the Lebanese economy in the short run. Yet, as Lebanon becomes an oil-exporting country, the Lebanese economy would be receiving less oil revenue. Moreover, to avoid the Dutch disease, the government should put in place an effective development strategy that not only fosters intergenerational equity but also promotes the development of the infrastructure and the quality of education.

Notes

1 However, other researchers observe there are positive effects associated with the influx of the Syrian refugees to Lebanon. For instance, Turner (2015) argues that the upper and upper middle class in Lebanon benefitted from the low labour cost of the Syrian workforce.
2 Net inflow of remittances is equal to remittance inflows minus remittance outflows.
3 The data provided by the Central Administration of Statistics (2018) implies that the nominal GDP growth between 2014 and 2015 was 3.41 per cent, whereas real GDP growth for that same period was only 0.41 per cent. There is a concern here in terms of the calculation of real GDP because, given that there was an important deflation in 2015, one would suspect real GDP growth to be greater than the growth rate in nominal GDP.

References

Abdelnour, Z. (2003), 'The corruption behind Lebanon's electricity crisis', *The Middle East Intelligence Bulletin*, 5:8–9, August–September. https://www.meforum.org/meib/articles/0308_l1.htm (accessed 20 June 2019).

Al-Jack, S. (2018), 'Lebanon's state electricity company: A pawn for political corruption', *Asharq al-Awsat*, 23 July. https://bit.ly/3tbByOh (accessed 16 June 2019).

Azar, G. (2017), 'Two bids received for the oil and gas exploration', *An-Nahar*, 13 October. https://bit.ly/30x7nou (accessed 10 June 2019).

Banque du Liban (2019), 'Statistics and research'. www.bdl.gov.lb/statistics-and-research.html (accessed 3 March 2019).

Barakat, M., J. Naayem, S. Saad Baba, F. Kanso, G. Arabian, and F. Nahlawi (2018), 'Lebanon economic report for a re-alignment of growth with its long-term potential', Beirut: Bank Audi.

Baumeister, C., and L. Kilian (2016), 'Lower oil prices and the US economy: Is this time different?', *Brookings Papers on Economic Activity*, 2, 287–357.

Baumeister, C., L. Kilian, and X. Zhou (2018), 'Is the discretionary income effect of oil price shocks a hoax?', *The Energy Journal*, 39:SI2, 117–38.

Bergmann, P. (2019), 'Oil price shocks and GDP growth: Do energy shares amplify causal effects?', *Energy Economics*, 80, 1010–40.

Bernanke, B. (2006), 'The economic outlook', Speech delivered before the National Italian American Foundation, 28 November. https://www.federalreserve.gov/newsevents/speech/bernanke20061128a.htm (accessed 3 March 2019).

Bordo, M. D., J. L. Lane, and A. Redish (2004), 'Good versus bad deflation: Lessons from the gold standard era', National Bureau of Economic Research, W10329.

Calderón, C., and L. Servén (2004), 'The effects of infrastructure development on growth and income distribution', Policy Research Working Paper, 3400. Washington, DC: World Bank.

CBC (2015), 'Economy better able to withstand oil shock, Bank of Canada says', 11 June. https://bit.ly/38s0okQ (accessed 2 May 2019).

CEIC (2019), 'Lebanon motor vehicles sales', CEIC data. https://www.ceicdata.com/en/indicator/lebanon/motor-vehicles-sales (accessed 4 May 2019).

Central Administration of Statistics (2018), 'Annual national accounts 2004–2018'. www.cas.gov.lb/index.php/national-accounts-en (accessed 18 March 2019).

Central Administration of Statistics (2019), 'Economic statistics'. www.cas.gov.lb/index.php/economic-statistics-en (accessed 24 March 2019).

Chaya, J. (2019), 'Breaking down Banque du Liban's financial engineering', Beirut Today, 27 August. http://beirut-today.com/2019/08/27/banque-du-libans-engineering/ (accessed 10 September 2019).

Credit Libanais (2015), 'Oil & gas sector: A new economic pillar for Lebanon', Credit Libanais Economic Research Unit, January. https://bit.ly/3eglLJs (accessed 1 June 2019).

Credit Libanais (2017), 'The impact of remittances on economic growth', Credit Libanais Economic Research Unit, May. https://bit.ly/2PIk8dS (accessed 24 May 2019).

Cust, J., and D. Mihalyi (2017), 'The presource curse', *Finance and Development*, 54:4, 37–40.

Davis, S. J., and J. Haltiwanger (2001), 'Sectoral job creation and destruction responses to oil price changes', *Journal of Monetary Economics*, 48:3, 465–512.

Dibeh, G. (2002), 'The political economy of inflation and currency depreciation in Lebanon, 1984–92', *Middle Eastern Studies*, 38:1, 33–52.

Edelstein, P., and L. Kilian (2009), 'How sensitive are consumer expenditures to retail energy prices?', *Journal of Monetary Economics*, 56:6, 766–79.

Energy Information Administration (2013), 'Overview of oil and natural gas in the Eastern Mediterranean region', U.S. Energy Information Administration, 15 August. https://bit.ly/3l3V6kK (accessed 1 July 2019).

Energy Information Administration (2020), 'Lebanon', U.S. Energy Information Administration. www.eia.gov/international/overview/country/LBN (accessed 15 March 2020).

Hamilton, J. D. (1983), 'Oil and the macroeconomy since World War II', *Journal of Political Economy*, 91:2, 228–48.

Hamilton, J. D. (1996), 'This is what happened to the oil price-macroeconomy relationship', *Journal of Monetary Economics*, 38:2, 215–20.

Hamilton, J. D. (2003), 'What is an oil shock?', *Journal of Econometrics*, 113:2, 363–98.

Hanushek, E. A. (2013), 'Economic growth in developing countries: The role of human capital', *Economics of Education Review*, 37, 204–12.

Herrera, A. M. (2018), 'Oil price shocks, inventories, and macroeconomic dynamics', *Macroeconomic Dynamics*, 22:3, 620–39.

Herrera, A. M., M. B. Karaki, and S. K. Rangaraju (2017), 'Where do jobs go when oil prices drop?', *Energy Economics*, 64, 469–82.

Herrera, A. M., M. B. Karaki, and S. K. Rangaraju (2019), 'Oil price shocks and US economic activity', *Energy Policy*, 129, 89–99.

Institute for Economics and Peace (2019), 'Global terrorism index', Vision of Humanity. https://bit.ly/3u9Ynmh (accessed 10 May 2019).

IPT (2019), 'Variation of Lebanese fuel prices'. www.iptgroup.com.lb/ipt/en/our-stations/fuel-prices (accessed 5 May 2019).

Iskandarani, A. (2019), 'Garbage crisis continues to heap misery on Lebanese people', *The National*, 7 May. https://bit.ly/3rAeecA (accessed 15 September 2019).

Kilian, L. (2014), 'Oil price shocks: Causes and consequences', *Annual Review of Resource Economics*, 6:1, 133–54.

Kilian, L. (2017), 'The impact of the fracking boom on Arab oil producers', *The Energy Journal*, 38:6, 137–60.

Kilian, L., and R. J. Vigfusson (2011), 'Are the responses of the US economy asymmetric in energy price increases and decreases?', *Quantitative Economics*, 2:3, 419–53.

LCPS (2016), 'The repercussions of the Syrian refugee crisis on Lebanon: The challenges of providing services and creating jobs', The Lebanese Center for Policy Studies, Roundtable Report Series, January. https://bit.ly/2OaGy71 (accessed 10 May 2019).

Lebanon Gas and Oil (2019), 'Minister Nada Boustani for a regional cooperation and a transparent sector', 24 July. https://bit.ly/3ugCyl8 (accessed 3 May 2021).

LOGI (2018), 'Is Lebanon preordained to become a "presource" curse country?', 15 November. https://bit.ly/3l4LTsA (accessed 12 May 2019).

Loo, B., and J. Magaziner (2017), 'Education in Lebanon', World Education News and Reviews, 2 May. https://wenr.wes.org/2017/05/education-in-lebanon (accessed 3 September 2019).

McDowall, A. (2019), 'Fixing Lebanon's ruinous electricity crisis', Reuters, 29 March. https://reut.rs/3qyM8NE (accessed 20 September 2019).

Mork, K. A. (1989), 'Oil and the macroeconomy when prices go up and down: An extension of Hamilton's results', *Journal of Political Economy*, 97:3, 740–4.

Mourad, A. (2014), 'Caught between constitution and politics: The presidential vacuum in Lebanon', Heinrich Böll Stiftung Middle East, 4 July. https://bit.ly/30wpKKc (accessed 12 June 2019).

Neaime, S. (2004), 'Sustainability of budget deficits and public debt in Lebanon: A stationarity and co-integration analysis', *Review of Middle East Economics and Finance*, 2:1, 43–61.

Obeid, J. (2018), 'Lebanon's electricity crisis', *Executive Magazine*, 17 December. www.executive-magazine.com/economics-policy/lebanons-electricity-crisis (accessed 12 September 2019).

Republic of Lebanon Ministry of Energy and Water (2019), 'Updated policy paper for the electricity sector', March. https://bit.ly/3eTjouY (accessed 25 September 2019).

Republic of Lebanon Ministry of Finance (2014), 'Trade brief: The international trade monthly bulletin', December. https://bit.ly/3cjqO9F (accessed 12 May 2019).

Republic of Lebanon Ministry of Finance (2015), 'Trade brief: The international trade monthly bulletin', December. https://bit.ly/2PUsjUp (accessed 12 May 2019).

Republic of Lebanon Ministry of Finance (2019), 'Fiscal performance', Ministry of Finance. www.finance.gov.lb/en-us/Finance/EDS/FP (accessed 2 March 2019).

Republic of Lebanon Ministry of Finance Public Debt Directorate (2019), 'Gross and net public debt as percentage of GDP', Ministry of Finance. https://bit.ly/3qA2Iwg (accessed 4 March 2019).

Reserve Bank of Australia (2015), 'Statement on monetary policy', February. www.rba.gov.au/publications/smp/2015/feb/box-c.html (accessed 3 May 2019).

Sachs, J. D. (2007). 'How to handle the macroeconomics of oil wealth?', in M. Humphreys, J. Sachs, and J. Stiglitz (eds), *Escaping the resource curse*. New York: Columbia University Press, pp. 193–213.

Saghir, J. (2019), 'Infrastructure investments as a driver of economic growth: What can Lebanon learn?', Issam Fares Institute, 16 August. https://bit.ly/3t3YJde (accessed 28 September 2019).

Sandler, T., and W. Enders (2008), 'Economic consequences of terrorism in developed and developing countries: An overview', in P. Keefer and N. Loayza (eds), *Terrorism, economic development, and political openness*. Cambridge: Cambridge University Press, pp. 17–47.

The Economist (2017), 'Why Lebanon produces no oil', 4 May. https://econ.st/2N3AX1A (accessed 30 November 2019).

Turner, L. (2015), 'Explaining the (non-)encampment of Syrian refugees: Security, class and the labour market in Lebanon and Jordan', *Mediterranean Politics*, 20:3, 386–404.

UNHCR (2019), 'Registered Syrian refugees by date', Operational Portal Refugee Situations, 31 January. https://data2.unhcr.org/en/situations/syria/location/71 (accessed 18 March 2019).

Union of Arab Banks (2016), 'Remittances to the Arab region: Economic, financial, and developmental impact', Union of Arab Banks, June. https://bit.ly/3ufavCH (accessed 12 June 2019).

Villanueva, P. A. G. (2019), 'Why civil society cannot battle it all alone: The roles of civil society environment, transparent laws and quality of public administration in political corruption mitigation', *International Journal of Public Administration*, 43:6, 552–61.

World Bank (2017), 'Migration and remittances data'. https://bit.ly/3l2b92t (accessed 4 June 2019).

World Bank (2019), 'Personal remittances, received (% GDP) – Lebanon', World Bank Data. https://bit.ly/3vdXnhO (accessed 4 June 2019).

Yellen, J. (2016), 'The outlook, uncertainty, and monetary policy', Speech delivered at the Economic Club of New York, 29 March. https://bit.ly/3entHcd (accessed 10 March 2019).

11

Oil and the political economy in the Middle East: Overcoming rentierism?

Martin Beck and Thomas Richter

Introduction

The final chapter of this volume focuses on the changing political economy in the Middle East. It initially examines what we believe are the most evident consequences of the post-2014 oil price decline. Although by no means in a comprehensive way, we also touch upon some of the most pertinent effects of the shock induced by the economic slowdown as triggered by the COVID-19 pandemic in 2020. First, migrant workers in the Arab Gulf are the main social losers of policy adjustments post-2014. They are the majority of the population in the oil rentiers, but their social and political positions are weak. Second, citizens who are predominantly employed in the well-paid public sector proved to be capable of repelling burdensome adjustments. This was, on the one hand, the result of social resistance and, on the other, the outcome of deeply entrenched informal ties to the ruling families. Third, adjustment policies post-2014 show some of the institutional weaknesses characteristic of rentier states. These weaknesses often hindered the policy coordination necessary for proper formulation and implementation of adjustment policies. Fourth, for the three semi-rentier states dealt with in this volume – Egypt, Jordan, and Lebanon – the expectation that they could profit from the oil price decline in 2014 has not been fulfilled. The short-term positive effect of a reduced energy bill that lowered the budget deficit quickly evaporated with no significant fiscal reform being initiated.

Beyond some of the empirical consequences of the post-2014 oil price decline, this conclusion also highlights three conceptual dimensions that we consider relevant for the theoretical advancement of rentierism. First, we argue that it is important to bring state–class relations back into the discussion. We emphasise that a theoretical revision of rentierism on the basis of the state–class concept might help to sharpen our understanding of adjustment policies within the rentier and semi-rentier states in the Middle East. Second, we stress the importance of institutions during periods of policy adjustment within countries shaped by the inflow of hydrocarbon resources. As this is an issue which has been rather neglected in the broader theoretical debate on rentier states, more attention should be given to the country-specific institutional set-ups in order to better comprehend the dissimilarities in adjustment policies between the different rentier and semi-rentier states. Third, we explore the issue of rentier state autonomy by calling for a more nuanced understanding in that regard. By recognising several scope conditions, we suggest a context-specific understanding of state autonomy within rentier states. In the final section of the conclusion, we reflect on our results by discussing the appropriateness of rentierism in light of the oil price declines of 2014 and 2020. We argue that rentierism is still highly relevant with regard to both empirical dynamics in the Middle East and academic discussions on its political economy.

Social losers of adjustment policies in the Arab Gulf

Labour migrants are the primary losers of the domestic adjustment policies implemented in the Arab Gulf since 2014. This was again confirmed in the immediate aftermath of the COVID-19-initiated economic slowdown. As emphasised in the contributions of Gray (2021) on Qatar, Hoetjes (2021) on Kuwait, and Young (2021) on the United Arab Emirates (UAE), and confirmed by the findings in this volume's chapters on the other Gulf Cooperation Council (GCC) members, rentier states in the Arab Gulf have treated migrant workers as a pivotal source of government revenue and increasingly imposed fees and taxes, many of which are exclusively geared towards foreigners.

Migrant workers represent the majority of the labour force throughout the GCC countries: well over 90 per cent in Qatar, at least 80 per cent in the Emirates and Kuwait, and about 75 per cent in Bahrain and Oman. Only in Saudi Arabia does the local population make up almost half of the workforce (Diop *et al.*, 2018: 36). Migrant workers do not have organisational power in any of these countries: employers of migrant workers issue temporary contracts only and must personally vouch for them. The *kafāla* (sponsorship) system weakens the position of migrant workers further by restricting or even excluding the possibility of changing employers within the country, let alone between the GCC countries. Even leaving the country before the fulfilment of a contract is usually not viable because – except for the relatively few expatriate workers coming mostly from the Organisation for Economic Co-operation and Development countries – income-generating opportunities and working conditions in their African, Asian, and Arab home countries are significantly worse than in the Gulf.

The *kafāla* system is highly controversial, not only for ethical but also for economic reasons. The system has a negative impact on productivity. Due to the limited duration of contracts by default, neither employers nor employees have an interest in primary and continuous vocational training. Furthermore, the inflexibility of the *kafāla* system has created a layer of local intermediaries who, by trading visas for a fee, help to cushion its negative effects, thereby, however, creating another cost factor of an inefficient system stuck in rentierism. Prior to the oil price decline of 2014, some of the regimes in the Arabian Gulf were considering reforms of the *kafāla* system. Yet, with the partial exceptions of Bahrain and the UAE, serious attempts to implement reforms were only made after 2014 in Kuwait, Saudi Arabia, and Qatar (Diop *et al.*, 2018: 38–42). However, all reform projects have, for the time being, either failed or have been watered down to such an extent that the *kafāla* system continues to exist with no substantial alterations.

Migrant workers have paid by far the highest price for post-2014 adjustment policies in the Arab Gulf. They again did so as an immediate consequence of the COVID-19-induced oil price decline, which the regimes in the Arab Gulf used to aggressively promote nationalisation of the workforce. This confronted migrant workers with unemployment without social cushioning and the loss of residency (ILO, 2020: 11, 15–16). At the same time, the GCC countries failed to implement sustainable labour market reforms. Instead, several policies were initiated to more systematically subsidise wages of citizens employed in the private sector (Mogielnicki, 2020).

Beyond migrant workers, who have been most struck by the fallout, other segments of society have also been affected. As the oil price drop in 2014 has constrained the Arab Gulf states to further expand their public sectors, youth unemployment rates have become a more pressing issue (Aftandilian, 2017). In the present volume, Ennis and Al-Saqri (2021) highlight high unemployment rates among the youth in Oman, which is particularly a problem for women. Moreover, austerity measures such as subsidy cuts for electricity, water, and fuel, as described in detail by Ennis and Al-Saqri (2021) on Oman, Young (2021) on the UAE, and AlJazeeri (2021) on Bahrain, are felt harder by the poorer segments of the local population because they spend relatively more of their income on everyday goods. In the immediate aftermath of the COVID-19 economic slowdown, it became apparent, for instance, that women, who are over-represented in both paid and unpaid care work, were disproportionally negatively affected (ILO, 2020: 6).

Social constraints of adjustment policies

The existing literature has portrayed the influence of private-sector actors with a specific focus on business associations as particularly influential in shaping governments' adjustment policies during fiscal crisis within rentier states (e.g. Chaudhry, 1989; Lawson, 1991; Moore, 2002). Across these contributions, only very little reference is made to the impact of other social groups upon the choice of government reactions. The period of the post-2014 oil price drop, however, provides complementing evidence on some of the social constraints which exist for adjustment policies in the Middle Eastern oil rentiers.

Although since 2014 the Gulf states have taken a number of policy measures to cushion the budget deficits caused by the drop of oil revenues, they have not yet formulated, let alone implemented, truly structural reform measures, as discussed in the introductory chapter of this book (Beck and Richter, 2021). Also, the immediate response of the regimes to the crisis triggered by the COVID-19 pandemic does not bear witness to significant structural reform initiatives and was largely in line with previous policy initiatives (Mogielnicki, 2020). In the immediate years after the 2014 price drop, privatisation of public enterprises was not high on the agenda, while attempts at making the necessary structural reforms in the labour market and the public sector failed, in particular where these reforms would have directly affected citizens. Public protest or the threat of protest was central to this failure.

A prominent example relates to attempts to reduce salaries in the public sector, a segment which is largely reserved to citizens. In autumn 2016, for instance, the Saudi cabinet decided to cut the lavish bonuses in the public sector but reversed this decision in April 2017 after protests were called for in several cities across the country via social media (Reuters, 2017). As Mason (2021) points out in his contribution to this volume, parts of the Saudi Vision 2030 were revised due to the overarching importance of public-sector wages and social benefits for sociopolitical stability in Saudi Arabia. In Kuwait, where public wages make up more than half of current government expenditures (Hoetjes, 2021), the first oil workers' strike in twenty years took place in April 2016 as a response to government plans to reform the public-sector wage system and privatise parts of the state-owned oil company. After three days of protests, the reform measures were withdrawn, and shortly afterwards the government even announced a pay rise of 7.5 per cent (Arabian Business, 2016). No other serious effort in the immediate years after the 2014 oil price decline was made by the Kuwaiti government to reduce the public-sector wage bill (Hoetjes, 2021). As for Bahrain, AlJazeeri (2021) points to a close correlation between public unrest and the rise in current expenditures on wages in the public sector. Hence, she highlights the political nature of government expenditures during fiscal crisis as a key structural component of rentierism.

The case of Oman exemplifies the constraints of rentier states in the face of shrinking oil income per capita and simultaneous population growth. Muscat increased fees for government services, cut subsidies, and took a number of measures to increase the participation of nationals in the private sector. Yet, later in the autumn of 2017, when demands for the creation of new jobs attracted nationwide attention with more than 600,000 tweets and retweets on Twitter (Mukrashi, 2017a), the Omani cabinet responded immediately by announcing the creation of 25,000 new jobs for citizens (Ennis and Al-Saqri, 2021). However, when implementation turned out to be slow, in January 2018 young graduates took to the streets in several Omani cities. The previously made commitments were then hastily affirmed.

Policy adjustments and institutional weaknesses

A common feature cutting across the literature of rentierism is the argument that states built upon hydrocarbon income experience institutional dysfunctionalities at the domestic level. In other words, oil-rich countries suffer from an 'institutional curse' (e.g. Selim and Zaki, 2014; Blanco *et al.*, 2015: 239; see also Ross, 2015: 248–50). Since state institutions in rentier states are built and maintained for patronage, they tend to be vertically segmented. In other words, the state bureaucracy is organised along extended clientelist networks. Thus, as Hertog (2006) argues, rentier states in the Gulf developed structures of 'segmented clientelism', which constitutes a polity based on parallel, impermeable institutions with little communication between them. During fiscal crises, these institutions are then largely unable to coordinate horizontally in order to solve conflicts (Yamada, 2020). Country studies of this volume provide notable evidence of this institutional weakness.

In Bahrain, inefficient government institutions are among the key characteristics of one of the oldest rentier states in the Middle East. During the 1950s and 1960s, while still under protection of the United Kingdom, oil revenues were as high as 75 per cent of total government revenues (Lawson, 1991: 45). As AlJazeeri (2021) highlights in this volume, official government rhetoric has claimed that post-2014 adjustment policies initiated structural reform. Yet, this was actually not the case, since the adjustment policies implemented were contradictory and not suitable to fix the distorted institutional structure of governing institutions. Kuwait is another example of how existing institutional structures may hinder the implementation of adjustment policies within rentier states. Against the backdrop of conflicts over succession in the ruling family, the emancipation of tribal segments from Kuwait's ruling family, and the formation of a newly emerging youth movement apt to organise social protest, Hoetjes (2021) argues that, following the 2014 oil price decline, the Kuwaiti parliament evolved as the centre of opposition to almost all of the government's austerity measures. This institutionalised resistance has not only led to the withdrawal of plans to cut subsidies for citizens but also created a context in which most of the austerity measures were imposed on non-citizens only. In Saudi Arabia, policy adjustment based on the grand strategy of Vision 2030 was headed by the young prince and new strongman Muhammad bin Salman (MBS) (Mason, 2021). This plan emphasises the development of new economic sectors like entertainment, leisure, and tourism in order to reduce the dependence on oil. Since 2015, Saudi adjustment policies have been complemented by a centralisation of the decision-making process within the royal family. This was going hand in hand with an unprecedented marginalisation of the religious sector, especially a deprivation of the Saudi religious police (Bashraheel, 2019). However, Saudi post-2014 adjustments also provide examples of institutional weaknesses with regard to the implementation of reform policies. For instance, as highlighted by Al-Sulayman (2020) as 'reform dissonance', the period since 2014 is characterised by contradictory policy outputs due to the lack of communication and coordination between different institutional bodies within the Saudi bureaucracy.

Qatar's and the UAE's management of post-2014 policy adjustments differ from the other GCC members. As demonstrated by Gray (2021), policy adjustment policies in Qatar have shown a relatively high degree of dexterity and efficacy despite being embedded in institutions shaped by rentierism. This is all the more remarkable since policymaking was complicated due to the conflict escalation between Saudi Arabia, the UAE, and Bahrain versus Qatar, which led to a complete blockade of Qatar by its neighbouring GCC members. A major reason for this relative success is that Qatari leaders lean towards promoting economic growth and market confidence, thereby possibly avoiding some of the institutional weaknesses characteristic of rentier states. In the UAE, too, post-2014 adjustments worked rather smoothly as new stimuli for increasing government revenues beyond oil and economic diversification were set. Key to that, as pointed out by Young (2021), is the competitive policy environment within the UAE's federal system. Traceable to the pre-2014 period with Dubai as an accelerator of diversification, this created a policy flexibility, which took centre stage post-2014 and helped the central government to implement adjustment policies rather successfully in various areas such as fiscal policy and social development. However, it must be emphasised that neither Qatar nor the UAE have yet tackled necessary deep structural reforms, for instance in the fields of the labour market and income taxation policies.

Besides the domestic level, Arab Gulf states have also responded to the post-2014 oil price decline through mainly two intergovernmental organisations. These are the Organization of the Petroleum Exporting Countries (OPEC) and the GCC.[1] These institutions' contributions to absorbing the consequences of the oil price erosion differ widely: in the immediate aftermath of the oil price decline in 2014, GCC members failed to find a joint strategy to address the impacts of the oil price decline (Hassan, 2015). Yet, as highlighted by Young (2021), in late 2015, GCC members resumed an older initiative of implementing a value-added tax (VAT) of 5 per cent and agreed on a plan to introduce VAT by January 2018. However, despite its rather low ambition in terms of increasing government revenues, the reform so far has only been implemented by Saudi Arabia and the UAE in 2018, and Bahrain a year later (Mogielnicki, 2019: 2). That Saudi Arabia unilaterally decided to raise its VAT in summer 2020 confirms our observation of the diminishing role of the GCC as a policy-coordinating regional institution.

OPEC needed some time to absorb the shock of the 2014 oil price decline, too. However, after a failed attempt to bring in its major competitor, Russia, in April 2016 (Beck, 2016), the organisation finally managed to do so later in that year, thereby realising the ambitious project of creating OPEC+, which responded to the shock of the 2014 price decline by implementing production quota (Richter, 2017: 5; Beck, 2019). The resulting upsurge on oil prices was a notable – albeit limited – success of this first effective coordination among oil producers after the oil price drop of 2014. Strikingly, OPEC+ managed to repeat this relative success in the immediate aftermath of the price crisis triggered by the COVID-19 pandemic. Buttressed and pressed by then-US President Donald Trump, who intended to save the American oil industry from bankruptcy, OPEC+ reached an agreement, effective in May 2020, that set up a record-high reduction in oil production of 9.7 million barrels per day from May till July (Bremmer, 2020). After a partial recovery of oil prices from a historic low of less than USD 15 in late April 2020 to around USD 40 in the period from May to July, OPEC+ was able to continue its cooperation by slightly reducing cuts from 9.7 to 7.7 million barrels per day (El Gamal *et al.*, 2020).

A first conclusion to be drawn from these findings is that in oil-rentier states those institutions entrusted with generating rents tend to have stronger incentives to provide effective solutions. Second, the difference in ambition and implementation between policy adjustment initiatives coordinated by GCC and OPEC+ is also remarkable because classic rationalist approaches based on game theory would expect a different outcome: OPEC – and this is true for OPEC+ too – faces what Martin (1992: 769–70) terms a 'collaboration game' in which all actors have a strong incentive to defect, as this behaviour would lead to immediate benefits (see also Mason, 2021): if a member of OPEC exceeds its quota, its individual oil-rent income will increase. The GCC, however, plays a 'coordination game' (Martin, 1992: 775) which only appears difficult to solve if there are major distributional conflicts between the engaged parties. Yet, a growing competitiveness among GCC members notwithstanding, this does not apply to the present case, as the benefit of introducing VAT by a single GCC member is not highly affected by the decisions of others introducing the same taxation level.

However, Putnam's (1988) concept of two-level games hints at a solution to this puzzle. He argues that in international negotiations states not only face other states as opponents with whom they must find an agreement, but also need backing from pertinent domestic actors. Vis-à-vis their societies, OPEC+ members were in a more favourable position to collaborate on production controls than GCC members regarding the introduction of VAT. The reason is that, contrary to the imposition of a new source of domestic government revenue such as VAT, no relevant domestic interest against absorbing the country's fall of rent income existed. Thus, even though the GCC was much quicker than OPEC+ in finding an agreement, the former could not easily prevail against vested interests at the domestic level.

In fact, contributions in this volume – AlJazeeri (2021) on Bahrain, Ennis and Al-Saqri (2021) on Oman, and Hoetjes (2021) on Kuwait – provide evidence that these three non-complying GCC members have refrained for the time being from introducing VAT because societal actors successfully lobbied against its implementation. Arguably, the introduction of VAT in Bahrain, Saudi Arabia, and the UAE correlates with a higher degree of authoritarianism there, which provides ruling regimes with more leverage to push through potentially contested policy reforms. The ability of the Saudi regime to raise its VAT from 5 to 15 per cent in the immediate aftermath of the crisis induced by the COVID-19 pandemic points in the same direction. Findings from Freedom House (2020) support this understanding by showing that political rights and civil liberties are curtailed to a higher degree in Saudi Arabia, Bahrain, and the UAE than in Oman, Qatar, and Kuwait, the latter country being ranked as the only 'partly free' GCC member states.

The semi-rentier states: Failed opportunities

For several reasons, the 2014 drop in oil prices raised expectations that the economies and state budgets of the semi-rentier states Egypt, Jordan, and Lebanon would benefit from the oil price drop. First, as the decline in oil prices reduced production costs in all net oil-importing countries, this should have led to stimulating effects within the domestic economy with potentially rising tax revenues. Moreover, the lower energy bill should have enabled governments to abolish subsidies with positive effects for other spending items or the repayment of debt. Second, employment for Egyptian, Jordanian, and Lebanese migrant workers in the Gulf states was expected to shrink, which would have had selected positive effects for the domestic economies in a twofold way. The brain drain from skilled-worker migration could have been reduced. Moreover, lower amounts of inflowing remittances would have mitigated the Dutch-disease effect, which in turn could have contributed to a reduction in the overvaluation of local currencies, thereby encouraging local productivity. Third, the expected cutbacks of budget-to-budget transfers from the Arab Gulf states to the semi-rentiers would have created an incentive to cut subsidies and expand state capacities in extracting taxes. However, the empirical findings from the chapters on Egypt, Jordan, and Lebanon presented in this volume differ from these presumptions.

Both the Egyptian and the Jordanian regimes tried to use the drops in their energy bills as leverage to cut subsidies. However, the Egyptian approach was rather limited, and the more ambitious Jordanian attempt faced significant social opposition. At the same time, neither Amman nor Cairo made a substantial effort to launch fiscal reforms. This is linked to the fact that the development in remittances and foreign aid inflows took different directions than assumed. In both cases, foreign aid payments remained at a high level. In Egypt, the decline since 2015 was much less than to be expected after the previous year's boom of aid payments by Saudi Arabia and the UAE in support of Field Marshal Abdel Fattah el-Sisi's usurpation. While foreign aid from the Gulf to Jordan halved during the period 2014 to 2018, overall foreign aid to Jordan remained on a high level, also because the West maintained its support for the regime in Amman. In both countries remittances also persisted on a surprisingly high level: they remained constant in Jordan, whereas in Egypt – after a slight decrease in the immediate aftermath of the 2014 oil price decline – they even skyrocketed in 2016–17 (Adly, 2021; al Khouri and Silcock, 2021).

As demonstrated by the contributions on Egypt (Adly, 2021) and Jordan (al Khouri and Silcock, 2021), regional and domestic features of Middle East rentierism contribute to a better understanding of these puzzling findings. First, Saudi Arabia and the UAE have a strong interest in stabilising allied authoritarian regimes in Jordan and Egypt because they contribute to the containment of Riyadh's and Abu Dhabi's major regional political opponent: the Muslim Brotherhood. Therefore, both Gulf states are ready to bear high costs, which means maintaining high foreign aid payments as well as privileging Jordanian and, even more so, Egyptian nationals by granting them working permits. Second, the regimes of the net oil importers are experienced rent-seekers: they often succeed in converting their geopolitical importance into rent inflows (for the case of Jordan, see Beck and Hüser, 2015), whereas they frequently fail to implement structural reforms to overcome rentierism.

Rentierism in Lebanon is less state centric than in Jordan and Egypt: while Lebanon's reception of official development assistance (ODA) per capita is not much lower than Jordan's (World Bank, 2019), the country – in contrast to oil-producing Egypt and phosphate-exporting Jordan – does not yet extract natural resources whose revenues directly accrue in the coffers of the government. Additionally, the Egyptian and Jordanian regimes enjoy much higher state-controlled location rents than Lebanon: Egypt earns fees from the Suez Canal (Adly, 2021) and Jordan from the West for sustaining secure borders to Israel and the Israeli-occupied West Bank (Beck and Hüser, 2015: 94–5). On the other hand, Lebanon is well ahead of Jordan and Egypt with regard to the per capita inflow of remittances (Global Knowledge Partnership on Migration and Development, 2018: 20). Thus, labour remittances are crucial for the Lebanese economy; they bypass state institutions, however, and go directly to society.

In contrast to Egypt and Jordan, whose labour migration flows are largely confined to the Middle East, Lebanese labour migration is global. Thus, as Karaki (2021) makes plausible, Lebanon could overcompensate for its losses from Gulf remittances by increasing those from Western countries whose economies were boosted in the wake of the 2014 oil price decline. Karaki (2021) additionally highlights how and why remittances are of extraordinary relevance for the Lebanese economy: more than one quarter of them are stored by the Lebanese banking sector, a factor which is crucial for the nation's economy.

The unique structure of Lebanon's rent dependence contributes to an explanation for the country's socio-economic crisis in 2019–20. Structurally speaking, the Lebanese economy is in need of a high inflow of US dollars in order to balance its large national public debt with a ratio to gross domestic product (GDP) of above 150 per cent. Yet, starting already from the early 2010s, the private Lebanese banking system was only able to attract remittances to be deposited in US dollar accounts by offering increasingly high interest rates. Thus, from an already relatively high level of over 3 per cent in 2011, interest rates for US dollar accounts steadily went up further until they reached their peak of over 7 per cent in late 2019 (Bloominvest Bank S.A.L., 2020). As these high interest rates were credited in US dollars, the Lebanese banking system accumulated huge amounts of 'lollars' (Azzi, 2019), which is a Lebanese virtual currency that does not correspond to actually existing amounts of US dollars. When in 2019 the well-informed 1 per cent of top depositors lost trust in this unsustainable system, they withdrew about USD 28 billion. This resulted in a broad bank run in the second half of the year (Diwan, 2020), which left the Lebanese economy largely with virtual US dollars. Therefore, Lebanon faced the risk of comprehensive collapse, because its banks were drained of convertible currency that both the highly import-dependent economy and the extremely indebted government were in desperate need of.

The Arab Gulf: Bringing state–class relations back in

The drop in oil prices post-2014 has made it impressively clear that relations between state and social class have gained in relevance – or rather regained it – in the Arab Gulf. Analyses on the relations between state and class – defined by what economic resources a social group controls (Grant, 2001: 161) – commonly start with discussing society. According to liberal economics, this is so because in the 'production states' (Luciani, 1987) of the Global North, the society is prior to state, as the latter relies on the extraction of surplus from it. For instance, in a capitalist society, workers contribute with their labour and entrepreneurs with investment, and they nourish the state by paying taxes. What Marxist-inspired economic structuralism shares with liberal economics is the premise that social classes are prior to the state, but this outcome is understood as the result of the dominant class's ability to establish the state as an instrument to control subordinate classes. However, the 'distributive state' (Delacroix, 1980) of the Arab Gulf is not the product of decades-long and tense class conflicts and therefore differs fundamentally from the state in the Global North.

Delacroix's classic contribution concerning state–class relations in the Gulf monarchies was path-breaking, as it revealed for the first time the structural differences between the production state in the Global North and the rentier state in terms of state–class relations. Yet, in order to sharpen our conceptual understanding that the fiscal crisis caused by the 2014 oil price decline has threatened to shake the foundations of the rentier state, the approach needs elaboration in four dimensions. First, it should be spelt out more clearly that by appropriating oil rents as a fiscal income of external origin, in a historic process the ruling families managed to transform themselves into state classes. These state classes have, second, managed to consolidate their superior power position by systematically and successfully developing external rent-seeking activities. Third, in contrast to the prevailing understanding, an exploited class does exist within the Arab Gulf states: low-skilled migrant workers. This exploitation, which was aggravated in the wake of the 2014 oil price decline, is crucial for the maintenance of the political economies of the Arab Gulf. Finally, the state class uses its externally generated rents not only for benevolent distribution but also, in the face of drained resources, increasingly for coercion.

First, as pointed out by Delacroix (1980: 8), 'to be of capitalism is not the same as being a capitalist society.' On the one hand, modern Arab Gulf states and societies are indeed a product of the first truly globalised capitalist branch: the transnational oil industry (Beck, 2012). On the other hand, state and society in the Arab Gulf are *not* capitalist as is known from the Global North. The dominant class in the Arab Gulf is not a capital-controlling bourgeoisie independent from the state as it emerged from intense class struggle within Europe during the nineteenth century. Rather, the evolution of a new dominant class in the Arab Gulf rentiers was enabled by the reception of oil revenues that allowed for 'the emergence of new and ... large bureaucratic structures' (Crystal, 1990: 188). This new state bureaucracy, created and headed by key players from within each of the ruling families, constitutes a class of its own: the 'state class' (Elsenhans, 1996: 1–28; for a different perspective, see Hanieh, 2011: 2–14). The resources that this class controls differ from the means of production in the Global North as controlled by the capitalist class: the state class in the Arab Gulf relies on organisational and managerial skills that enable it both to externally accrue rents and to distribute them to maintain its power position.

Second, mainstream rentierism takes it as given that rentier states receive oil rents from external sources: rents are typically treated as a gift of nature (Beblawi, 1987: 49). However, only in the early 1950s did oil rents virtually drop into the lap of the Arab Gulf states. Thereafter, when the Gulf regimes aimed at enlarging their share of the rent vis-à-vis the oil companies, they had to develop effective rent-seeking strategies and tools, which were largely outward oriented.

Yet, before the emerging state classes were able to develop rent-seeking activities, they received support from the rising hegemon of the Western world, the USA. Since the US administration was highly interested in stabilising the then-poor and weak oil-producing states of the Arab Gulf, in 1950 it convinced its major transnational oil companies to grant a fifty-fifty rent-sharing formula in exchange for generous tax exemptions in the USA (Schneider, 1983: 27–31; Mommer, 2002: 125–6). By virtue of their own strength, the regimes in the Gulf at that time would not have been capable of reaching an agreement that copied the fifty-fifty formula that Venezuela's much more advanced oil ministry had reached after years-long tough negotiations with the transnational oil companies (Mommer, 2002: 107–18). Yet, already in the mid-1950s, ruling regimes in the Gulf had begun to understand that the fifty-fifty agreement was – its wording notwithstanding – not fair, because the bulk of the oil rents still went into the pockets of the transnational oil companies (Mommer, 2002: 118–33). At least equally annoying from the Arab Gulf regimes' point of view was the fact that these companies and the consortia they had built were in full control of decision-making regarding production volumes across all Gulf countries. Due to their oligopolist position in the world market, the consortia were furthermore able to manipulate prices and therewith also rent payments to the respective Arab Gulf regimes. When, by co-founding OPEC in 1960, the ruling royal regimes in Saudi Arabia and Kuwait – joined by Qatar in 1961 and the UAE in 1967 – became proactive rent-seekers, they faced the antagonism of the transnational oil industry. At the same time, by thus acquiring high rent-seeking competences, the Gulf states became prepared for a unique historical moment in the early 1970s, when favourable circumstances – in particular growing competition among Western oil companies and an upsurge in the global demand for oil – enabled them to triumph in their external rent-seeking policy by nationalising the hydrocarbon sector. This shift in ownership of the oil resources from the transnational oil companies to the rentier states became manifest in an 'oil price revolution' (Schneider, 1983: 101); however, it was much more than this, namely, nothing less than a full-fledged 'oil revolution' (Tétreault, 1985: 47).

Third, the perspective of the Arab Gulf state as a socio-economic system that is largely free of exploitation is widespread in the literature on the rentier state and across the growing community of Gulf studies scholars alike: Beblawi's (1987: 53) note of the rentier state as an '*état providence*, distributing favours and benefits to its population' has often been reiterated and has sometimes lured scholars into overly positive appraisals of the benevolent state class. El-Katiri (2014: 27), for instance, depicts 'the existence of a generous, modern welfare system, unparalleled in the developing world' as a characteristic feature of what she calls the 'guardian state'. Rentier states in the Arab Gulf appear to be benign towards society as a whole. Although the rentier state is discriminatory by distributing wealth extremely unequally, even the poorest get something from the state, as everyone benefits from subsidies, while in return no one has to pay taxes. There is some truth in this portrayal: the citizenry is indeed not burdened, let alone exploited. Their primary economic function is consuming the oil rent. However, it is thereby often neglected that large parts of the populations – the labour force of unskilled (and semi-skilled) migrant workers – are subject to substantial exploitation.

This exploitation has been organised by the state class from the top by elaborating the *kafāla* system inherited from British imperialism (AlShehabi, 2019), which is executed mostly by private entrepreneurs but also utilised throughout all segments of the citizenry. The *kafāla* system externalises the costs of economically dysfunctional relations between the state and its citizens. This externalisation comes at the expense of migrant workers, since most of them earn only a fraction of the payment received by GCC citizens employed in the largely unproductive und inefficient public sector. At the same time, Arab Gulf governments have very little leeway to reform the *kafāla* system: due to widespread expectations among citizens to receive salaries as high as in the public sector, the Arab Gulf's private companies, which are exposed to somewhat more competition, largely depend on cheap labour provided by migrant workers. Moreover, many private households employ at least one foreign domestic worker. In the case of Qatar, survey research has revealed that employers have a strong interest in further strengthening the bond with the foreign employee, which leaves the latter with little protection from exploitative practices (Diop *et al.*, 2018).

Finally, according to Delacroix (1980: 12), statehood in the Arab Gulf is singular insofar as the role of distribution prevails over coercion, while the latter is considered constitutive to the capitalist production state in the North both in the Weberian and Marxist tradition. However, as Smith (2017: 599) emphasises, oil-rent income constitutes 'flexibility of spending': rents can be used not only for patronage but also for coercion. Crystal shows that the state classes in the Arab Gulf responded to the challenges of the Arab uprisings and the oil price decline in 2014 by increasingly 'policing the people' (Crystal, 2018: 83). A policy of securitisation, which constructed a nexus between oil and terrorism, has been applied by all Gulf regimes to constrain different oppositional groups and activists who raised their voice through new social media. As Crystal (2018: 83–90) highlights using the example of the Arab Gulf regimes' responses to the Arab uprisings, this new kind of repression has been implemented on trend to a higher degree by those regimes for whom the era of oil income abundance has come to an end: Bahrain, Oman, and Saudi Arabia (Human Rights Watch, 2016, 2017; Amnesty International, 2018).

Policy adjustment and the importance of institutions

Based on the notion that oil income contains a distinct character which has often been pointed to as a 'curse' (Ross, 2013), the concept of rentierism tends to assume that oil rents make governments alike by providing incentives to create similar institutional structures, which then ultimately instigate similar policies. This conception traces back to rentierism as an inductively built approach based upon the detailed assessment of single cases from only one world region – the Middle East – thereby having largely ignored studies on the impact of oil income on countries in world regions with different institutional settings, for instance Venezuela (Karl, 1987). Strikingly, the idea of oil rents as having shaped governments uniformly is echoed in the 1990s by literature in comparative political economy, which argues that dependency upon one leading economic sector determines adjustment policies during periods of economic crisis (e.g. Frieden, 1991; Shafer, 1994).

One example of the misleading understanding of oil rents as uniformly shaping oil-exporting countries is provided by the reception of Chaudhry's works in some of the subsequent academic discussions. As Chaudry (1989, 1997) highlights using the example of emerging rentierism in Saudi Arabia and Yemen during the 1970s and 1980s, the 'role of domestic social contexts in mediating the effects of capital flows to developing countries' (Chaudhry, 1989: 145) is crucial. Strikingly, Chaudhry's widely cited research is often referenced in the opposite way to support the idea that oil states tend to be alike and therefore fabricate uniformly weak institutions, which eventually cause the same poor policy responses.[2]

As already noted by Okruhlik (1999: 309), 'oil enters into an ongoing process of development', and as reiterated by Moore (2002: 35), 'historical and institutional trajectories are crucial to explaining crisis politics'. Thus, the legacy of institutional structures prior to the oil era and subsequent changes in them as a factor shaping government responses during situations of fiscal crisis are worth revisiting. A recent contribution on rentierism in the Arab Gulf advances the conjunction between oil and institutions by specifying that the historical impact of oil wealth upon institutional developments created an 'institutional stasis' (Kamrava, 2018). In other words, at that moment when oil income started to flow into state coffers, state institutions began to freeze (for a more general version of this argument, see Smith, 2006). While this seems to be a welcome recognition of the relevance of social and political institutions within rentier states, it neglects some of the institutional dynamics and developments which occur under conditions of rent scarcity.

This volume has provided plenty of evidence underlining the importance of the dynamics of social and political institutions and their various influences across countries in the post-2014 period of oil price decline. As all contributors to this book make clear, oil income flowing into state coffers during the latest oil price peak in the early 2000s has not levelled out historically grown institutional differences by making institutions and public authorities alike. Particular institutional features within each of the Gulf states also indicate country-specific development paths after 2014.

When comparing some recent policy adjustments made by Kuwait and Saudi Arabia, the specific impact of different domestic institutions comes to the forefront. Kuwait is the only one of the Gulf monarchies to have produced a strong parliament, which has the right to withdraw confidence in any minister and even the prime minister by a simple majority vote (Herb, 2014: 50–1). Although the composition of the parliament was pro-government due to the opposition's boycott of the 2013 elections, a serious dispute arose between parliament and government over the latter's plans to cut fuel-price subsidies. When the majority of the deputies threatened to exercise their right to question ministers and the prime minister in plenary, Amir Sabah Al-Ahmad Al-Sabah dissolved parliament in October 2016. This backfired, however, as the opposition gained a clear majority in the new parliament elected in December 2016 and was able to block all further attempts to reduce government spending, not to mention structural reforms (Hoetjes, 2021).

Also, in the Saudi case, historically grown institutional features shaped the planning and implementation of adjustment policies. The most important elements of the reform projects under King Fahd in the 1980s were not very ambitious from the outset and failed or were severely diluted by the informal influence of the private sector (Chaudhry, 1997: 277–82). In contrast, not only is the Saudi reform programme that was launched in 2016 as Vision 2030 much more ambitious, but its first policy measures have also actually been implemented (Mason, 2021). For example, since summer 2017, every Saudi citizen who employs a migrant worker has to pay an annually increasing fee. VAT was introduced in January 2018, too. Although these measures are limited in scope, it should be noted that comparable projects failed in the 1980s due to resistance from the private sector. For example, neither the employers' share of social security costs planned in 1985 nor a tax reform announced in 1988 were implemented (Chaudhry, 1997: 274–5). The differences between the 1980s and 2010s can best be explained by the political centralisation that has taken place since 2015, which breaks with the tradition of Saudi rule. Since the death of the state founder, Ibn Saud, in 1953, all of his sons who succeeded him on the throne have been involved in a consensual system with the other senior princes. When King Salman, who was the last of these senior princes, ascended to the throne, he appointed his favourite son, MBS, as a leading figure among the remaining key decision makers. MBS then quickly became Saudi Arabia's de facto sole ruler and used his newly won power to purge the political kernel elite of all potential competitors. He also marginalised some of the most conservative segments of the Wahhabi 'ulama' that otherwise would have been able to block parts of the post-2014 adjustment policies, such as the opening of cinemas, the performance of concerts, or the development of Saudi Arabia as a tourist destination. This centralisation of Saudi rule paved the way for a comparatively successful implementation of adjustment policies during the immediate years after the 2014 oil price decline.

Rentier state autonomy and post-2014 policy adjustments

During the pioneering phase of rentierism, a deep conviction developed that, due to the external origin of oil income, governments may gain a degree of autonomy vis-à-vis their society previously unseen. As Mahdavy points out, a 'government that can expand its services without resorting to heavy taxation acquires an independence from the people seldom found in other countries' (1970: 466–7), while Luciani highlights that the allocation state 'being independent ... of the domestic economy, does not need to formulate ... economic policy: all it needs is an expenditure policy' (1987: 74). Anderson paints it more broadly by stressing that '[t]he availability of revenues generated outside the domestic economy ... has substantially lessened the reliance of many Middle Eastern governments on their own population, and the corollary ... appears to be domestic state autonomy' (L. Anderson, 1987: 9). Similarly, Crystal finds that 'oil-based states are unusual in ... their higher degree of autonomy from other social groupings' (1990: 6).

The idea that oil income makes governments more autonomous prevails and is widely acknowledged in the literature discussing the rentier state. For instance, the standard textbook of Middle East political economy notes: '[O]il can foster the rise of autonomous states that are little influenced by their society' (Cammett *et al.*, 2015: 322). Also, the view that rentier leaders are autonomous is still widespread within recent revisions of the rentier state: 'Once rent revenues started flowing into the state ... they also enhanced the autonomy of state leaders (Kamrava, 2018: 5–6). Among others, Hanieh (2011: 12) uses the autonomy assumption to justify his departure from rentierism, as he assesses the rentier state theory (RST) notion of relative autonomy as 'highly misleading'. Last but not least, the domestic autonomy of oil states is particularly popular in the quantitative studies testing for the impact of oil wealth upon democratisation (e.g. Ross, 2001; Aslaksen, 2010) and authoritarian survival (e.g. Ulfelder, 2007; Wright *et al.*, 2015).

Although the assumption that oil income makes governments more autonomous from social forces has been challenged before (e.g. Chaudhry, 1989, 1997; Hertog, 2010; Gray, 2011), it persists in rentierism without being critically discussed. Therefore, some points of clarification with regard to the autonomy of the rentier state vis-à-vis its society are in order. State autonomy is closely associated with the idea of state capacity, which can be defined as 'the ability of the government to formulate and implement policies' (Chaudhry, 1989: 102, footnote 2). An autonomous state, then, is considered to be an entity which, first, formulates policies unimpeded by social actors and, second, is also able to implement these policies even against the will of some of them. A rewarding perspective at the intersection of rentierism and policy adjustment should therefore address two elements:

first, the conditions under which the government of the rentier state is able to act without being constrained by social actors and, second, against which social actors the government is capable of doing so.

A first aspect which sheds some light on how autonomous a rentier state is concerns the relations between migrant workers and the state in the Arab Gulf. As several of the contributions on the Gulf countries in this volume have demonstrated, this social group is the key loser of post-2014 policy adjustments. In addition to imposing poor labour conditions on migrant workers, governments also burdened them with additional economic costs by increasing fees, diminishing subsidies, and introducing taxation. The few pockets of resistance formed by migrant workers after 2014 have been met with heavy repression by the respective regimes. Rentier states' autonomy from migrant workers can therefore be considered to be consistently high with regard to policy formulation and implementation. This is quite remarkable, as very few other states in the world enjoy that level of autonomy from the majority of their population.

A second element which needs to be stressed in the context of discussing the autonomy of a rentier state concerns the close ties between the rentier state and its citizens. Since their foundation, Arab Gulf states have distributed large amounts of their oil income towards comprehensive public employment schemes. This led to one of the highest proportions of citizens employed in the public sector worldwide. As exemplified by the large public wage gain during the Arab uprisings, political crises tend to fortify this structural feature. Thus, in all Arab Gulf countries, government expenditure on wages and salaries takes up by far the biggest share of the budget and does not adjust as steeply during contradictions in the oil market as they increase during booms. In the immediate years after the 2014 oil price decline, none of the countries has been able to reform its public sector. The few attempts failed due to considerable social resistance and the state classes' fear of emerging social instability. At the dawn of the oil age, the state classes were autonomous and thus free to spend their oil income by co-opting social groups through state building. Over time, however, this conduct created obligations towards the beneficiaries of rent distribution. The history of government spending on public employment is therefore an important scope condition for a rentier's autonomy regarding the implementation of structural reforms in its public sector. In other words, the autonomy of the rentier state has 'declined precipitously' since its foundation (Hertog, 2010: 267).

A final dimension of partially constrained state autonomy after the oil price decline in 2014 became relevant in those Gulf rentiers in which parliaments exist that are elected by all adult citizens, these being Bahrain, Kuwait, and Oman.[3] Within these three countries, parliamentarian debates and some formal rights of parliamentarian approval allow social groups and coalitions of social groups to more visibly limit the autonomy of the state class. The role of the Kuwaiti parliament as the veto-player for governmental reform initiatives following the 2014 oil price decline is the most striking example, but there is also evidence of interferences by other parliaments in the Arab Gulf. In Bahrain, considerable resistance emerged from within the elected lower house of the parliament – the *majlis al-nuwwāb* – against some of the government's austerity measures after 2014. Although the Bahraini *majlis al-nuwwāb* is less powerful than Kuwait's *majlis al-umma*, it can exert some influence due to its formal right to approve both the annual budget and all laws. This led to postponement of new taxation (Barbuscia, 2018) and the termination of measures to reform the country's bloated subsidy system (Yaakoubi, 2019). The *majlis al-shūrā* in Oman has seen some enhancement of status after demands were expressed by citizens to widen political participation during the Arab uprisings. Among the changes implemented since then has been the right to question so-called service ministers and to discuss the annual budget. Yet, after one of the most vocal deputies was sentenced to prison in 2013 and three similarly engaged incumbent members were barred from elections in 2015, these formal options were rarely applied (Bertelsmann Stiftung, 2020a: 15). Nevertheless, in 2016 some of the government's austerity measures were delayed due to opposition from within the majlis (R. Anderson, 2016), and in 2017 the fuel-price reform was revised after members of the majlis publicly urged for better consideration of lower income groups (Mukrashi, 2017b).

Arab rent-dependent countries beyond the Gulf: Egypt, Jordan, and Lebanon

Through petrolism, the development of net oil importers like Egypt, Jordan, and Lebanon has also been deeply shaped by hydrocarbon rents. Petrolism started interweaving the Arab Gulf states – under the leadership of Saudi Arabia – with all of the Arab net oil importers in the wake of the oil revolution in a twofold way (Korany, 1986). The Arab Gulf states used part of their newly acquired oil-rent abundance to stabilise the political regimes of resource-poor Arab countries, particularly in the Mashriq, by transferring political rents, mostly in the form of budget-to-budget transfers. At the same time, the prospering Gulf states were opening their labour markets for Arab migrant workers. These migrant workers sent back home a good deal of their salaries as remittances, which also constitutes a rent income because it is not balanced by investment or the labour of the recipient. Petrolism has certainly constituted Egyptian, Jordanian, and Lebanese dependence on the Arab Gulf countries. This, however, does not imply that the former are passive objects of the latter's policies. Indeed, semi-rentier states often act as rent-seekers. As the influx of rents in Egypt, Jordan, and Lebanon encountered different kinds of socio-economic structures and political institutions, the evolution of rentierism through petrolism within these countries has manifested in particular ways.

Ever since the British initiative to establish the state of Jordan in the early 1920s, the regime in Amman has been dependent on rents. At the beginning, it was the United Kingdom that had to grant King Abdullah I an appanage. After World War Two, the USA started to become Jordan's major donor. From the 1970s onwards, the Arab Gulf states developed into the main financial supporters, yet Jordan managed to also maintain political rent payments from the West (Beck and Hüser, 2015: 89–90).

Similar to the development of state classes in the Arab Gulf, the Hashimite rulers of Jordan created a state bureaucracy largely based on successful external rent-seeking. Yet, due to a much smaller volume of accrued amounts, two important differences exist. First, the Hashimite regime was unable to be as benign to its citizens as state classes were throughout the Arab Gulf. Notwithstanding a similar pattern – providing jobs in the public sector and launching subsidy policies – distribution had to be more moderate. Moreover, entry to the public sector was largely confined to Transjordanians, whereas the majority of Jordanian citizens – Palestinian refugees from the 1947–49 Palestine War – were mostly excluded. Second, members of the royal family were not put into key positions within the regime. In other words, Jordan is a so-called linchpin monarchy (Bank *et al.*, 2014: 166), which is a political system that exclusively centres around the king.

The strategic backbone of Jordanian rentierism has been the Hashimite Kingdom's shared border with Israel. Jordan swiftly turned this into an only rarely interrupted asset for receiving political rents from actors who are interested in Israel remaining securely within its borders. This strategic rent was supplemented by phosphate rents. Thus, starting in the 1970s, petrolism fostered the already consolidated Jordanian semi-rentier state by increasing and diversifying rent income with new donors – the Arab Gulf states – and new sorts of rents – remittances.

In contrast to Jordan, Egypt looks back on a history of state institutions prior to colonialism and rentierism. In the nineteenth century, under the reign of Muhammad Ali, attempts to modernise and industrialise had already been launched taking Europe as a model (Lustick, 1997). Decades later, in the 1950s, state bureaucrats in uniform under the leadership of Gamal Abdel Nasser conducted a 'revolution from above' (Trimberger, 1978), thereby paving the way for the emergence of another state class in the Middle East. However, in contrast to the Arab Gulf and Jordan, the primary means of consolidating Egypt's rule was originally not external rent-seeking but class struggle. The state bureaucracy deprived the then-dominant class – the big landowners and a handful of industrialists and the banking sector – of their power through a land reform and a nationalisation policy, respectively. A state-managed import substitution industrialisation policy was launched, which, however, turned out to be a failure at the end of the 1960s. Still, the Egyptian state bureaucracy remained in a superior position vis-à-vis society (Richards and Waterbury, 2008: 188–90). The rise of petrolism had a double effect for Egypt. On the one hand, external rents provided the regime with urgently needed income to deal with the socio-economic crisis that was fuelled by the disastrous defeat against Israel in the June War of 1967. External rent-seeking policies became a key strategy of Egypt's foreign policies from the 1970s onwards (Springborg, 2014: 397). On the other hand, petrolism constituted the turning point in the hierarchy of Middle Eastern regional powers: Egypt, which had threatened the Arab Gulf states with exporting its republican revolution and which had waged a proxy war against Saudi Arabia in Yemen during the 1960s, started to become dependent on and, in the wake of the Arab uprisings, even subordinated to Saudi Arabia and the UAE with regard to regional affairs.

Lebanon, as an independent state separated from Syria, does not have a contemporary history. Almost as much as the Jordanian state was designed in London, the state of Lebanon was formed in Paris. However, the Lebanese history of societal development differs fundamentally from Jordan's. Around 1900, attempts were launched to promote a silk industry (Khater, 1996). Yet, what proved to have an even deeper impact on the socio-economic development of Lebanon was the globalisation of its labour force, which started in the mid-nineteenth century (Tabar, 2009). This process was facilitated by Western cultural imperialism, which enabled Christian Lebanese to migrate to the Global North and – as quasi-Western bridgeheads – to the Global South alike, for instance to Nigeria, from the late nineteenth century (Ndukwe, 2017). Soon after petrolism kicked off, Lebanon was hit hard by a devastating civil war (1975–90). Two developments occurred during that period. On the one hand, Lebanon participated in petrolism insofar as many Lebanese citizens migrated en masse to all destinations abroad, including the Gulf. On the other, the political economy of the Lebanese Civil War was deeply shaped by the influx of huge amounts of petrodollars, mainly associated with the financial activities of Lebanese-Saudi billionaire Rafiq Hariri, who later served two terms as Lebanese prime minister, from 1992 to 1998 and from 2000 to 2004 (Hourani, 2015). After the war, Saudi Arabia significantly contributed to the reconstruction of Lebanon, thereby establishing a further distorted version of the pre-war sectarian state (Al-Tamimi, 2018).

As a recipient of large amounts of aid, budget support, and remittances, Lebanon shares similarities with the two other semi-rentier states, Egypt and Jordan. Moreover, the political class of Lebanon is as elitist, and possibly similarly would-be authoritarian, vis-à-vis its society as the ruling regimes in Egypt and Jordan. However, there are two major differences that set the Lebanese case apart from the other state class regimes discussed in this book: the co-existence of several elite groups built on institutionalised sectarianism and the relative strength of Lebanese society vis-à-vis the state (Saouli, 2019), which is reinforced by a comparatively high societal control of rent sources, namely, remittances.

Egypt, Jordan, and Lebanon are all participants in labour migration to the Arab Gulf, which is based on the *kafāla* system. However, for two reasons the Lebanese, Jordanian, and, albeit to a much lesser degree, Egyptian societies are not only exploited, but are exploitative themselves because they import cheap labour from African and Asian countries. First, the highly skilled labour force of Egypt, Jordan, and, due to its advanced private educational system, especially Lebanon has been migrating to the Gulf. Like expatriate workers from the Global North, they are usually not over-exploited. Second, all three semi-rentiers also import cheap foreign labour: Jordan and Lebanon on the basis of *kafāla* and Egypt in the form of irregular foreign workers, mainly from the Horn of Africa and Sudan (Thomas, 2010).

The Lebanese capital provides a notable example. In 2010, 11.6 per cent of all households in Beirut employed domestic workers, most of whom lived in the employer's house (Fakih and Marrouch, 2014: 343, 348–9). Nearly all of these household helpers – more traditionally referred to as 'maids' – are women from Asian or sub-Saharan African countries. After the end of the civil war in the early 1990s, they replaced mostly Syrian-Arab and (Syrian-)Kurdish domestic workers (Jureidini and Moukarbel, 2004: 590). The beneficial participation of the Beirut upper and middle class in the exploitative characteristics of the Lebanese *kafāla* system is not only reflected by very low salaries – Jureidini and Moukarbel (2004: 587) note a salary range between USD 100 and 350 per month (which correlates to the ethnic heritage and different educational levels of the employees) – but is also manifested in (extremely) long working days, few (if any) days off, and not least in the precarious legal status of the persons who work as 'maids'.

As is shown by Adly (2021) for Egypt and by al Khouri and Silcock (2021) for Jordan, rentierism in semi-rentier states has also left its traces on the institutional arrangements within these countries. In particular, although not as extreme as in the Arab Gulf, institutions necessary to tax citizens and the economy are under-developed, and existing state organisations lack capacity to coordinate reform policies. Thus, from this perspective, unsurprisingly both regimes failed to use the opportunities of the oil price decline post-2014 to launch structural reforms. Lebanon is a different case, as the institutional weaknesses of its sectarian system are not primarily the result of externally acquired rents and related rent-seeking activities but the outcome of non-violent and violent conflict management between sectarian groups that were initially politicised by imperialist forces. However, there is plenty of evidence that Lebanon's dysfunctional consociational system (Salloukh *et al.*, 2015; Mabon, 2019) is reinforced by rent-seeking activities. For the latter, Saudi political rents for neoliberal reconstruction since the early 1990s managed by the Hariri family and its supporters, on the one hand, and Iranian payments to its Mediterranean ally Hizbullah, on the other, are the most striking examples. Evidence for the former aspect can be drawn from research showing that public funding in the realms of health, education, and infrastructure in Lebanon follows sectarian lines rather than imperatives of equality and poverty alleviation (Salti and Chaaban, 2010; see also Cammett, 2014).

In general terms, autonomy of the state vis-à-vis the society in the three semi-rentier states is lower than in the Arab Gulf. This difference is mainly due to the fact that the share of foreign migrant workers (from whom both the rentier and semi-rentier states enjoy a very high degree of autonomy) is small in Egypt, Jordan, and Lebanon. Moreover, particularly in Egypt and Jordan, decades-long policies of co-opting segments of the population that are crucial for maintaining political stability by way of employment in the public sector constrains the regimes' autonomy towards these groups. The narrow limits of state autonomy in Lebanon are mainly the result of sectarianisation, which enabled the political parties to corrupt the state and its agencies.

With the possible exception of Bahrain, social resistance to economic adjustment policies in the form of street riots and mass demonstrations have a much stronger tradition in the semi-rentier states than in the rentiers in the Arab Gulf (Brand, 1992; Salevurakis and Abdel-Haleim, 2008). Yet, similar to the Gulf states, some of the more effective ways to dilute austerity measures in the 1990s were grounded in informal linkages, like affiliations of Transjordanian tribal groups with the royal palace in Amman (Peters and Moore, 2009).

In the years after the oil price drop of 2014, the three semi-rentiers have taken different paths with regards to state–societal dynamics. As shown by al Khoury and Silcock (2021) using the case of the 2018 social protests against the Jordanian government, discontent seems to lead to social resistance, but deep political change is not in sight. However, Egyptian and Lebanese developments differ. After el-Sisi seized power in July 2013, Egyptian authoritarianism tightened, as evidenced in, for instance, opposition groups being silenced by the jailing of many of their members, the blocking of independent media, and the work of independent non-governmental organisations being criminalised (Momani, 2018). In Lebanon, however, the *You Stink!* movement challenged the consocational system (Beck, 2015), and the broad, resolutely non-sectarian social movement of 2019–20 brought it to the brink of collapse. These striking differences in state–societal dynamics in the three semi-rentier states complement our finding for the Arab Gulf countries that country-specific domestic institutions are much more important than outlined in orthodox rentierism.

Still prevailing: Rentierism

Some scholars suggested overcoming rentierism by either leaving behind some of its main assumptions (e.g. Gray, 2011) or declaring its relevance as obsolete (e.g. Hanieh, 2011). Gray observes that Arab Gulf rentier states started to spend their oil wealth 'more intelligently' (2011: 2) and finds that the 'depth and impact of ... reforms were startling' (2011: 15). He concludes that since the 1990s all Arab Gulf states, albeit to different degrees, have entered the stage of 'late rentierism' which makes the rentier state 'more responsive, globalized, and strategic in its thinking' (Gray, 2011: 24). A more dismissive perspective is represented by Hanieh, who departs from rentierism by pointing out that the 'class character of the Gulf economy is seldom tackled with any theoretical sophistication' (2011: 12).

On the one hand, we agree with Gray (2011) and Hanieh (2011) that rentierism as formulated during the 1980s needs refreshing to fully grasp the dynamics of contemporary Middle Eastern political economies. On the other, we believe in a more gradual and intrinsic advancement of the rentier state approach. Gray (2011) rightly argues that challenges from globalisation, social change, new technologies, and the global free-trade regime have created a context for a new major bargain taking place within the countries of the Arab Gulf. However, this is not a recent development only and should not be solely viewed as an act of enlightenment. As we outlined above, starting already in the 1960s, institutions crucial for external rent-seeking developed remarkable competences. Moreover, exploitation of migrant workers never diminished, while repression increased. We share with Hanieh (2011) the conviction that approaches of rentierism – except that of Delacroix (1980) – have largely failed to address state–class relations. However, as we attempted to show above, there are beneficial ways from within rentierism to overcome this deficit. In contrast to Hanieh (2011: 1–28, 2018: 64–7) and Gray (2011: 32), who portray the political economies of Arab Gulf as 'capitalist' and 'entrepreneurial state capitalist', respectively, we claim that the dominant class in the Arab Gulf, as well as the semi-rentier states of Egypt, Jordan, and Lebanon, is not genuinely guided by interests in economic development or capital accumulation but by the overarching goal of maintaining its socially and politically superior position. Economic development is thus rather a tool for achieving the latter. If the nature of the Arab Gulf regimes were genuinely (state) capitalist, they would have responded to the oil price drop in 2014 – and potentially also to the one in 2020 – by launching much deeper structural adjustment policies to make their economies fit for competition in global capitalist markets.

However, based on the above highlighted findings of the country studies in this volume, there is little evidence that ambitious structural reforms, as outlined in the heuristic framework of Beck and Richter (2021: 16–26), were launched after 2014. Also, initial summaries of the Arab Gulf regimes' immediate responses to the COVID-19-related oil price decline did not find any evidence that the 2020 socio-economic crisis had triggered more far-reaching structural reforms (e.g. Mogielnicki, 2020). Driven by a logic of maintaining their dominant social and political position, state classes in the Middle East have tended to act very carefully to avoid grievances among citizens, from whom they enjoy little autonomy with regard to implementing reforms. It is crucial to the state classes not to alienate their citizens, even if this implies sacrificing economic development goals. Therefore, due to anticipated and sometimes real social resistance, which is primarily based on informal institutions but sometimes also manifests itself in formal processes, attempts at structural reforms that would hit the citizens' welfare were only half-heartedly launched, withdrawn, or heavily diluted during the implementation phase. As we have attempted to show above, rentierism as a scholarly approach is capable of addressing these ongoing dynamics in a productive way by paying close attention to state–class relations and related issues of exploitation and repression, by highlighting the importance of intervening institutions, and by employing the idea of state autonomy in a more nuanced way.

In 1985, in the midst of a first deep fiscal crisis in the Middle East, Tétreault (1985: 189) expressed the fear that a breakdown of rentierism might drive the region back into a state prevalent before the export of oil by quoting a refrain attributed to Ahmad Yamani, who had then been in office as Saudi oil minister for twenty-three years:

> My grandfather rode a camel.
> My father drove a car.
> I fly in jet planes.
> My son will drive a car.
> My grandson will ride a camel.

On the one hand, such a scenario being the long-term result of lastingly low oil prices should not be readily dismissed, because the structural resilience of rentierism as a system of governance could mean that oil dependency will not be overcome. If global energy transition away from hydrocarbons becomes a reality – and there are some indicators that after decades of procrastination the COVID-19 pandemic might finally facilitate its acceleration (BP, 2020) – the Arab Gulf would then indeed be ill-prepared.

On the other hand, the resilience of rentierism also implies that the Arab Gulf states have gradually become virtuosic in rent-seeking since the 1950s. There can be hardly any doubt that the COVID-19 pandemic crisis further shrank the room for manoeuvre of the Arab Gulf regimes. For instance, tourism might not become a viable rent-seeking option in the foreseeable future. This also applies to preferred investments in sectors like aviation. However, at least in the short run, the Arab Gulf's by far most effective response to the oil price crash in the spring of 2020 was an impeccable rent-seeking activity: the April 2020 OPEC+ deal. Thus, our claim that rentierism prevails not only applies to the immediate years after the oil price decline of 2014 and 2020 but may also fit for the more distant future well into the twenty-first century.

Notes

1 There has been no effective initiative of the Arab League to cushion the oil price drop of 2014.

2 To give just a few examples: Kennedy and Tiede reference Chaudhry (1989) as claiming that oil 'incontrovertibly' inhibits the creation of functioning institutions (Kennedy and Tiede, 2013: 761). Su *et al.* cite Chaudry's research for showing that a rentier state 'cannot be expected to use the windfalls wisely, such as investing in appropriate infrastructure and managing the boom and bust. Instead, the government is typically fiscally irresponsible and engages in wasteful spending and pays for expensive security apparatus' (Su *et al.*, 2016: 22). A notable exception to this problematic view on Chaudhry's work is provided by Young (2020).

3 All deputies of the *majlis al-shūrā* in Saudi Arabia are appointed by the king (Bertelsmann Stiftung, 2020b: 8). The Qatari constitution prescribes that the majority of the members of parliament be elected by the people; however, the regime has kept postponing elections since the early 2000s (Freer, 2019: 12). In the UAE, the members of the *majlis al-watani al-ittihadi* are elected by only a small number of citizens. Moreover, this body holds advisory power only (Freer, 2019: 16). Gray (2021), Mason (2021), and Young (2021) did not find any evidence that the parliamentarian bodies in Qatar, Saudi Arabia, and the UAE, respectively, played a role in adjustment policies after the oil price decline in 2014.

References

Adly, A. (2021), 'Egypt's twisted hydrocarbon dependency: A case of persistent semi-rentierism', in M. Beck and T. Richter (eds), *Oil and the political economy in the Middle East: Post-2014 adjustment policies of the Arab Gulf and beyond*. Manchester: Manchester University Press, pp. 164–91.

Aftandilian, G. (2017), 'Youth unemployment remains the main challenge in the Gulf states', Arab Center Washington DC, 11 July. https://bit.ly/3esGha4 (accessed 26 March 2020).

AlJazeeri, S. (2021), 'Upgrading towards neoclassical rentier governance: Bahrain's post-2014 oil price decline adjustment', in M. Beck and T. Richter (eds), *Oil and the political economy in the Middle East: Post-2014 adjustment policies of the Arab Gulf and beyond*. Manchester: Manchester University Press, pp. 36–57.

al Khouri, R., and E. Silcock (2021), 'Oil and turmoil: Jordan's adjustment challenges amid local and regional change', in M. Beck and T. Richter (eds), *Oil and the political economy in the Middle East: Post-2014 adjustment policies of the Arab Gulf and beyond*. Manchester: Manchester University Press, pp. 192–212.

AlShehabi, O. H. (2019), 'Policing labour in empire: The modern origins of the kafala sponsorship system in the Gulf Arab States', *British Journal of Middle Eastern Studies*. https://doi.org/10.1080/13530194.2019.1580183 (accessed 26 March 2020).

Al-Sulayman, F. (2020), '"Reform dissonance" in the modern rentier state: How are divergent economic agendas affecting state-business relations in Saudi Arabia?', *British Journal of Middle Eastern Studies*, 47:1, 62–76.

Al-Tamimi, N. (2018), 'Saudi policy in Lebanon: No easy option for Riyadh', Italian Institute for International Political Studies, 4 May. https://bit.ly/3eBlEbE (accessed 1 April 2020).

Amnesty International (2018), 'GCC summit: Human rights in the Gulf under renewed scrutiny', 7 December. https://bit.ly/38qQb8k (accessed 9 May 2020).

Anderson, L. (1987), 'The state in the Middle East and North Africa', *Comparative Politics*, 20:1, 1–18.

Anderson, R. (2016), 'Oman's majlis al shura accused of a lack of urgency for austerity', Gulf Business, 5 July. https://bit.ly/3qFbf14 (accessed 18 May 2020).

Arabian Business (2016), 'Oil workers in Kuwait get 7.5% salary raise after negotiations', 25 May. https://bit.ly/3uOr5sU (accessed 28 October 2019).

Aslaksen, S. (2010), 'Oil and democracy: More than a cross-country correlation?', *Journal of Peace Research*, 47:4, 421–31.

Azzi, D. (2019), 'Lollar', Twitter, 22 December. https://twitter.com/dan_azzi/status/1208647074713092096 (accessed 31 March 2020).

Bank, A., T. Richter, and A. Sunik (2014), 'Durable, yet different: Monarchies in the Arab Spring', *Journal of Arabian Studies*, 4:2, 163–79.

Barbuscia, D. (2018), 'UPDATE 1-Bahrain will go ahead with value-added tax, finance minister says', Reuters, 21 February. https://reut.rs/2OGrON5 (accessed 19 May 2020).

Bashraheel, A. (2019), 'Rise and fall of the Saudi religious police', *Arab News*, 22 September. www.arabnews.com/node/1558176/saudi-arabia (accessed 12 May 2020).

Beblawi, H. (1987), 'The rentier state in the Arab world', in H. Beblawi and G. Luciani (eds), *The rentier state*. London: Croom Helm, pp. 49–62.

Beck, M. (2012), 'Dynasties', in H. K. Anheier, M. Juergensmeyer, and V. Faessel (eds), *Encyclopedia of global studies*. Thousand Oaks: Sage Publications, pp. 436–9.

Beck, M. (2015), 'Contextualizing the current social protest movement in Lebanon', E-International Relations, 10 October. https://bit.ly/30LCflj (accessed 11 May 2020).

Beck, M. (2016), 'Saudi Arabia's foreign policy and the failure of the Doha oil negotiations', E-International Relations, 21 June. https://bit.ly/3l4qpvP (accessed 29 June 2019).

Beck, M. (2019), 'OPEC+ and beyond: How and why oil prices are high', E-International Relations, 24 January. https://bit.ly/2OJm1Gg (accessed 24 October 2019).

Beck, M., and S. Hüser (2015), 'Jordan and the "Arab Spring": No challenge, no change?', *Middle East Critique*, 24:1, 83–97.

Beck, M., and T. Richter (2021), 'Pressured by the decreased price of oil: Post-2014 adjustment policies in the Arab Gulf and beyond', in M. Beck and T. Richter (eds), *Oil and the political economy in the Middle East: Post-2014 adjustment policies of the Arab Gulf and beyond*. Manchester: Manchester University Press, pp. 1–35.

Bertelsmann Stiftung (2020a), 'BTI 2020 country report – Oman'. https://bit.ly/2ODWmim (accessed 18 May 2020).

Bertelsmann Stiftung (2020b), 'BTI 2020 country report – Saudi Arabia'. https://bit.ly/3esMJ0J (accessed 18 May 2020).

Blanco, L. R., J. B. Nugent, and K. J. O'Connor (2015), 'Oil curse and institutional changes: Which institutions are most vulnerable to the curse and under what circumstances?', *Contemporary Economic Policy*, 33:2, 229–49.

Bloominvest Bank S.A.L. (2020), 'Interest rates on USD term savings & deposits', BRITE. https://bit.ly/2OmgeXt (accessed 26 March 2020).

BP (2020), 'Energy outlook 2020'. https://on.bp.com/3bAYwby (accessed 29 September 2020).

Brand, L. A. (1992), 'Economic and political liberalization in a rentier economy: The case of the Hashemite Kingdom of Jordan', in I. F. Harik and D. J. Sullivan (eds), *Privatization and liberalization in the Middle East*. Bloomington: Indiana University Press, pp. 167–87.

Bremmer, I. (2020), 'If there is a winner, it is Trump', *Time*, 16 April. https://time.com/5821593/opec-deal-trump/ (accessed 1 October 2020).

Cammett, M. (2014), *Compassionate communalism: Welfare and sectarianism in Lebanon*. Ithaca: Cornell University Press.

Cammett, M., I. Diwan, A. Richards, and J. Waterbury (2015), *A political economy of the Middle East*. Boulder: Westview Press.

Chaudhry, K. A. (1989), 'The price of wealth: Business and state in labor remittance and oil economies', *International Organization*, 43:1, 101–45.

Chaudhry, K. A. (1997), *The price of wealth: Economies and institutions in the Middle East*. Ithaca: Cornell University Press.

Crystal, J. (1990), *Oil and politics in the Gulf: Rulers and merchants in Kuwait and Qatar*. Cambridge: Cambridge University Press.

Crystal, J. (2018), 'The securitization of oil and its ramifications in the Gulf', in H. Verhoeven (ed.), *Environmental politics in the Middle East: Local struggles, global connections*. London: C Hurst & Co Publishers Ltd, pp. 75–97.

Delacroix, J. (1980), 'The distributive state in the world system', *Studies in Comparative International Development*, 15:3, 3–21.

Diop, A., T. Johnston, and K. T. Le (2018), 'Migration policies across the GCC: Challenges in reforming the kafala', in P. Fargues and N. M. Shah (eds), *Migration to the Gulf: Policies in sending and receiving countries*. Cambridge: Gulf Research Center, pp. 33–60. https://bit.ly/3nURBPb (accessed 26 March 2020).

Diwan, I. (2020), 'Why Lebanon's debt problem is super hard to sort out', The Lebanese Center for Policy Studies, April. http://lcps-lebanon.org/featuredArticle.php?id=282 (accessed 8 April 2020).

El Gamal, R., A. Lawler, and O. Astakhova (2020), 'OPEC+ presses for compliance with oil cuts', Reuters, 19 August. www.reuters.com/article/us-oil-opec-idUSKCN25F0Y0 (accessed 29 September 2020).

El-Katiri, L. (2014), 'The guardian state and its economic development model', *The Journal of Development Studies*, 50:1, 22–34.

Elsenhans, H. (1996), *State, class and development*. London: Sangam Books Limited.

Ennis, C. A., and S. Al-Saqri (2021), 'Oil price collapse and the political economy of the post-2014 economic adjustment in the Sultanate of Oman', in M. Beck and T. Richter (eds), *Oil and the political economy in the Middle East: Post-2014 adjustment policies of the Arab Gulf and beyond*. Manchester: Manchester University Press, pp. 79–101.

Fakih, A., and W. Marrouch (2014), 'Who hires foreign domestic workers? Evidence from Lebanon', *The Journal of Developing Areas*, 48:3, 339–52.

Freedom House (2020), 'Freedom in the world 2019: Countries and territories'. https://freedomhouse.org/countries/freedom-world/scores (accessed 9 May 2020).

Freer, C. (2019), 'Clients or challengers? Tribal constituents in Kuwait, Qatar, and the UAE', *British Journal of Middle Eastern Studies*. https://doi.org/10.1080/13530194.2019.1605881 (accessed 9 May 2020).

Frieden, J. (1991), *Debt, development, and democracy*. Princeton: Princeton University Press.

Global Knowledge Partnership on Migration and Development (2018), 'Migration and remittances: Recent developments and outlook', World Bank, Migration and Development Brief, 30. https://bit.ly/3eeakBS (accessed 16 March 2020).

Grant, J. A. (2001), 'Class, definition of', in R. J. B. Jones (ed.), *Routledge encyclopedia of international political economy*. New York: Routledge, pp. 161–3.

Gray, M. (2011), 'A theory of "late rentierism" in the Arab states of the Gulf', Georgetown University Center for International and Regional Studies, Occasional Paper, 7. https://bit.ly/3qRlK1s (accessed 29 April 2020).

Gray, M. (2021), 'Qatar: Leadership transition, regional crisis, and the imperatives for reform', in M. Beck and T. Richter (eds), *Oil and the political economy in the Middle East: Post-2014 adjustment policies of the Arab Gulf and beyond*. Manchester: Manchester University Press, pp. 102–23.

Hanieh, A. (2011), *Capitalism and class in the Gulf Arab states*. New York: Palgrave Macmillan.

Hanieh, A. (2018), *Money, markets, and monarchies: The Gulf Cooperation Council and the political economy of the contemporary Middle East*. Cambridge: Cambridge University Press.

Hassan, I. K. (2015), 'GCC's 2014 crisis: Causes, issues and solutions', Al Jazeera Centre for Studies, 31 March. https://bit.ly/2N7lwp7 (accessed 26 March 2020).

Herb, M. (2014), *The wages of oil: Parliaments and economic development in Kuwait and the UAE*. Ithaca: Cornell University Press.

Hertog, S. (2006), 'Segmented clientelism: The political economy of Saudi economic reform efforts', in P. Aarts and G. Nonneman (eds), *Saudi Arabia in the balance: Political economy, society, foreign affairs*. New York: New York University Press, pp. 111–43.

Hertog, S. (2010), *Princes, brokers, and bureaucrats: Oil and the state in Saudi Arabia*. Ithaca: Cornell University Press.

Hoetjes, G. (2021), 'Stalled reform: The resilience of rentierism in Kuwait', in M. Beck and T. Richter (eds), *Oil and the political economy in the Middle East: Post-2014 adjustment policies of the Arab Gulf and beyond*. Manchester: Manchester University Press, pp. 58–78.

Hourani, N. (2015), 'Capitalists in conflict: The Lebanese Civil War reconsidered', *Middle East Critique*, 24:2, 137–60.

Human Rights Watch (2016), 'Arab Gulf states: Attempts to silence 140 characters', 1 November. https://bit.ly/3la28V1 (accessed 10 May 2020).

Human Rights Watch (2017), 'Arab Gulf states: Assault on online activists', 12 July. https://bit.ly/2NapHRa (accessed 10 May 2020).

ILO (2020), 'COVID-19: Labour market impact and policy response in the Arab states', May. https://bit.ly/3qBnbRD (accessed 7 October 2020).

Jureidini, R., and N. Moukarbel (2004), 'Female Sri Lankan domestic workers in Lebanon: A case of "contract slavery"?', *Journal of Ethnic and Migration Studies*, 30:4, 581–607.

Kamrava, M. (2018), 'Oil and institutional stasis in the Persian Gulf', *Journal of Arabian Studies*, 8:1, 1–12.

Karaki, M. B. (2021), 'Lower oil prices since 2014: Good news or bad news for the Lebanese economy?', in M. Beck and T. Richter (eds), *Oil and the political economy in the Middle East: Post-2014 adjustment policies of the Arab Gulf and beyond*. Manchester: Manchester University Press, pp. 213–36.

Karl, T. L. (1987), 'Petroleum and political pacts: The transition to democracy in Venezuela', *Latin American Research Review*, 22:1, 63–94.

Kennedy, R., and L. Tiede (2013), 'Economic development assumptions and the elusive curse of oil', *International Studies Quarterly*, 57:4, 760–71.

Khater, A. F. (1996), '"House" to "goddess of the house": Gender, class, and silk in 19th-century Mount Lebanon', *International Journal of Middle East Studies*, 28:3, 325–48.

Korany, B. (1986), 'Political petrolism and contemporary Arab politics, 1967–1983', *Journal of Asian and African Studies*, 21:1–2, 66–80.

Lawson, F. H. (1991), 'Managing economic crises: The role of the state in Bahrain and Kuwait', *Studies in Comparative International Development*, 26:1, 43–67.

Luciani, G. (1987), 'Allocation vs. production state', in H. Beblawi and G. Luciani (eds), *The rentier state*. London: Croom Helm, pp. 63–84.

Lustick, I. S. (1997), 'The absence of Middle Eastern great powers: Political "backwardness" in historical perspective', *International Organization*, 51:4, 653–83.

Mabon, S. (2019), 'Desectarianization: Looking beyond the sectarianization of Middle Eastern politics', *The Review of Faith & International Affairs*, 17:4, 23–35.

Mahdavy, H. (1970), 'Patterns and problems of economic development in rentier states: The case of Iran', in M. A. Cook (ed.), *Studies in the economic history of the Middle East from the rise of Islam to the present day*. New York: Oxford University Press, pp. 428–67.

Martin, L. L. (1992), 'Interests, power, and multilateralism', *International Organization*, 46:4, 765–92.

Mason, R. (2021), 'The nexus between state-led economic reform programmes, security, and reputation damage in the Kingdom of Saudi Arabia', in M. Beck and T. Richter (eds), *Oil and the political economy in the Middle East: Post-2014 adjustment policies of the Arab Gulf and beyond*. Manchester: Manchester University Press, pp. 124–44.

Mogielnicki, R. (2019), 'Value-added tax in Gulf Arab states: Balancing domestic, regional, and international interests', The Arab Gulf States Institute in Washington, 26 August. https://bit.ly/3qBnXhv (accessed 31 March 2020).

Mogielnicki, R. (2020), 'Is this time different? The Gulf's early economic policy response to the crises of 2020', The Arab Gulf States Institute in Washington, 19 August. https://bit.ly/3ceqrwU (accessed 1 October 2020).

Momani, B. (2018), 'Egypt's IMF program: Assessing the political economy challenges', Brookings, 30 January. https://brook.gs/3l7P5Dq (accessed 12 January 2019).

Mommer, B. (2002), *Global oil and the nation state*. Oxford: Oxford University Press.

Moore, P. W. (2002), 'Rentier fiscal crisis and regime stability in the Middle East: Business and state in the Gulf', *Studies in Comparative International Development*, 37:1, 34–56.

Mukrashi, F. A. (2017a), 'Oman to create 25,000 jobs for nationals', *Gulf News*, 5 October. https://bit.ly/3qwuN7S (accessed 29 October 2019).

Mukrashi, F. A. (2017b), 'Oman shura council calls for ceiling for fuel prices', *Gulf News*, 2 February. https://bit.ly/2OkRnmV (accessed 20 May 2020).

Ndukwe, I. (2017), '"Everyone is hustling here": The Lebanese of Nigeria', Al Jazeera, 28 January. https://bit.ly/3rEbOd1 (accessed 8 April 2020).

Okruhlik, G. (1999), 'Rentier wealth, unruly law, and the rise of opposition: The political economy of oil states', *Comparative Politics*, 31:3, 295–315.

Peters, A. M., and P. W. Moore (2009), 'Beyond boom and bust: External rents, durable authoritarianism, and institutional adaptation in the Hashemite Kingdom of Jordan', *Studies in Comparative International Development*, 44:3, 256–85.

Putnam, R. D. (1988), 'Diplomacy and domestic politics: The logic of two-level games', *International Organization*, 42:3, 427–60.

Reuters (2017), 'Saudi Arabia restores perks to state employees, boosting markets', 22 April. https://reut.rs/2OkRQWd (accessed 2 October 2019).

Richards, A., and J. Waterbury (2008), *A political economy of the Middle East*. Boulder: Westview Press.

Richter, T. (2017), 'Structural reform in the Arab Gulf states – Limited influence of the G20', GIGA Focus Middle East, 3. https://bit.ly/3co55x9 (accessed 12 November 2019).

Ross, M. L. (2001), 'Does oil hinder democracy?', *World Politics*, 53:3, 325–61.

Ross, M. L. (2013), *The oil curse: How petroleum wealth shapes the development of nations*. Princeton: Princeton University Press.

Ross, M. L. (2015), 'What have we learned about the resource curse?', *Annual Review of Political Science*, 18:1, 239–59.

Salevurakis, J. W., and S. M. Abdel-Haleim (2008), 'Bread subsidies in Egypt: Choosing social stability or fiscal responsibility', *Review of Radical Political Economics*, 40:1, 35–49.

Salloukh, B. F., R. Barakat, J. S. Al-Habbal, L. W. Khattab, and S. Mikaelian (2015), *The politics of sectarianism in postwar Lebanon*. London: Pluto Press.

Salti, N., and J. Chaaban (2010), 'The role of sectarianism in the allocation of public expenditure in postwar Lebanon', *International Journal of Middle East Studies*, 42:4, 637–55.

Saouli, A. (2019), 'Sectarianism and political order in Iraq and Lebanon', *Studies in Ethnicity and Nationalism*, 19:1, 67–87.

Schneider, S. A. (1983), *The oil price revolution*. Baltimore: Johns Hopkins University Press.

Selim, H., and C. Zaki (2014), 'The institutional curse of natural resources in the Arab world', Economic Research Forum December. https://bit.ly/3rDgMXq (accessed 6 April 2020).

Shafer, M. D. (1994), *Winners and losers: How sectors shape the developmental prospects of states*. Ithaca: Cornell University Press.

Smith, B. (2006), 'The wrong kind of crisis: Why oil booms and busts rarely lead to authoritarian breakdown', *Studies in Comparative International Development*, 40:4, 55–76.

Smith, B. (2017), 'Resource wealth as rent leverage: Rethinking the oil–stability nexus', *Conflict Management and Peace Science*, 34:6, 597–617.

Springborg, R. (2014), 'Too "big" to succeed?', in R. Looney (ed.), *Handbook of emerging economies*. London: Routledge, pp. 397–415.

Su, F., G. Wei, and R. Tao (2016), 'China and natural resource curse in developing countries: Empirical evidence from a cross-country study', *China & World Economy*, 24:1, 18–40.

Tabar, P. (2009), 'Immigration and human development: Evidence from Lebanon', United Nations Development Programme, Human Development Research Paper 2009/35, 1 August. https://bit.ly/2PPmrLT (accessed 2 February 2020).

Tétreault, M. A. (1985), *Revolution in the world petroleum market*. Westport: Praeger.

Thomas, C. (2010), 'Migrant domestic workers in Egypt: A case study of the economic family in global context', *American Journal of Comparative Law*, 58:4, 987–1022.

Trimberger, E. K. (1978), *Revolution from above: Military bureaucrats and development in Japan, Turkey, Egypt and Peru*. New Brunswick: Transaction Books.

Ulfelder, J. (2007), 'Natural-resource wealth and the survival of autocracy', *Comparative Political Studies*, 40:8, 995–1018.

World Bank (2019), 'Net ODA received per capita (current US$)', World Bank Data. https://data.worldbank.org/indicator/dt.oda.odat.pc.zs (accessed 26 March 2020).

Wright, J., E. Frantz, and B. Geddes (2015), 'Oil and autocratic regime survival', *British Journal of Political Science*, 45:2, 287–306.

Yaakoubi, A. E. (2019), 'Bahrain ditches subsidy reform plan as political tensions simmer', Reuters, 7 May. https://reut.rs/3v94qbL (accessed 2 October 2019).

Yamada, M. (2020), 'Can a rentier state evolve to a production state? An "institutional upgrading" approach', *British Journal of Middle Eastern Studies*, 47:1, 24–41.

Young, K. E. (2020), 'Sovereign risk: Gulf sovereign wealth funds as engines of growth and political resource', *British Journal of Middle Eastern Studies*, 47:1, 96–116.

Young, K. E. (2021), 'Federal benefits: How federalism encourages economic diversification in the United Arab Emirates', in M. Beck and T. Richter (eds), *Oil and the political economy in the Middle East: Post-2014 adjustment policies of the Arab Gulf and beyond*. Manchester: Manchester University Press, pp. 145–63.

Index

EU authorised representative for GPSR:
Easy Access System Europe, Mustamäe tee 50,
10621 Tallinn, Estonia
gpsr.requests@easproject.com

www.ingramcontent.com/pod-product-compliance
Lightning Source LLC
Chambersburg PA
CBHW051954270326
41929CB00015B/2652